History of Rotating Machinery Dynamics

HISTORY OF MECHANISM AND MACHINE SCIENCE

Volume 20

Series Editor

MARCO CECCARELLI

Aims and Scope of the Series

This book series aims to establish a well defined forum for Monographs and Proceedings on the History of Mechanism and Machine Science (MMS). The series publishes works that give an overview of the historical developments, from the earliest times up to and including the recent past, of MMS in all its technical aspects.

This technical approach is an essential characteristic of the series. By discussing technical details and formulations and even reformulating those in terms of modern formalisms the possibility is created not only to track the historical technical developments but also to use past experiences in technical teaching and research today. In order to do so, the emphasis must be on technical aspects rather than a purely historical focus, although the latter has its place too.

Furthermore, the series will consider the republication of out-of-print older works with English translation and comments.

The book series is intended to collect technical views on historical developments of the broad field of MMS in a unique frame that can be seen in its totality as an Encyclopaedia of the History of MMS but with the additional purpose of archiving and teaching the History of MMS. Therefore the book series is intended not only for researchers of the History of Engineering but also for professionals and students who are interested in obtaining a clear perspective of the past for their future technical works. The books will be written in general by engineers but not only for engineers.

Prospective authors and editors can contact the series editor, Professor M. Ceccarelli, about future publications within the series at:

LARM: Laboratory of Robotics and Mechatronics
DiMSAT – University of Cassino
Via Di Biasio 43, 03043 Cassino (Fr)
Italy
E-mail: ceccarelli@unicas.it

For other titles published in this series, go to
www.springer.com/series/7481

J.S. Rao

History of Rotating Machinery Dynamics

Springer

J.S. Rao
Altair Engineering
Chief Science Officer
Outer Ring Road
560103 Bangalore
India
E-mail: js.rao@altair.com

ISBN 978-94-007-1164-8 e-ISBN 978-94-007-1165-5

DOI 10.1007/978-94-007-1165-5

Typesetting: Data supplied by the authors

Cover Design: Scientific Publishing Services Pvt. Ltd., Chennai, India

Printed on acid-free paper

9 8 7 6 5 4 3 2 1

springer.com

Dedicated to the memory of my parents

Jammi Chikka Rao
Jammi Ramanamma

Contents

Foreword

Human history may be built on the development of technology, but the history of rotordynamics to date has been a blank page of that history. Dr. Rao's book will do much to fill this empty space. Engineers are not historians, but it seems that wherever the history of a specialty is written, it is a specialist – Dr. Rao in this case – who must rise and address this need. Clearly, the history of any specialized area presents a very difficult task for anyone other than the expert himself.

Fortunately, Dr. Rao has chosen to address this task, a task it must be said that he is well-equipped to undertake.

Recognizing the blank page which stood before him, Rao has done well to first put his efforts into a broad historical perspective before attempting the specialized areas that make up the complex science of rotor-dynamics. For this is a subject which extends from shaft dynamics, through the static and dynamic properties of various types of bearings and seals, and often involves both blade and disk dynamics. A well-established basis of knowledge in many technologies is therefore needed before one could think of considering the history of all these subjects in any depth. Small wonder then that this task has lain unaddressed for years, waiting for a specialist like Dr. Rao, who has spent a lifetime of scholarship and preparation before attempting a history of a subject as broad as this.

Rao begins with chapters on the wheel, the water wheel and on the windmill. Recognizing the contributions that Aristotle and later Archimedes made to the science which underlies the technology of rotating machinery, he discusses the contributions of the great engineers of ancient times, and he mentions the fallacies that occasionally grew from such work, even from the teachings of the great Aristotle and Ptolemy. In this, Rao subtly poses an interesting question: is it right to expect that even from birth that science shall be perfect? Evidently not: rather, we come to realize through our own work the value of such a book as this, that the creation of science and technology, is an on-going task, and one for which we must all assume some responsibility through our own publications. From the Ancients Rao next moves through the aridity of the Dark Ages in the West, through the Renaissance to the great re-generation of science that began with the work of certain inspired scientists, and on to the emergence of the new technology of the Industrial Revolution, and to its flowering in the service of man.

The pace quickens as Rao finds himself in familiar territory, and he begins to address the foundations of the modern science of rotating machinery dynamics. The many methods that the author himself has used and explored in his work on blade vibrations are now laid out in historical terms, thus making clear to us how the many clever innovators that we have to thank for our firm grip on the technology of our subject achieved their results. The contributions of Euler, Lagrange, Rayleigh, and Dunkerley to the history of this subject are now set before us in fresh, historical terms, rather than in the familiar terms with which we are familiar. Here, these authors appear as real people for the first time, in the perspective of their subject. Euler with his thirteen children, going blind in St. Petersburg; Lagrange, too worried after publishing his timeless *Mechanique Analytique* to even open the copy on his desk for two years, amid the revolution which swirled below in the streets of Paris; and Rayleigh, in the blissful isolation of a honeymoon cruise up the Nile River, writing the *Theory of Sound*. Now at last for us these people have real, human faces.

In quick succession Rao now moves through the graphical methods of the late 19th and early 20th century until need again outstrips ability. Here it is the hands of Holzer which introduce numerical methods to address situations where formulas only lead to intractable mathematics. We begin to see that future the world of technology will exists in digital terms, and not formula terms. Jeffcott studied the vexed "critical speed" problem of shaft dynamics, by making the first rigorous experimental study, and then by using these results during his analysis of the problem. He found that this was a forced vibration problem, not an instability problem as some before him had claimed. Rao next considers the development and the contributions of the computer. He considers in parallel the birth of matrix methods, and of the Prohl–Myklestad method, as preparations through design for the solution of problems relating to the unbalance response and balancing of real rotors. Thus this fundamental question of rotordynamics had waited sixty years before it was explained by Jeffcott's contribution, and then it waited a further forty years, through the innovations and squabbles of the 1970s, until a full and accepted solution for unbalance response was finally reached.

This of course says nothing about the other great problem of rotordynamics which emerged in the 1920s, that of rotor instability. Rao's book treats both bearing instability and shaft hysteretic instability, and considers the impact of this and other questions of shaft dynamics, such as misalignment and asymmetry of the rotor. He makes it clear why almost a hundred years had to elapse before a complete solution for the shaft problem was reached. To complete his work, Rao addresses the much-awaited introduction of the Finite Element method into both rotor-dynamics and bearing dynamics, and then turns to another topic for which he is well prepared: the problems of blade and disk dynamics. Rao describes from first-hand knowledge how the blade vibration problem with its many geometrical complexities was investigated. He describes the evolution of methods for single airfoil analysis with discrete bending-bending-torsion approaches, and eventually through to finite element approaches, where with the evolution of greater computer capabilities it became possible to study both the steady stresses and the dynamic stresses in blades, blade groups, and today in complete blade-disk assemblies.

We may be confident that with this history, as part of the larger history of the Theory of Machines and Mechanisms that Professor Marco Ceccarelli has so worthily undertaken as editor, will do much to set the record straight on the origins of rotor-dynamics within the historical perspective of machinery development. This book, and other works in this series, will be received and read with relish by savants and students alike, that is, by all who seek knowledge of the background to their art. With the writing of this book, Dr Rao has secured for himself yet again a place in the history of engineering, as much as his technical work in this subject has earned him a place in its science. All who practice this fine art must now feel indebted to Dr Rao for this impressive work of documentation and perspective.

Neville F. Rieger, PhD, DSc, Fellow ASME
Chief Scientist, STI Technologies, Inc.
Rochester, NY, USA
Former Gleason Professor Mechanical Engineering
Rochester Institute of Technology, Rochester, NY, USA
Founder President
STI Technologies, Inc, Rochester, NY, USA
Chairman IFToMM Rotor Dynamics Technical Committee 1998-2006 and
Emeritus Chairman since 2008

Preface

It has been an eventual 20th century that transformed the mankind in the rapid development and usage of rotating machinery in various industrial sectors. Its all due to de Laval and Parsons inventions of impulse and reaction turbines who could bring the dream of Hero of Alexandria alive in 2100 years. This book is written to retrace the steps of history.

While the book traces the events leading to Laval and Parsons Turbines, the emphasis is on rotor and blade dynamics aspects that pushed these turbines to limits in the last century. The tabular and graphical methods developed in precomputer era have taken different form in the last 50 years through finite element methods. The methods evolved in the last century are discussed in detail to help the modern day designers and researchers.

Man has become mobile with the ending of last ice age 15000 years ago and has since been looking for doing daily chores in an easier manner. He invented the wheel for this purpose and put it use nearly 5000 years ago in the form of a Potter's wheel, then to transportation and as a grinding wheel in about 3000 years of time. This wheel is fundamental to rotating machinery as briefly presented in Chapter 1.

There was no science to take this development further, the first organized thinking before medieval period came from Aristotle about 2350 years ago. Archimedes in Alexandria about 2250 years ago has put forward some realistic and sound ideas on science; in fact in his period, Hero made the first workable reaction steam turbine. This was the first machine man attempted but could not take it to real usage because of lack of science. Chapter 2 briefly discusses these.

Without much science man has used wheel in the form of Water wheels and Wind mills to reduce the burden of grinding food grains and smithy. These are discussed briefly in Chapters 3 and 4.

The need for science was felt more and more in understanding structures and despite strict religious practices; there was renaissance in scientific thinking as by Leonardo da Vinci that led to scientific revolution with Nicholas Copernicus announcing heliocentric theory. Events moved fast with the invention of Calculus by Newton and Leibniz and beam structures received maximum attention three centuries ago. The rotor owes its knowledge base from the beam theories in this scientific revolution period. These aspects were discussed in Chapter 5.

While the science is taking good shape, engineers were looking forward to develop machines particularly removing water from mines. Otto von Guericke discovered vacuum in this period and renaissance engineers beginning from Denis Papin developed "atmospheric engines" or "fire engines" for this purpose. These developments were briefly discussed in Chapter 6.

With the discovery of Latent Heat by Joseph Black, the stage has been set to James Watt for producing a reciprocating steam engine which ushered in industrial revolution more than 200 years ago. Chapter 7 briefly discusses these aspects.

The reciprocating engine was never the dream of mankind and with Laval and Parsons Inventions, the reciprocating steam engine gave way to turbomachines which has rotors and mounted blades; this engine was considered "Vibration Free". This is the beginning of rotor dynamics just over a century ago. Chapter 8 briefly discusses these aspects.

The rotor and blade dynamics depended heavily on science and the fundamentals of elasticity are first discussed in Chapter 9. The elasticity equations are not amenable for a solution that involves coupled partial differential equations with 15 unknowns. Therefore energy methods were developed using Calculus of Variations from Newton which are fundamental to derivation of modern finite element methods. The energy methods were discussed in Chapter 10. We also discuss here some design examples adopted from Rayleigh's principle as adopted over a century ago in industry. Hamilton's principle which is the most general principle in Dynamics is discussed in this chapter with examples that cannot be addressed with finite elements even today.

When the turbomachinery were invented and their application became ubiquitous with the dynamo, the industry needed methods to determine critical speeds of rotors and blades. Since the methods of elasticity were not suitable without computers, initially strength of materials approach was developed and numerical methods using graphical or tabular form were adopted. These methods were discussed in Chapter 11.

Matrix methods were adopted in the mid 20th century and discussed in Chapter 12. With the advent of computer era the finite element method made rapid strides and they are discussed in Chapter 13 for beams, plates and shells. Nonlinear contact element methods required for analysis of bladed-disks were also discussed here. Typical application of commercial finite element codes for a turbomachine blade as practiced today is also discussed.

Rotor dynamics deals with rotating structures and the way in which these principles were developed from Jeffcott analysis in 1918 to handle various special aspects on fluid film supports, instabilities, gyroscopic effects, etc., are presented in Chapter 14. These methods in transfer matrix form are discussed in Chapter 15. The application of finite element methods for rotor dynamics is discussed in Chapter 16. Solid model rotor dynamics a recent development is also discussed in this chapter.

Chapter 1
Beginnings of the Wheel

The dates of the Stone Age vary considerably for different parts of the world. It began about 2 million years ago and ended in different parts from 6000 to 2500 BC. Throughout the immense time span of the Stone Age, see Scarre [2] and Schick and Nicholas [3], vast changes occurred in climate and in other conditions affecting human culture. Humans themselves evolved into their modern form during the latter part of it. The Stone Age has been divided accordingly into three periods: the Paleolithic, Mesolithic and Neolithic, see *Technology of Man* [5].

The Paleolithic, or Old Stone Age, was the longest period. It began about 2 million years ago, when stone tools were first used by humanoid creatures, and ended with the close of the last ice age about 13,000 BC. After 13,000 BC more clement weather patterns resulted in the greater availability of food. In tropical and temperate forest regions, Paleolithic tools, still chipped from stone, were adapted to the new conditions. This period is known as the Mesolithic, or Middle Stone Age. In this period, sometime many thousands of years ago, man found that a section of a tree trunk could be moved under the force of gravity because it was round. If the branches and twigs of the trunk were removed, the speed of the rolling log improved, see Figure 1.1.

Early men began to place runners under a heavy load, which they discovered would make it easier for the load to be dragged. This was the invention of the sledge. Men then began to combine the roller and the sledge. As the sledge moved forward over the first roller, a second roller was placed under the front end to carry the load when it moved off the first roller. It was discovered that the rollers that carried the sledge became grooved with use and that these deep grooves actually allowed the sledge to advance a greater distance before the next roller was needed. Thus, the rollers developed into wheels. In this process sections of wood between the grooves of the roller were cut away to form an axle and wooden pegs were fastened to the runners on each side of the axle. A slight improvement was made to the cart. This time, instead of using pegs to join the wheels to the axle, holes for the axle were drilled through the frame of the cart. Axle and wheels were now made separately [1,4].

Fig. 1.1 Earliest movement achieved by pulling and pushing under log rollers. Artist impression by Lakshmi.

Fig. 1.2 Potter's wheel of 3500 BC. Artist impression by Lakshmi.

The wheel is probably the most important mechanical invention of all time. Nearly every machine built since the beginning of the industrial revolution involves a single, basic principle embodied in one of mankind's truly significant inventions. It is hard to imagine any mechanized system that would be possible without the wheel or the idea of a symmetrical component moving in a circular motion on an axis. From tiny watch gears to automobiles, jet engines and computer disk drives, the principle is the same.

Agricultural villages had begun to develop by 8000 BC. This is known as the Neolithic period, or New Stone Age. Stone tools became highly polished and varied. By 6000 BC pottery appeared and copper was used for the first time in some regions followed by bronze. Before the introduction of iron, most tools and weapons were made of bronze, so this period was called the Bronze Age. During this time the slow potter's wheel was invented, see Figure 1.2. In about 3000 BC, the Egyptians developed the fast wheel, a completely mobile, carefully balanced apparatus of stone. Based on diagrams on ancient clay tablets, the earliest known use of this essential invention was a potter's wheel that was used at *Ur* in Mesopotamia (part of modern day Iraq) as early as 3500 BC. The first use of the wheel for transportation was probably on Mesopotamian chariots in 3200 BC (Figure 1.3). It is interesting to

Fig. 1.3 The wheel put in use for movement. Artist impression by Lakshmi.

Fig. 1.4 Indian spoked chariot. Artist impression by Lakshmi.

Fig. 1.5 Middle age grinding wheel. Artist impression by Lakshmi.

note that wheels may have had industrial or manufacturing applications before they were used on vehicles.

The wheel was furthered improved on later by the Egyptians, who made wheels with spokes, which could be found on Egyptian chariots of around 2000 BC. Over in Ancient India, chariots with spoked wheels dating back to around 1500 B.C. were also discovered, see Figure 1.4. The Greeks too adopted the idea of wheel-making from the Egyptians and made further improvements to it. Later, during the time of the Roman Empire, the Romans too engaged themselves in wheel-making and produced the greatest variety of wheeled vehicles. They had chariots for war, hunting, and racing, two-wheeled farm carts, covered carriages, heavy four-wheeled freight wagons and passenger coaches.

With the collapse of the Roman Empire in AD 476, the wheel became widely used for war machines across the old empire. The grinding wheel (see Figure 1.5) was introduced from Arabia to Europe in the middle ages, greatly improving the effect of bladed combat weapons.

References

1. Locke, I. (1995) *The Wheel and How It Changed the World*, Facts on File Inc., New York.
2. Scarre, C. (Ed.) (1988) *Past Worlds: The Times Atlas of Archaeology*, Times Books.
3. Schick, K.D. and Nicholas, T. (1993) *Making Silent Stones Speak: Human Evolution and the Dawn of Technology*, Simon & Schuster.
4. Strandth, S. (1979) *A History of Machine*, A&W Publishers, Inc., New York.
5. *Technology of Man* (1979) United Technologies Corporation.

Chapter 2
Science before the Medieval Period

Greeks took the elementary mathematics of the Egyptians and developed it into tools to serve the physicist and the engineer of the day. About 600 BC geometry from Egypt was imported and Greeks began to develop it and arithmetic into separate branches of mathematical science. In the next several hundred years, Hippocrates (460- BC), Aristotle (384–322 BC), Euclid (323–285 BC), and others systematized what was then known about geometry and arithmetic [2].

Aristotle was probably the first to conceive the idea of organized research. From the standpoint of the physical sciences, early Greek philosophers produced no fundamentally sound principles which are now accepted, but they gave for centuries their authority to many false premises. Early Greeks such as Aristotle had devoted considerable time to the study of objects in motion instead of bodies at rest. The world of science might then have benefited materially from their speculations, instead of being left with a lot of vague statements about *natural* and *unnatural* motion which actually served to confuse and retard the successful development of the fundamental truths of Mechanics.

Some of the Aristotle school teachings include false premises such as:

- Substances are divided into corruptible and incorruptible.
- Bodies are classified as absolute heavy bodies and absolute light bodies and seek their places, the light bodies on top.
- Motions are classified as natural motions and violent motions.
- Large bodies fall quicker than small ones,

These teachings retarded progress in Mechanics.

Archimedes (287–212 BC) of Syracuse in Sicily is believed by many to be the first mathematical genius the world had so far produced. He is also believed by many to have put his accomplishments into written form. He systematized simple machines and propounded the theory of their functions [1, 3]. It was probably he who invented the compound pulley, a device for increasing traction or lifting power and he propounded the theory of lever, both one and two armed. He regarded the wheel as the circular figure described by a rotating one-armed lever, and the screw as the circular analogy of the inclined plane. The Archimedean screw is illustrated

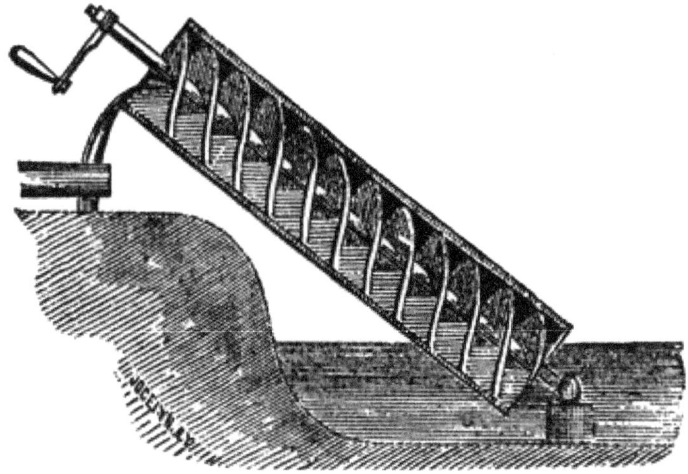

Fig. 2.1 As the handle a is turned, a certain amount of water is brought into the helical screw, which then brings water up to a reservoir or trough. (Courtesy Ianmacm of Wikipedia)

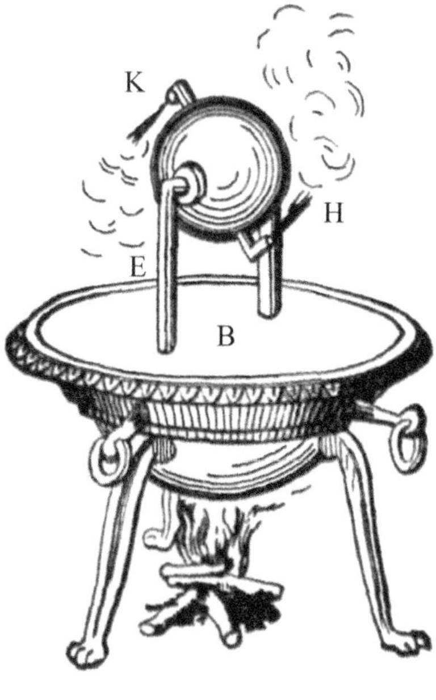

Fig. 2.2 Hero's aelopile. (Courtesy Wikimedia from Hero's "Spiritalia", edited by Woodcroft, of London)

in Figure 2.1. One of his famous sayings is "Give me a place to stand and I will move the earth".

Archimedes received his education at the University of Alexandria, where groups of mathematicians and scientists worked, devoting themselves to the construction of numerous fascinating machines [4]. The greatest and most colorful of what is known as the Alexandrian school of engineers was undoubtedly Hero who lived sometime during the second century BC. His best invention is the *aelopile*, the first reaction turbine, which converted heat into mechanical energy through the medium of steam, see Figure 2.2.

Hero's *aelopile*, the first reaction turbine could not produce useful work, as its speed was not sufficient to create the required high head of steam. As shown in Figure 2.2 – The lower container "B" was partly filled with water and was heated. The steam that was then produced was led via pipes "E" to a metal sphere, which could turn on its axle and had exhaust pipes "K" and "H", whose openings were directed at right angles to the axle. As steam escaped, the sphere rotated because of the steam's reaction.

In the 1780s James Watt worked on the theoretical operating conditions of a reaction turbine and concluded that such a turbine could not be built given the state of contemporary technology.

References

1. Crombie, A.C. (1979) *Augustine to Galileo*, Harvard University Press.
2. Girvin, H.F. (1948) *A Historical Appraisal of Mechanics*, International Textbook Company, Scranton, PA.
3. Heath, T.L. (1897) *The Works of Archimedes*, edited in modern notation, with introductory chapters, Cambridge University Press.
4. Pedersen, O. (1993) *Early Physics and Astronomy: A Historical Introduction*, 2nd. ed., Cambridge University Press.

Chapter 3
Water Wheels

In all likelihood, the earliest tools employed by humankind for crushing or grinding seeds, nuts, and other food-stuffs consisted of little more than a flat rock, upon which the material was crushed by pounding with a stone or tree branch. The archaeological records show that as early as 30,000 years ago, Cro-Magnon artists employed the mortar and pestle to grind and mix the pigments they used to create their magnificent "cave-art".

Far more efficient than the flat rock or even the mortar and pestle was the handmill, which appears to have long pre-dated the agricultural revolution. The handmill consists of a flat rock, often hollowed or concave, on which the grain, seeds, or other materials is placed, and a grinding stone, which is rolled across the grain, thus reducing the grain to flour. Although the handmill is still, today, in use in many parts of the world, approximately 2,000 years ago humankind began to harness water-power

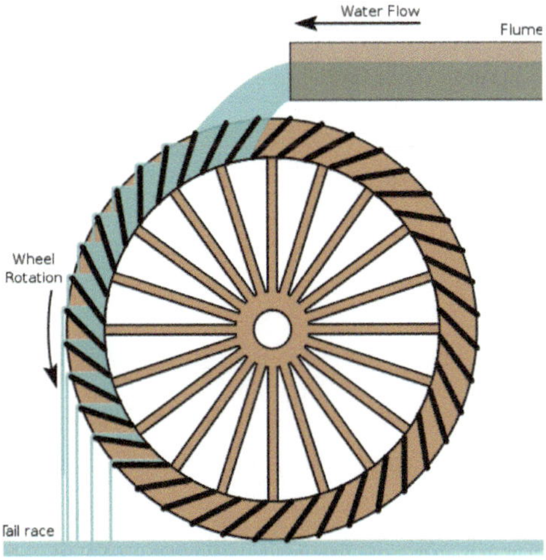

Fig. 3.1 An overshot wheel. (Courtesy Daniel M. Short at Wikimedia)

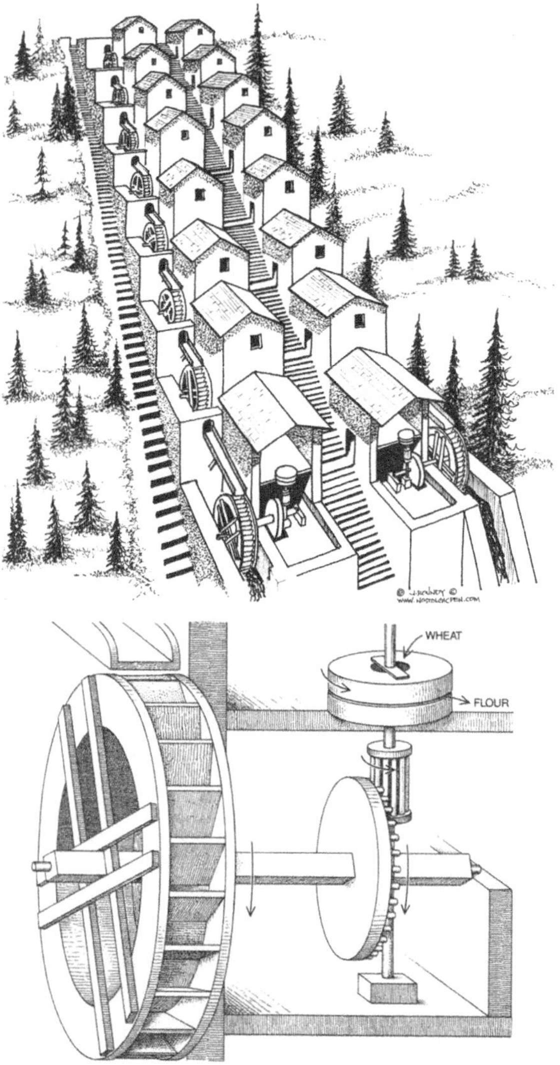

Fig. 3.2 Flour Mill at Barbegal near Arles in southern France dating from 4th century AD. (Courtesy Jay Roundy)

to turn the stones that ground its grain. It was probably the first tool for creating mechanical energy to replace direct human and animal power [3].

The first description of a water wheel is from Vitruvius, a Roman engineer (31 BC–14 AD), who composed a 10 volume treatise on all aspects of Roman engineering. From classical times, there have existed three general varieties of water wheels: the horizontal wheel and two variations of the vertical wheel see Hansen (on line website). Figure 3.1 shows an overshot wheel commonly employed to grind grain.

Fig. 3.3 Water Mill employed in China for iron casting. (Courtesy Wang Zhen from Wikipedia)

One of the most remarkable Roman applications of a waterwheel was at Barbegal near Arles in southern France. Dating from the 4th century AD, the factory was an immense flour mill which employed 16 overshot water wheels, see Figure 3.2.

Waterpower was an important source of energy in ancient Chinese civilization. One of the most intriguing applications was for iron casting; see Figure 3.3. According to an ancient text, in 31 AD the engineer Tu Shih invented a water-powered reciprocator for the casting of [iron] agricultural implements. Waterpower was also applied at an early date to the grinding of grain. Large rotary mill appeared in China about the same time as in Europe (2nd century BC). But while for centuries Europe relied heavily on slave and donkey-powered mills, in China the waterwheel was a critical power supplier.

Renaissance engineers studied the waterwheel and realized that the action of water on a wheel with blades would be much more effective if the entire wheel were somehow enclosed in a kind of chamber. They knew very well that only a small amount of the water pushing or falling on a wheel blade or paddle actually strikes it, and that much of the energy contained in the onrushing water is lost or never actually captured. Enclosing the wheel and channeling the water through this chamber would result in a machine of greater efficiency and power. However, they were hampered by lack of any theoretical understanding of hydraulics. Both of these problems were resolved to some degree in the eighteenth century, with one of the earliest examples of a reaction turbine being built in 1750 by the German mathematician and naturalist Johann Andres von Segner (1704–1777) [4]. In his system, the moving water entered a cylindrical box containing the shaft of a runner or rotor and flowed out through tangential openings, acting with its weight on the inclined vanes of the wheel.

These developments occurred during a significant period of scientific revolution and renaissance. Euler founded by this period Scienta Navalis, a new branch of science "rational mechanics" [1], treats ideal fluids in the first volume and on the seafaring and ship engineering in the second volume. To a great degree, we owe to Euler the principles of the impeller drive and the screw. In his time these were considered only as theory; however, Euler's experiments on Segner's water-powered machine and the related theory of water turbines are well known.

References

1. Fellmann, E.A. (2007) *Leonhard Euler*, translated by E. Gautschi and W. Gautschi, Birkhäuser Verlag, Berlin.
2. Hansen, R.D. Water wheels, http://www.waterhistory.org/histories/waterwheels and http://www.waterhistory.org/gallery/histories/barbegal/aap.jpg.html?g2_imageViewsIndex=1.
3. Hodge, A.T. (1990) A Roman factory, *Scientific American*, November Issue, p. 106.
4. Kaiser, W. (1977) *Johann Andreas Segner: Der 'Vater der Turbine'*, Leipzig.

Chapter 4
Wind Mills

Over 5,000 years ago, the ancient Egyptians used wind to sail ships on the Nile River. While the proliferation of water mills was in full swing, windmills appeared to harness more inanimate energy by employing wind sails. The wind wheel of Heron of Alexandria marks one of the first known instances in history of wind powering a machine [1]. The first practical windmills were the vertical axle windmills invented in eastern Persia, as recorded by the Persian geographer Estakhri in the 9th century, see Hassan and Hill [2]. Prototypes of windmills were probably known in Persia (present day Iran) as early as the 7th century AD with their sails mounted on a vertical axis, see Figure 4.1. Towards the end of the 12th century, windmills with sails mounted on a horizontal axis appeared in Europe; the first of this kind probably appeared in Normandy, England. These are post mills, where the sails and machinery are mounted on a stout post and the entire apparatus has to be rotated to face the wind.

Two centuries later the tower mill was introduced, enclosing the machinery in a stationary tower so that only the cap carrying the sails needed to be turned to the wind.

In 1854 Daniel Halliday obtained the first American windmill patent. His windmill had four wooden blades that pivoted and would self adjust according to wind speed. It had a tail which caused it to turn into the wind (Figure 4.2).

Thus far we have seen how a rotor evolved from ancient devices. Rotating (or reciprocating) machinery could not be meaningfully developed until sufficient scientific background was laid. Though Archimedes paved the way in the 3rd century BC for scientific thinking, the stronghold of the Church in the western world during the medieval period had impeded further developments. In mid-16th century a revival began but with slow progress as the Church continued its insistence on the belief that the Earth is the center of the universe as proscribed by the Bible. Ptolemy, who lived around 90 to 168 AD in the medieval period, subscribed to and widely promoted Plato's 4th century BC geocentric model [3].

It is this author's belief that various religions evolved in different parts of the world in an independent manner at different times, all of them with a common purpose and approach, viz, inculcation of good behavior in humans that had been

Fig. 4.1 An early Persian vertical axis windmill. (Artist impression by Lakshmi)

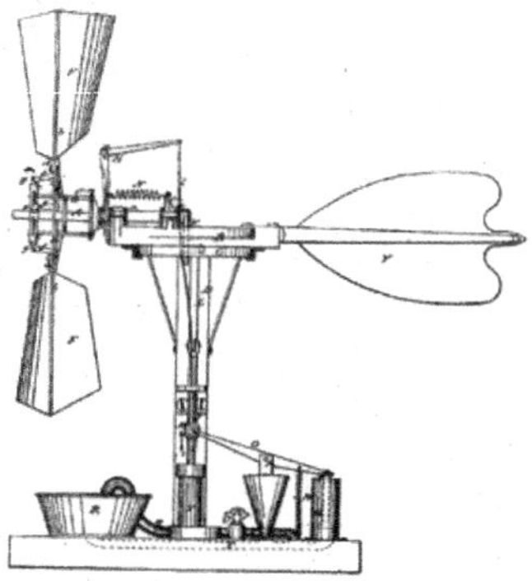

Fig. 4.2 Early American windmill mounted on a horizontal axle. (Coursey of United States Patent and Trademark Office)

elevated from the mammal kingdom. Any belief preached in these religions was vehemently opposed if a thinker expressed a different opinion based on scientific evidence. Thus there were delays in the ascendancy of scientific thinking, but eventually it surfaced as a scientific revolution which, in turn, paved the way to the industrial revolution. While scientific theories explained such phenomena as beam bending, which was a precursor for understanding of rotors, the industrial revolution provided the need for development of high-speed rotors. These aspects are discussed in the next section.

References

1. Drachmann, A.G. (1961) Heron's windmill, *Centaurus*, 7, p. 145.
2. Hassan, A.Y. and Hill, D.R. (1986) *Islamic Technology: An Illustrated History*, Cambridge University Press.
3. Pedersen, O., *A Survey of the Almagest*, Odense University Press, 1974.

Chapter 5
Renaissance and Scientific Revolution

Technology has been traditionally the realm of craftsmen working by rough rules of trial and error. The existing knowledge base was a mass of confusion in the absence of a unified understanding of the behavioral motion of solids and fluids [7, 31, 35]. The man of knowledge was a natural philosopher rather than a scientist.

The reawakening of scientific thought was brought about during the Renaissance Period (1400–1600) and carried into the period of the scientific revolution. Leonardo da Vinci (1452–1519) has recently been credited for some fundamental contributions to solid mechanics, fluid mechanics and mechanical design much before the scientific revolution. His contributions appear in Codex Madrid I, one of two remarkable notebooks that were discovered in 1967 in the National Library of Spain (Madrid), after being misplaced for nearly 500 years, see [1, 45]. He correctly concluded that, in bending of beams due to transverse loads, plane cross-sections remain plane before and after bending and rotate as shown in Figure 5.1. Da Vinci lacked Hooke's law and calculus to complete the theory; we had to wait for Galileo to improve this further before Euler and Bernoulli formed correct equations for simple bending.

The prevailing consensus otherwise is that Galileo Galilei (1564–1642) made the first attempts at developing a theory of beams [54]. He recognized the Principle of Virtual Work as a general law. Galileo however made an incorrect assumption in the development of the theory of beams; he did not recognize that at any section of the beam there was equilibrium of the tensile and compressive stresses. This equilibrium of the tensile and compressive stresses was not discovered until years later, by Mariotte (1686) in Paris [36].

The event which most historians of science call the scientific revolution can be dated roughly as having begun in 1543, the year in which Nicolaus Copernicus published his *De revolutionibus orbium coelestium* [5, 10, 20, 21, 26, 39]. It is widely accepted that Copernicus's *De revolutionibus* followed the outline and method set by Egyptian mathematician and astronomer Ptolemy in his *Almagest* in the second century [37].

Ancient Empedoclean division of the "elements" into Air, Water, Fire and Earth amounts to a little more than a geographical ordering of the familiar around us.

Fig. 5.1 Leonardo da Vinci's understanding of bending of beams

The Chinese divided the five elements as, viz., wood, fire, earth, metal, and water. The *Pancha Mahabhuta*, or "five great elements", of Hinduism are *Prithvi* or *Bhumi* (Earth), *Ap* or *Jala* (Water), *Agni* or *Tejas* (Fire), *Vayu* or *Pavan* (Air or Wind), and *Akasha* (Aether). Fire or Heat is fundamental to machines where energy is converted into useful mechanical work. Galileo (1564–1642) was amongst the first to argue in *Il saggitore* and in *Discorsi* that the sensation of heat is caused by the rapid motion of certain specific atoms (Galilean atomic model). Rene Descartes (1596–1650) was more explicit than Galileo and tried to explain all physical phenomena in terms of extension and motion only. Isaac Newton (1642–1727) built upon the work of Kepler and Galilei, see Galileo Galilei (1638, 1974), Galileo's mathematical treatment of acceleration and his concept of inertia both reflect earlier medieval analyses of motion [9]. His development of the calculus opened up new applications of the methods of mathematics to science. He showed that an inverse square law for gravity explained the elliptical orbits of the planets, and advanced the theory of Universal Gravitation [32, 48, 62].

Robert Boyle (1627–1691) was a notable disciple of Descartes. His major contribution to heat remains his association with chemistry and his famous law relating the pressure to the volume of an elastic fluid, or gas. Generally speaking, 17th century theories of heat combined the idea of a subtle fluid with that of motion of its constituent corpuscles or atoms. But the atoms remained scientifically inscrutable until Daltonian chemistry in 1808 was accepted and the unknown function of their motions was not understood until the concept of energy was established in the 19th century.

Scientific revolution made rapid strides beginning with Newton (1786). Gottfried Leibniz (1646–1716) [30], Pierre Varignon (1654–1722) [58], Jacob Bernoulli (1654–1705) [2], Johann Bernoulli (1667–1748) [3], Daniel Bernoulli (1700–1782) [4],

Leonhard Euler (1707–1783) [16], a Swiss mathematician, proposed the Euler equations, which describe conservation of momentum for an inviscid fluid, and conservation of mass. Claude Louis Marie Henry Navier (1785–1836) [38] and George Gabriel Stokes (1819–1903) [51] introduced viscous transport into the Euler equations, which resulted in the Navier–Stokes equations.

Lateral bending of simple long slender beams was correctly explained by Euler and Bernoulli in 1750. Euler made a number of contributions to Mechanics. The better known of these are in Mechanics (Strength of Materials). During his tenure as a Professor of Mathematics at St. Petersburg, he published in 1736 what has since been generally recognized as being the first book on Analytical Mechanics.

It was 65 years after the publication of Newton's Principia (1687) that Jean le Rond D'Alembert (1717–1783) published his Traite de Dynamique (1743). The Euler–Bernoulli model includes the strain energy due to bending and the kinetic energy due to lateral displacement. The Euler–Bernoulli model dates back to the 18th century. Jacob Bernoulli (1654–1705) first discovered that the curvature of an elastic beam at any point is proportional to the bending moment at that point. Daniel Bernoulli (1700–1782), nephew of Jacob, was the first to formulate the differential equation of motion of a vibrating beam. Later, Jacob Bernoulli's theory was accepted by Leonhard Euler (1707–1783) in his investigation of the shape of elastic beams under various loading conditions. Many advances in understanding of the elastic curves were made by Euler. The Euler–Bernoulli beam theory, sometimes called the classical beam theory, Euler beam theory, Bernoulli beam theory, or Bernoulli–Euler beam theory, is the most commonly used because it is simple and provides reasonable engineering approximations for many problems. However, the Euler-Bernoulli model tends to slightly overestimate the natural frequencies. This problem is exacerbated for the natural frequencies of the higher modes. Also, the prediction is better for slender beams than non-slender beams (see also [23]).

For a rotor dynamist, beam theory is the backbone of all analysis; all rotors were modeled as beams throughout the 20th century and only recently have we begun to understand solid model rotor dynamics. Thus beam models play a significant role in the history of rotor dynamics. The Rayleigh beam theory (1877) [44] provides a marginal improvement on the Euler–Bernoulli theory by including the effect of rotation of the cross-section. The kinetic energy due to rotation is accounted for in the analysis; it decreases the natural frequencies in the Euler–Bernoulli model. The effect of shear on the Euler–Bernoulli beam was accounted for by Timoshenko (1921, 1922) for high-frequency responses where shear or rotary effects are not negligible. This beam model is valid even today in the 21st century.

Joseph Louis Lagrange (1788) who lived from 1736 to 1813 made fundamental contributions during the scientific revolution and transformed Newtonian mechanics into a branch of analysis, Lagrangian mechanics as it is now called, and exhibited the so-called mechanical "principles" as simple results of the variational calculus. Hamilton (1834) who lived from 1805-1865 announced subsequently his famous principle of dynamics which is fundamental to the derivation of all vibration problems today.

In 1808 Ernst Florenz Friedrich Chladini (1756–1827) a German musician and scientist conducted experiments on vibrating plates, exhibiting the so-called Chladini figures [57]. Chladini repeated the pioneering experiments of Robert Hooke of Oxford University who, on 8 July 1680, had observed the nodal patterns associated with the vibrations of glass plates. Hooke ran a bow along the edge of a plate covered with flour, and saw the nodal patterns emerge.

Chladini's technique, first published in 1787 in his book, consisted of drawing a bow over a piece of metal whose surface was lightly covered with sand. The plate was bowed until it reached resonance and the sand formed a pattern showing the

Fig. 5.2 Sand sprinkled on a vibrating plate drifts to the nodal lines, revealing the standing wave pattern produced by one of several frequencies. (Courtsey of The Video Encyclopedia of Physics Demonstrations, Montana State University, Bozeman, MT, USA)

nodal regions, see Figure 5.2 with modern experimentation: what we know now as rectangular plate mode shapes. The sand moved about until it reached the nodes. This was the first experimental demonstration of 2-D harmonic motion.

There was no mathematical explanation for the seemingly strange phenomenon. The French Emperor Napoleon was very much intrigued by the demonstration. At his instance, the French Academy of Sciences announced a big prize for finding a mathematical solution for the same. Most mathematicians did not attempt to solve the problem, mainly because Lagrange had said that the mathematical methods available at the time were inadequate for the purpose. It was believed that molecular structure, theorized for materials, had to be taken into consideration in order to find an explanation of this kind of phenomenon. Sophie Germain (1776–1831) attempted to solve the problem and was awarded the Prize for deriving a plate theory in 1815 – though it had some deficiencies corrected later by Kirchhoff in 1850 [6, 24, 34, 43]. It took 65 years for the community to begin to understand how to deal with a two-dimensional structure.

Charles-Augustin de Coulomb (1736–1806) was a French natural philosopher, and military engineer. In the Histoire de l'Academie for 1784, published in 1787 at Paris, a paper by Coulomb appeared entitled "Recherches Théoriques et Expérimentales sur la Force de Torsion et sur l'Élasticité des Fils de Metal" [11]. This paper seems to be the first recorded discussion of torsion and apparently is what Saint-Venant refers to on pages 331, 340, and 341 of his work *Torsion* when he mentions Coulomb's ancient theory of torsion. Coulomb does not express his torque in terms of the elastic rigidity but makes it proportional to the moment of inertia of the normal section about the longitudinal axis of the rod or wire. Adhémar Jean Claude Barré de Saint-Venant (1797–1886) was a French mechanician and mathematician who contributed to early stress analysis. It took us another 40 years from the formulation of plate theory to explain the Torsion problem of noncircular rods which was first solved by Saint Venant in the mid-1850s [54]. Torsion plays an important role in rotor dynamics in drive train vibration problems that is responsible for earlier fatigue failures.

The vibration problem of elastic systems was lucidly discussed by Lord Rayleigh (1842–1919) in 1877. Walter Ritz (1878–1909) made significant contributions to variational methods in deriving energy methods [46, 47].

Fundamentals of the Theory of Elasticity or physics of deformable bodies were established during the scientific revolution. Engineers would refer to these deformable bodies as structures. Augustus Edward Hough Love was a British geophysicist and mathematician. Love's book, *A Treatise on the Mathematical Theory of Elasticity*, 2 vols. (1892–1893) [33], is an indispensable reference work for engineers, mathematicians, and physicists; it is the most complete and authoritative treatment of classical elasticity in existence. There also exist several excellent textbooks among which books by Dym and Shames [14], Sokolnikoff [50], Timoshenko and Goodier [55] and Todhunter [56].

Elasticians have evolved approximate methods through energy principles to solve problems posed by the theory of elasticity. Energy principles are fundamental to the solution of all elasticity problems including those where the displacement field is time dependent (vibration problems). Initially these problems happened to be beams, then plates and solid structures with intricate geometry. Again there are several texts that describe the energy methods in an elegant manner, some of them are Fox [17], Lanczos [28], Langhaar [29], Love [33], Prescott [41], Rao [42], Washizu [59, 60] and Weinstock [61].

The industrial revolution began with reciprocating steam engines devised by James Watt in 1780, and the 19th century witnessed a rapid expansion in various industrial sectors. Unfortunately, the reciprocating steam engine has several problems because of external combustion and excessive alternating load due to reciprocating masses that limit speeds and capacities. The industry was looking for non-reciprocating systems, purely rotating systems that could usher in an era of so-called "Vibration Free" engines. Rotor Dynamics is different from Structural Dynamics, as we deal with a rotating structure. Basically, all the vibration phenomena will be valid, however, there are several differences and we have to set up procedures for handling the rotors and their vibratory phenomena.

More immediately important were the developments in studies of atmosphere that revealed the existence of a finite ocean of air (the atmosphere) which exerts a considerable pressure on everything underneath it. (Clausius published the basic ideas of the second law of thermodynamics in 1850.)

Until the middle of the 19th century, the fruits of scientific revolution made little impact and the contribution of science to technology remained occasional. But the cultural and social revolution of the 16th and 17th centuries created conditions for the union of science with technology which was to become the basis of modern industrial development, see [15, 25, 49].

References

1. Ballarini, R. (2003) The Da Vinci–Euler–Bernoulli Beam Theory?, *ME Magazine*, http://www.memagazine.org/contents/current/webonly/webex418.html.
2. Bernoulli, Jacob (1694) *Curvatura Laminae Elasticae*, Acta Eruditorum, Lipsiae, June.
3. Bernoulli, Johann (1724) Discours sur les loix de la communication du mouvement, chapters 1–3, Prize Essay, Paris.
4. Bernoulli, Daniel (1751) De vibrationibus et sono laminarum elasticarum commentationes physico-geometricae, Commentari Academiae Scientiarum Imperialis Petropolitanae. T.13 ad annum 1741 p. 43. p. 105.
5. Bruce, S.E. (1982) Kepler as Historian of Science: Precursors of Copernican Heliocentrism According to De revolutionibus, *Proceedings of the American Philosophical Society*, 126, p. 367.
6. Bucciarelli, L.L. and Dworsky, N. (1980) *Sophie Germain: An Essay in the History of the Theory of Elasticity*, Springer.
7. Butterfield, H. (1965) *The Origins of Modern Science, 1300–1800*, Free Press.
8. Chladini, E.F.F. (1787) Entdeckungen über die Theorie des Klanges.
9. Clagett, M. (1961) *The Science of Mechanics in the Middle Ages*, University of Wisconsin Press.
10. Cohen, H.F. (1994) *The Scientific Revolution: A Historiographical Enquiry*, University of Chicago Press.
11. Coulomb, C.A. (1784) Recherches théoriques et expérimentales sur la force de torsion, & sur l'élasticité des fils de métal: Application de cette théorie à l'emploi des métaux dans les Arts & dans Jonathan A. Hill, Bookseller, Inc., USA.
12. D'Alembert, J.L. (1743) Traite de Dynamique.
13. Descartes, René (1984–1991) *The Philosophical Writings of Descartes*, 3 vols., trans. J. Cottingham, R. Stoothoff, D. Murdoch and A. Kenny, Cambridge University Press.
14. Dym, C.L. and Shames, I.H. (1973) *Solid Mechanics, A Variational Approach*, McGraw-Hill Book Co.
15. Dumas, M. (Ed.) (1962) *Histoire Generale des Techniques*, Vols. I–IV, Paris.
16. Euler, L. (1736–1770) Mechanica sive Motus Scienta Analytice Exposita, 1736, Sur la force des colonnes, *Berlin-Brandenburgischen Akademie der Wissenschaften – Memoires de l'Academie de Berlin*, Tom. XIII, 1759, p. 252, De motu vibratorio fili flexilis, corpusculis quotcunque onusti, *Novi Comentarii Academiae Scientarum Imperialis Petropolotanae*, Vol. IX, 1764, Genuina Principia Doctrinae de Statu aequilibri et motu corporum tam perfecte flexibilium quam elasticorum, *Novi Comentarii Academiae Scientarum Imperialis Petropolotanae*, Vol. XV, 1770.
17. Fox, C. (1950) *An Introduction to the Calculus of Variations*, Oxford University Press.
18. Galileo Galilei (1638) *Discorsi e Dimostrazioni mathematische*, Leiden.
19. Galileo Galilei (1974) *Two New Sciences*, trans. Stillman Drake, University. of Wisconsin Press,
20. Grant, E. (1996) *The Foundations of Modern Science in the Middle Ages: Their Religious, Institutional, and Intellectual Contexts*, Cambridge University Press.
21. Hall, A.R. (1983) *The Revolution in Science, 1500–1750*, 3rd ed., Longman.
22. Hamilton, W.R. (1834–1835) On a General Method in Dynamics, *Philosophical Transaction of the Royal Society*, Part I, 1834, pp. 247–308; Part II, 1835, pp. 95–144.
23. Han, S.M., Benaroya, H. and Wei, T. (1999) Dynamics of Transversely Vibrating Beams Using Four Engineering Theories, *Journal of Sound and Vibration*, 225(5), p. 935.
24. Kirchoff. G. (1876) *Vorlesungen über mathematisch Physik: Mechanik*, Leipzig.
25. Kranzberg, M. and Pursell, C.W. Jr. (1967) *Technology in Western Civilization*, Vols. I–II, New York.
26. Kuhn, T. (1957) *The Copernican Revolution*, Harvard University Press.
27. Lagrange, J.L. (1788) *Mécanique Analytique*, 2 vols., Gauthier-Villars et fils, Paris.
28. Lanczos, C. (1949) *The Variational Principle of Mechanics*, University of Toronto.

29. Langhaar, H.L. (1962) *Energy Methods in Applied Mechanics*, John Wiley & Sons.
30. Leibniz, G. (1684) *Demonstrationes novae de Resistentia solidorum*, Acta Eruditorum Lipsiae, p. 319.
31. Lindberg, D.C. (1992) *The Beginnings of Western Science: The European Scientific Tradition in Philosophical, Religious, and Institutional Context, 600 B.C. to A.D. 1450*, University. of Chicago Press.
32. Lindberg, D.C. and Westman, R.S. (Eds.) (1990) *Reappraisals of the Scientific Revolution*, Cambridge University Press.
33. Love, A.E.H. (1944) *Mathematical Theory of Elasticity*, Dover.
34. Mahanty, S. (2006) Marie-Sophie Germain, The Remarkable Woman Mathematicians of France, *Dream 2047*, 8(10), p. 38.
35. Maier, A. (1982) *On the Threshold of Exact Science: Selected Writings on Late Medieval Natural Philosophy*, University of Pennsylvania Press.
36. Mariotte, E. (1686) *Traits du mouvement des eaux*, Paris.
37. McGuire, J.E. and Rattansi P.M. (1966) Newton and the 'Pipes of Pan', *Notes and Records of the Royal Society of London*, Vol. 21, No. 2.
38. Navier Lois (1821) De l'équilibre et du mouvement des corps solides élastiques, Paper Read to the Académie des Sciences, May 14, 1821.
39. Neugebauer, O. (1968) On the Planetary Theory of Copernicus, *Vistas in Astronomy*, Vol. 10.
40. Newton, I. (1786) *Principia Mathematica*, Earl Gregg Swem Library, College of William & Mary.
41. Prescott, J. (1946) *Applied Elasticity*, Dover.
42. Rao, J.S. (1992) *Advanced Theory of Vibration*, John Wiley & Sons.
43. Rao, J.S. (1998) *Dynamics of Plates*, Marcel Dekker.
44. Rayleigh, J.W.S. (1945) *Theory of Sound*, Macmillan, London, 1877, Dover Publication.
45. Reti, L. (Ed.) (1974) *The Unknown Leonardo*, McGraw-Hill Co., New York.
46. Ritz, W. (1909) Über eine neue Methode zur Lösung gewisser Variationsprobleme der mathematischen Physik, *Journal für die Reine und Angewandte Mathematik*, Vol. 135, p. 1.
47. Ritz, W. (1911) *Gesammelte Werke*, Gauthier-Villars.
48. Shapin, S. (1996) *The Scientific Revolution*, University of Chicago Press.
49. Singer, C. et al. (Eds.) (1954) *A History of Technology*, Vols. I–V, New York.
50. Sokolnikoff, I.S. (1956) *Mathematical Theory of Elasticity*, McGraw-Hill Book Co.
51. Stokes, G.G. (1849) On the Theories of the Internal Friction of Fluids in Motion, and of the Equilibrium and Motion of ElSstic solids. *Cambridge Philosophical Society Transactions*, Vol. 8, p. 287.
52. Timoshenko, S.P. (1921) On the Correction for Shear of the Differential Equation for Transverse Vibrations of Prismatic Bars, *Philosophical Magazine*, Vol. 41, p. 744.
53. Timoshenko, S.P. (1922) On the Transverse Vibrations of Bars of Uniform Cross-Section, *Philosophical Magazine*, p. 125.
54. Timoshenko, S.P. (1955) *History of Strength of Materials*, McGraw-Hill Book Co.
55. Timoshenko, S.P. and Goodier, J.N. (1951) *Theory of Elasticity*, McGraw-Hill Book Co.
56. Todhunter, I. (1960) *A History of the Theory of Elasticity and of the Strength of Materials: From Galilei to Lord Kelvin*, Dover.
57. Ullmann, D. (2007) Life and Work of E.F.F. Chladini, *The European Physical Journal – Special Topics*, Vol. 145, No. 1, p. 25.
58. Varignon, P. (1702) De la Resistance des Solides en general pour toit ce qu'on peut faire d'hypothises touchant la force ou la tenacite des Fibres des Corps a rompre; Et en particulier pour les hypotheses de Galilee & de M. Mariotte, Memoires de l'Acadimie, Paris, p. 66.
59. Washizu, K. (1962) Variational Principles in Continuum Mechanics, Report 62-2, University of Washington, College of Engineering.
60. Washizu, K. (1982) *Variational Methods in Elasticity and Plasticity*, Pergamon Press.
61. Weinstock, R. (1952) *Calculus of Variations with Applications to Physics and Engineering*, McGraw-Hill Book Co.
62. Westfall, R.S. (1971) *The Construction of Modern Science*, John Wiley and Sons.

Chapter 6
Renaissance Engineers

Medieval and Renaissance Europe possessed only one effective heat engine, the combustion engine in the form of the cannon (Figure 6.1) [1].

The credit for making pressure exerted by the atmosphere entirely explicit belongs to Otto von Guericke (reprint 1963), who in 1672 published the famous book in which he described his air pump and the experiments that he made with it from the mid 1650s onwards. His famous demonstration is illustrated in Figure 6.2. Once it was understood that atmosphere exerts pressure, it was a matter of creating a vacuum and allowing the atmospheric pressure to move the piston in a cylinder.

Denis Papin (1647–1712) a French physicist, mathematician and inventor is best known for his pioneering invention of the steam digester, the forerunner of the steam engine [5]. He visited London in 1675, and worked with Robert Boyle from 1676 to 1679, publishing an account of his work in *Continuation of New Experiments* in 1680. During this period, Papin invented the *steam digester,* a type of pressure cooker. He first addressed the Royal Society in 1679 on the subject of his digester, and remained mostly in London until about 1687, when he left to take up an academic post in Germany. While in Leipzig in 1690, having observed the mechanical power of atmospheric pressure on his "digester", he built a model of a piston steam engine, the first of its kind, see Figure 6.3. The Papin experiment was a metal tube (closed at one end) with a piston inside. Under the piston there was a small quantity of water which, heated and transformed into steam, raised the piston which reached the edge of the cylinder where it was stopped by a flange. A stream of cold water was sprayed onto the cylinder. The steam inside condensed. This produced a partial vacuum and the outside air pressure forced the piston down (active stroke). The tube had three roles: boiler, cylinder and steam condenser. The steam engine will build pressure step by step, separating those three roles [3].

Thomas Savery (1650–1715) was an English military engineer and inventor who in 1698 patented the first crude steam engine, based on Denis Papin's Digester or pressure cooker of 1679. On 2 July 1698 Savery patented an early steam engine; he demonstrated it to the Royal Society on 14 June 1699. In 1702 Savery described the machine in his book *The Miner's Friend; or, An Engine to Raise Water by Fire*, in which he claimed that it could pump water out of mines. His machine (see Fig-

Fig. 6.1 A 17th century forge-welded iron cannon in Thanjavur (S. India). (Courtesy Wikipedia)

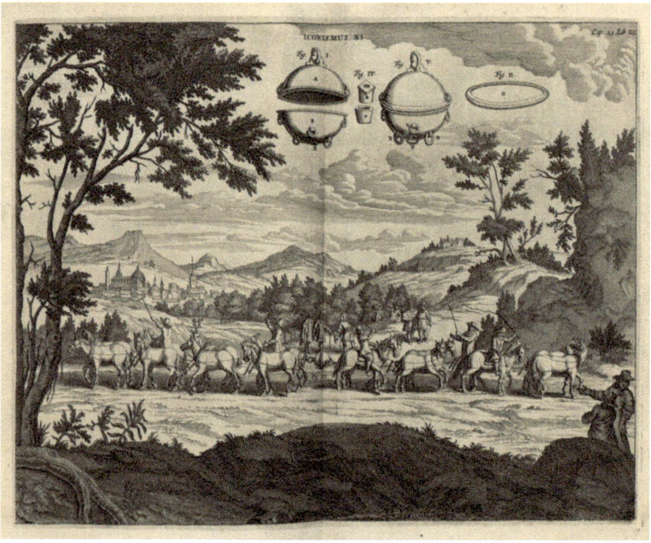

Fig. 6.2 Teams of horses trying unsuccessfully to pull apart vacuum-filled copper spheres in Magdeburg demonstration for Emperor Ferdinand III. (Courtesy Gaspar Schott of Wikipedia)

ure 6.4), consisted of a closed vessel filled with water into which steam under pressure was introduced. This forced the water upwards and out of the mine shaft. Then a cold water sprinkler was used to condense the steam. This created a vacuum which sucked more water out of the mine shaft through a bottom valve [7].

In 1705 Papin, with the help of Gottfried Leibniz, developed a second steam engine using steam pressure rather than atmospheric pressure. Papin's steam engine was the first breakthrough since Hero's reaction turbine of the 2nd century BC, which never functioned in reality. In the installation at Kassel (Figure 6.5), steam was fed from boiler "o" to a vessel "b" in which there was a float serving as a piston

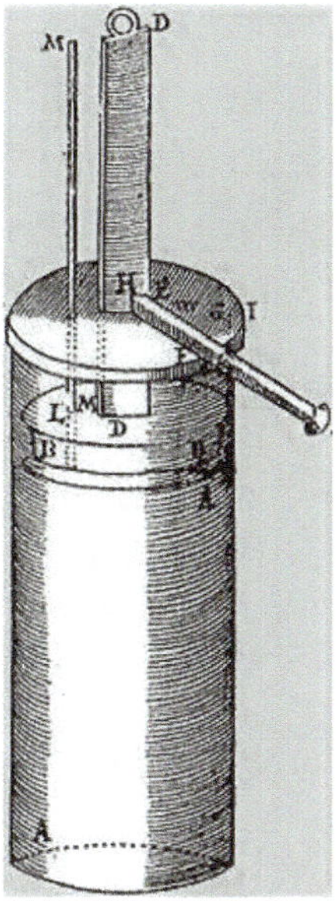

Fig. 6.3 The first steam engine 1690. (Courtesy Wikipedia)

Fig. 6.4 Thomas Savery (1698) vacuum pump. (Courtesy Wavesmikey at Wikipedia)

“c”. Papin planned to fill a container “d” in the piston with red-hot scrap iron in order to superheat the steam, but in reality this could not be done. When the steam pushed the piston downwards, the check valve “e” of the container for water to be pumped was closed. Simultaneously the check valve “f” of the ascending pipe “g” was opened and the water was pumped into a cistern “h”, from where it flowed to the

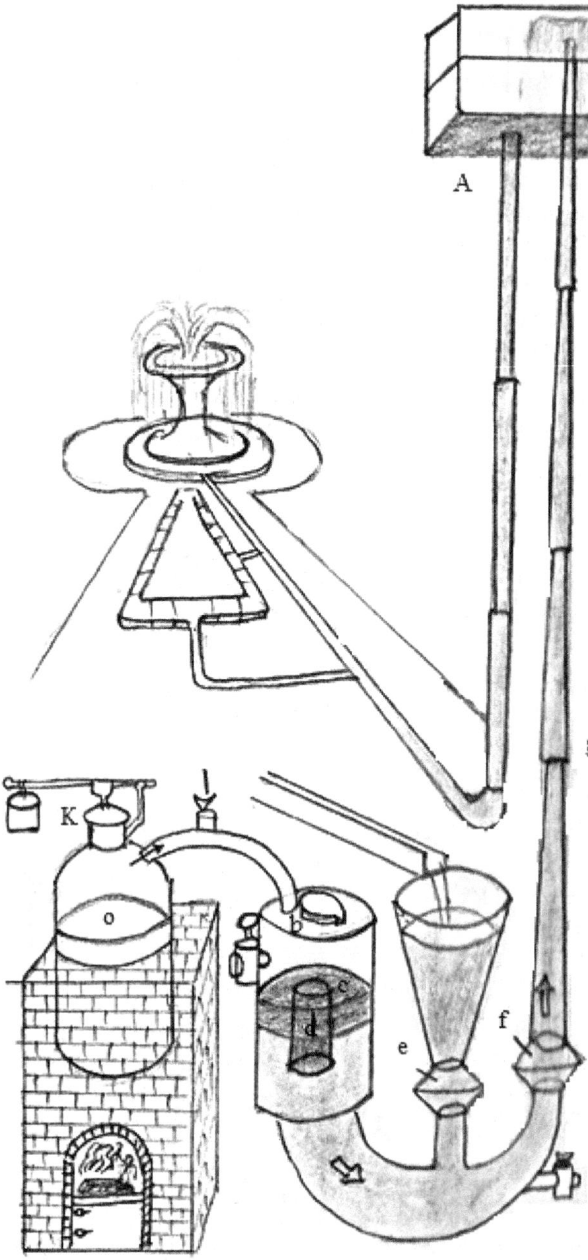

Fig. 6.5 Papin Steam Engine installed in Kassel gardens of the Duke of Hesse. (Artist impression by Lakshmi)

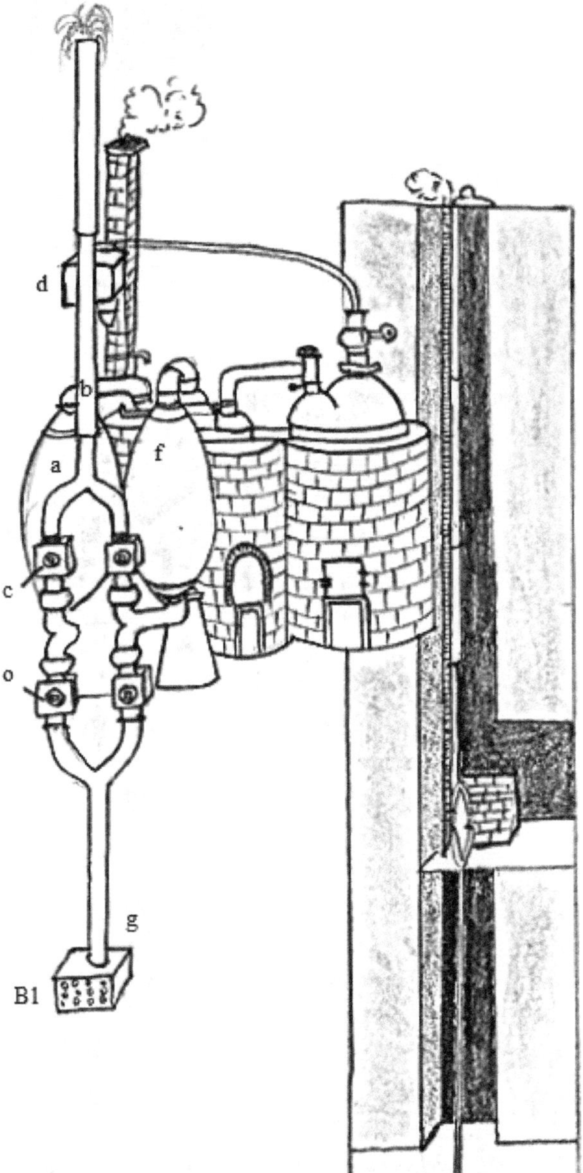

Fig. 6.6 Thomas Savery Steam Engine in action in a water-logged mine (Artist impression by Lakshmi)

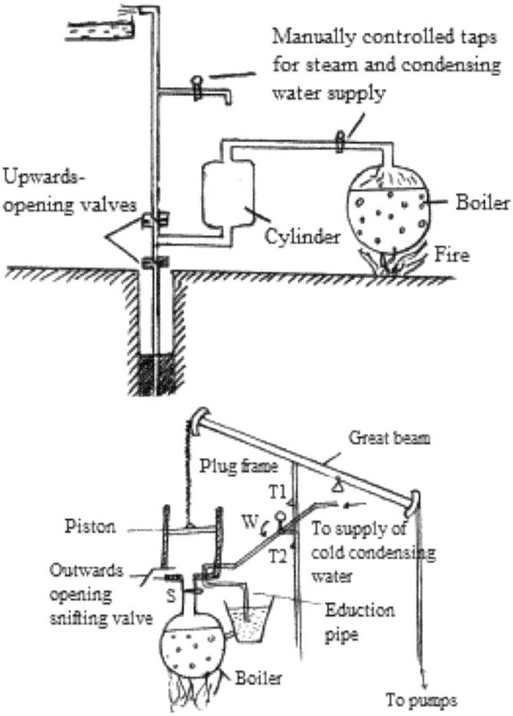

Fig. 6.7 Improvements in Newcomen Engine 1712 compared with Savery Engine (above). (Artist impression by Lakshmi)

fountains. When the piston reached the bottom of "b" a tap "j" on "c" was opened and the steam escaped. Because of the water pressure, "f" was then closed and "e" opened and more water poured in. "k" is a safety valve [5].

Savery's Vacuum pump installed in a water logged mine is illustrated in Figure 6.6. It had neither a piston nor a safety valve. Steam from the boiler was fed into a vessel "a" and the water in it was forced out through an ascending pipe "b" by way of check valve "c". When "a" had been emptied, the flow of steam was stopped and the vessel was cooled by means of cold water, which was sprayed over it from vessel "d". Since a vacuum was created when the steam was condensed, water was again sucked into "a" by way of a check valve "e". While "a" was being cooled, steam was fed into the water filled vessel "f", which was emptied, cooled and refilled with water "g".

Further developments, up until the appearance of James Watt are recently being questioned, e.g., Valenti [8] writes

> The early history of the invention of the steam engine shows without doubt that the British Royal Society, including Isaac Newton personally, deliberately prevented the industrial and naval applications of steam power for nearly 100 years. In fact, the Royal Society was so intent on burying Denis Papin's 1690 invention of a paddle-wheel-driven steamship, worked

Fig. 6.8 Newcomen Engine Installation, (Courtsey Dartmouth Directory Ltd)

> out in collaboration with Gottfried Wilhelm Leibniz, that it stole his work, and created a mythical story of how two British "Newtonian" heroes, Savery and Newcomen, invented the steam engine, for the sole purpose of raising water from coal mines – a myth that has persisted in the history books until today.

Be that as it may, we will proceed to discuss further developments in steam engines prior to the industrial revolution.

Thomas Newcomen (1663–1729) made explicit provision for the expulsion of air from the cylinder. The snifting valve in Figure 6.7 opened outwards in such a way that the rush of steam into the cylinder at the beginning of each cycle carried the accumulated air out with it through the valve. In this way once a cycle, the engine made a wheezing noise – like a man snifting with a cold – as it cleared itself of the air. The movement up and down of the plug frame, a long board hanging from the great beam, causes the tappet, or plug "T_1" set in it to trip the weight-operated valve, "W". When the plug frame moves in the opposite direction, another tappet, "T_2" resets the valve. A similar mechanism (not shown) controls the steam supply by means of the valve "S". The eduction pipe enables the condensed steam and the warmed condensing water to be returned via a well to the boiler, thus conserving heat (see also [4]).

Newcomen's steam engine was the first practical device to harness the power of steam to produce mechanical work. His first working engine was installed at a coal mine at Dudley Castle in Staffordshire in 1712. Such engines were used throughout England and Europe to pump water out of mines starting in the early 18th century, see Figure 6.8 and were the basis for James Watt's later improved versions [2]. Although Watt is far more famous today (largely due to Matthew Boulton's tireless salesmanship), Newcomen rightly deserves the majority of the credit for widespread introduction of steam power.

References

1. Balasubramaniam, R. et al. (2004) A Marvel of Medieval Indian Metallurgy: Thanjavur's Forge-Welded Iron Cannon, *Journal of Metallurgy*, January 1, p. 17.
2. Cardwell, D.S.L. (1971) *From Watt to Clausius*, Cornell University Press, Ithaca, NY.
3. Debus, A.G. (1978) *Man and Nature in the Renaissance*, Cambridge University Press, Cambridge.
4. Jenkins, Rhys (1936) Savery, Newcomen and the Early History of the Steam Engine, in *The Collected Papers of Rhys Jenkins*, Newcomen Society, Cambridge, p. 48.
5. McConnell, A. (2004) Papin, Denis (1647–1712?), *Oxford Dictionary of National Biography*, Oxford University Press.
6. von Guericke, Otto (1963) *Experimentanova Magdeburgica*, Amsterdam (1672), reprinted Otto Zeller, Aalen.
7. Savery, Thomas (1702) *The Miner's Friend: Or, an Engine to Raise Water by Fire*. S. Crouch.
8. Valenti, P. (1979) Leibniz, Papin and the Steam Engine: A Case Study of British Sabotage of Science, printed in *The American Almanac*, 1996; first version published in *Fusion Magazine*, December 1979.

Chapter 7
Industrial Revolution

Between 1780 and 1850, in a space of just seven decades, the face of England was changed by a far-reaching revolution, without precedent in the history of mankind.

Glasgow University had one of the Newcomen engines for its natural philosophy class. In 1763, one hundred years after the birth of Newcomen, this apparatus went out of order and Professor John Anderson gave James Watt (1736–1819) the opportunity to repair it. After the repair and while experimenting with it, Watt was struck by the enormous consumption of steam because, at every stroke, the cylinder and piston had to be heated to the temperature of boiling water and cooled again. This prevented the apparatus from making, with the available boiler capacity, more than a few strokes every minute. He quickly realized that wastage of steam was inherent in the design of the engine and became obsessed with the idea of finding some remedy. From the discovery of Joseph Black (1728–1799), he deduced that the loss of latent heat was the most serious defect in the Newcomen engine [2]. The work of James Watt [3] is thus the key application of science to engineering which led to the birth of the industrial revolution.

In 1765 he conceived the idea of a separate condensing chamber for the steam engine to separate the condensation system from the cylinder, injecting the cooling water spray in a second cylinder, connected to the main one [1]. When the piston had reached the top of the cylinder, the inlet valve was closed and the valve controlling the passage to the condenser was opened. External atmospheric pressure would then push the piston towards the condenser. Thus the condenser could be kept cold and under less than atmospheric pressure, while the cylinder remained hot. Important as the separate condenser idea was, in the fully developed version of 1775 that went into production, changes had to be more far-reaching. There was no spray, the condenser being immersed in a water tank and at each stroke the warm condensate was drawn off and sent up to a hot well by a vacuum pump which also helped to evacuate the steam from under the power cylinder. The still-warm condensate was recycled as feed water for the boiler.

James Watt's single acting pumping engine of 1788 is shown in Figure 7.1. This engine worked a crossbeam for pumping. The cylinder was closed (by a cap) and heated by a warm steam jacket. The condenser, positioned underground, was cooled

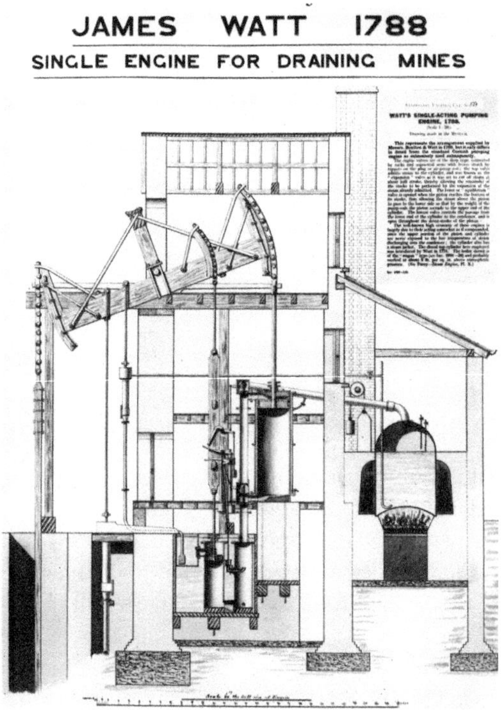

Fig. 7.1 James Watt's Steam Engines in 1788 - Pumping Engine. (Courtesy National Museum of Science & Industry)

and vacuum operated (by a pump). When the piston reached the top of its stroke the exhaust-valve opened and a partial vacuum was produced below the piston (inside the cylinder communicating to the condenser). Above the piston, at the same time, the entrance of steam helped the atmospheric pressure to drive the piston down. On this stroke the crossbeam raised water in the pump. When the piston reached the bottom of the stroke the inlet valve closed and an equilibrium valve opened to allow steam to pass from above to below the piston. The engine piston (now with the same pressure above and below) was driven up by the crossbeam and the descent of the very, very heavy pump piston and rod. Note the presence of the condenser and the warm steam jacket that surrounds the cylinder.

There was a heavy demand for an engine that could produce a rotary motion to drive factory machinery. The reciprocating engine that produced rotary motion revolutionalized the world; the first engine James Watt built in 1787–1788 is also shown in Figure 7.2. Reciprocating machinery has inherent disadvantages at high speeds, they have practically disappeared in the modern day world; there are still steam locomotives operating in a few places, e.g., Fairy Queen, the oldest running vintage steam locomotive in the world, built in the year 1855 by the British firm Kinston, Thompson & Hewitson for the British firm East India Railways, see Figure 7.3, and

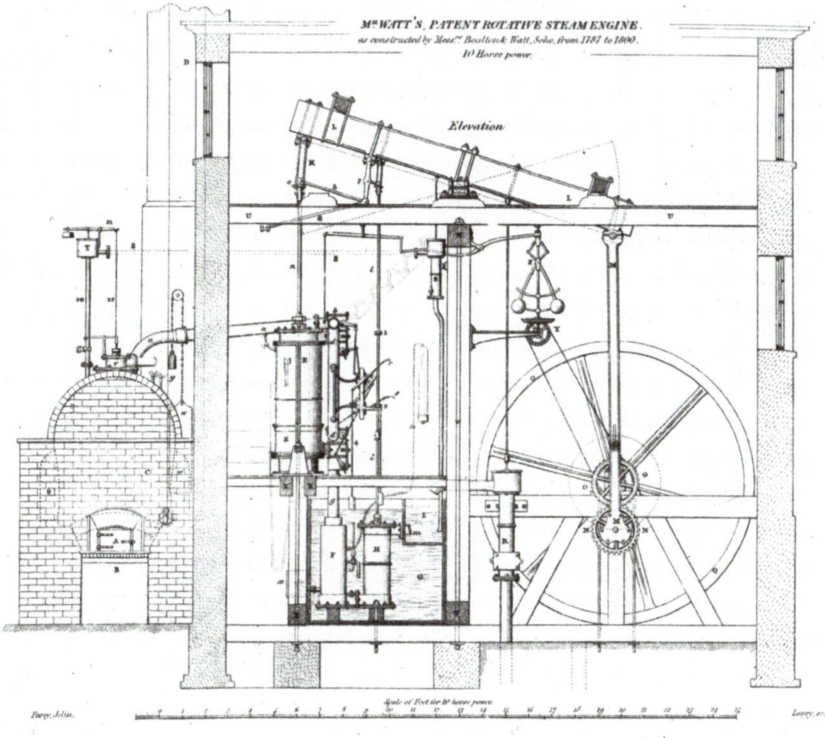

Fig. 7.2 James Watt's Steam Engines in 1788 – Engine with rotary motion for factories. (Courtesy National Museum of Science & Industry)

Fig. 7.3 Fairy Queen (http://www.visit-indya.com/tours/fairyqueen.html)

occasional reciprocating engines for producing small amounts of power in sugar mills, but they are gone. Internal combustion engines still thrive for transportation, power generation, and so on.

The internal combustion engines could run at higher speeds with a multi cylinder arrangement and reciprocating parts balancing. However, they are not vibration free engines. Though reciprocating machines are fascinating, they have no rotors directly and in the drive train too, the speeds are limited because of the reciprocating drive. Thus, not much attention was given to rotor dynamics during this era. We will study the growth of turbomachinery which necessitated the studies on rotor dynamics.

References

1. Dickinson, H.W. and Jenkins, R. (1989) *James Watt and the Steam Engine*, Encore Editions, London.
2. Ogg, David (1965) *Europe of the Ancient Regime: 1715–1783*. Harper & Row.
3. Rao, J.S. (1999) Watt – Two Hundred Years After His Retirement, in *10th World Congress on the Theory of Machines and Mechanisms*, Oulu, Finland, Vol. 1, p. 63.

Chapter 8
Turbomachines

During the 2nd century BC, Hero demonstrated the principle of a reaction turbine, but could not realize any useful work. Despite the scientific revolution followed by the industrial revolution, James Watt, while attempting to build a steam turbine, came to the conclusion that it could not be built given the state of contemporary technology.

About the year 1837 several reaction steam wheels were made by Avery at Syracuse, New York, and by Wilson at Greenock [3], for driving circular saws and cotton gins, see Figure 8.1. Steam was introduced into it through a hollow shaft, and, by the reaction of the jets at the extremities, caused rotation.

By the middle of the 19th century, there was a fair amount of understanding of the vibration characteristics of strings, beams, membranes and plates; however, nothing was known about a beam when it becomes a shaft and rotates. William John Macquorn Rankine (1820–1872) proposed in 1869 that a critical speed exists for a rotor which is the limit of speed for centrifugal whirling. It was not known whether a rotor can cross this limiting critical speed as Rankine proposed this as a limiting speed. Nearly two decades later in 1883 (about 100 years after Watt built his steam engine), De Laval of Stockholm undertook the problem with a considerable measure of success. Karl Gustaf Patrik de Laval (1845–1913) was a Swedish engineer and inventor who made important contributions to the design of steam turbines and dairy machinery. He built the first steam turbine (impulse turbine) [5], where high pressure steam was blown through nozzles whose inner shape "a" allowed the steam to expand to low pressure, see Figure 8.2. Its velocity was then greatly increased, and when the steam jets hit the turbine wheel's bent vanes "b", the wheel was set in motion. The slim, resilient turbine axle is mounted in its bearings 1 which were fixed in spherical segments 2 so that the axle could stand up to the whipping resonance vibrations.

Once a rotating machine was achieved with steam as motive force, there was a tremendous expansion in the capacity of power generation. Just one year after Laval's turbine, Charles Parsons [3] in 1884 came up with the first reaction turbine [5]. Sir Charles Algernon Parsons (1854–1931) was a British engineer, best known for his invention of the steam turbine (see Figure 8.3). Because of heavy vibrations

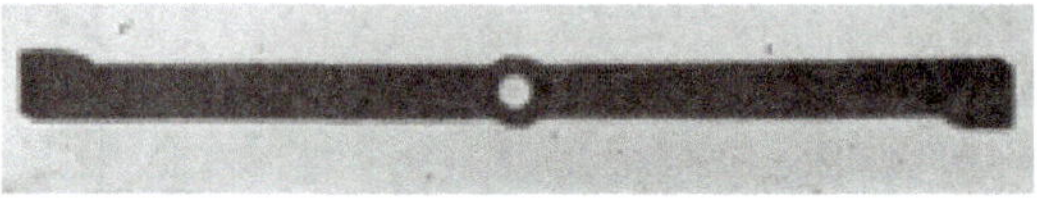

Fig. 8.1 Rotor of Avery's turbine. (Courtesy Cambridge University Press)

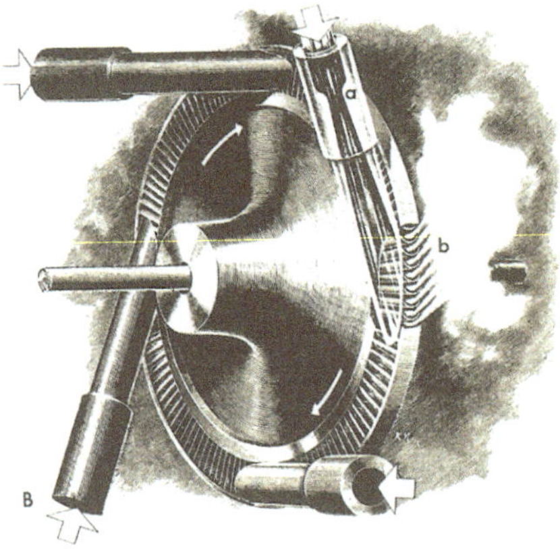

Fig. 8.2 Laval's turbine and critical speed. (Courtesy Cambridge University Press)

in reciprocating machines with severe torque and speed fluctuations, these turbines are hailed "vibration free engines" for that time at least.

Laval was also able to derive the whirl radius y in terms of shaft eccentricity δ, shaft stiffness F and weight W and explain how the rotor rotates smoothly beyond the critical speed, which Rankine thought 14 years ago that the critical speed is the limiting speed. These are the benefits from the Science revolution to speed up the industrial revolution.

With the invention of Dynamo in 1878 by Thomas Alva Edison and installation of Pearl Street Electric Power Station in 1882 (see Figure 8.4), the path has been cleared to produce electricity in an unprecedented scale which brought in a phenomenal expansion of the steam turbine; the early part of 20th century has seen 2 MW turbines, by 1920 the first 50 MW machine was made and by end of the II World War, 100 MW machines began to produce power. The capacity rose to 1000 MW by 1970 and in 1980 a single machine produced 1500 MW electricity (see Figure 8.5), see GE Steam Turbines, and *A Century of Progress* [1].

Heinkel He-178 was the world's first turbojet-powered aircraft in 1939, flying nearly two years before the British Gloster E28. It was powered by propellerless engine, a jet engine or gas turbine rather than a piston engine. Hans von Ohain (1911–1998) studied at University of Göttingen and when 22 years old he first conceived the idea of a continuous cycle combustion engine in 1933; he patented a jet propulsion engine design similar in concept to that of Frank Whittle (1907–1996) but different in internal arrangement in 1934. After receiving his degree in 1935, Ohain became the junior assistant of Robert Wichard Pohl, then director of the

Fig. 8.3 First Parson's turbine. (Courtesy Cambridge University Press)

Fig. 8.4 Edison's Jumbo Dynamo – a 27-ton machine that produced 100 kilowatts at Pearl Street Electric Power Station. (Courtesy Thomas Edison National Historical Park)

Physical Institute of the University. Hans von Ohain and Frank Whittle are both recognized as being the co-inventors of the jet engine. Each worked separately and knew nothing of the other's work. While Hans von Ohain is considered the designer of the first operational turbojet engine, Frank Whittle was the first to register a patent for the turbojet engine in 1930. Hans von Ohain was granted a patent for his turbojet engine later in 1936 [2]; however, Hans von Ohain's jet was the first to fly in 1939. Frank Whittle's jet first flew in 1941.

While working at the University, von Ohain met an automotive engineer, Max Hahn, and eventually arranged for him to build a model of his engine. When it was complete he took it to the University for testing, but ran into serious problems with combustion stability. Often the fuel would not burn inside the flame cans, and would instead be blown through the turbine where it would ignite in the air, shooting flames out the back and overheating the electric motor powering the compressor.

In February 1936, Pohl wrote to Ernst Heinkel telling him of the von Ohain design and its possibilities. Heinkel arranged a meeting where his engineers were

Fig. 8.5 GE's N series steam turbines for nuclear applications up to 1500 MW. (From Steam Brochure GE Energy, 4200 Wildwood Parkway, Atlanta, GA 30339)

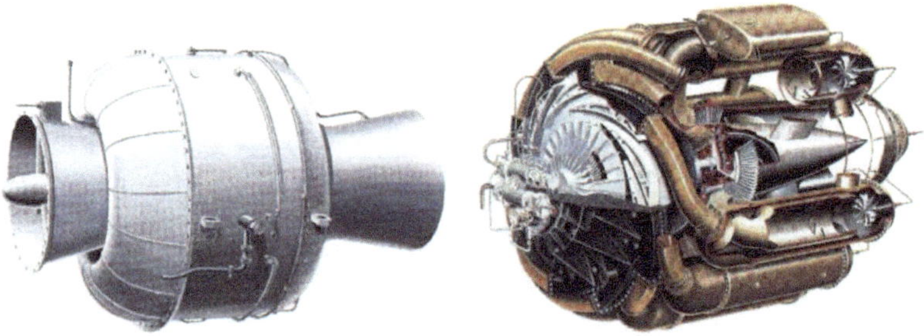

Fig. 8.6 Hans Von Ohain first jet engine (left) and Frank Whittle's jet engine (right)

able to grill von Ohain for hours, during which he flatly stated that the current "garage engine" would never work but there was nothing wrong with the concept as a whole. The engineers were convinced, and in April, von Ohain and Hahn were set up at Heinkel's works at the Marienehe airfield outside Rostock, Germany in Warnemünde.

Once moved, a study was made of the airflow in the engine, and several improvements made over a two month period. Much happier with the results, they decided to produce a completely new engine incorporating all of these changes, running on hydrogen gas. The resulting Heinkel-Strahltriebwerk 1 (HeS 1), German for Heinkel Jet Engine 1 (see Figure 8.6), was built by hand-picking some of the best machinists in the company, much to the chagrin of the shop-floor supervisors. Hahn, meanwhile, worked on the combustion problem, an area he had some experience in.

Frank Whittle was a Royal Air Force Officer, proposed in a thesis that planes would need to fly at high altitudes, where air resistance is much lower, in order to achieve long ranges and high speeds [6]. Piston engines and propellers were unsuitable for this purpose, so he concluded that rocket propulsion or gas turbines driving propellers would be required: jet propulsion was not in his thinking at this stage. In 1929, Whittle had considered using a fan enclosed in the fuselage to generate a fast

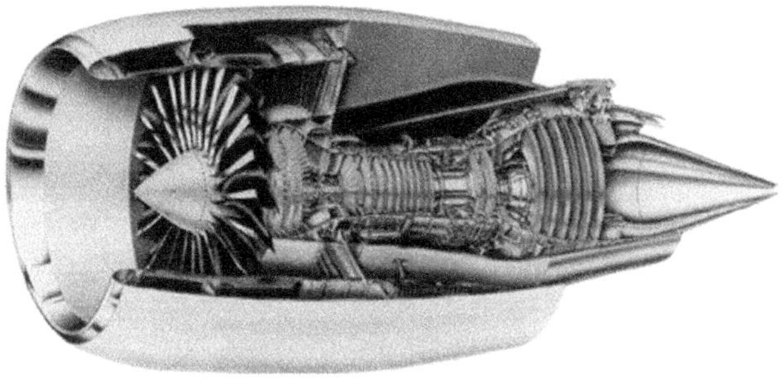

Fig. 8.7 GE 90 Aircraft Engine produced a record thrust of 120,316 pounds in 2001, 75 times that of W.2 in 1941. (Courtesy Dino of Turbokart)

flow of air to propel a plane at high altitude. A piston engine would use too much fuel, so he thought of using a gas turbine and patented his idea.

In 1935 Whittle secured financial backing and, with RAF approval, Power Jets Ltd was formed. They began constructing a test engine in July 1936, but it proved inconclusive. Whittle concluded that a complete rebuild was required, but lacked the necessary finances. Protracted negotiations with the Air Ministry followed and the project was secured in 1940. By April 1941 the engine W.2 (see Figure 8.6) was ready for tests and it produced 1600 lb thrust. The first flight Gloster E.28/39 took place on 15 May 1941. By October the Americans had heard of the project and asked for the details and an engine. A Power Jets team and the engine were flown to Washington to enable General Electric examine it and begin construction. The Americans worked quickly and their XP-59A Aircomet was airborne in October 1942, some time before the British Meteor, which became operational in 1944.

In just six decades later, the General Electric GE-90 115-B engine (see Figure 8.7) is designed for a thrust rating of 115,000 pounds (511 kN), making it the most powerful jet engine in the world [4]. It is produced by a tight-knit partnership of General Electric, Snecma Moteurs, FiatAvio and IHI, and is intended for Boeing's new longer-haul 777 versions, the 777-200LR (Long Range) and the 777-300ER (Extended Range).

Rolls-Royce developed its first three-shaft engine, the RB211, in the late 1960s/early 1970s for the Lockheed TriStar. It went on to power the Boeing 747 jumbo jet, Boeing 757 and Boeing 767 (see Figure 8.8). The Industrial version of the RB211 entered service in 1972.

There have been several challenges in the development of high speed rotating machinery, steam turbines, gas turbines, and radial and axial flow compressors, internal or external combustors the key elements that go in making the complete drive units during the 20th century. While slow speed machinery could be developed by design methodologies, with high-speed complex machinery precise estimation of several important state quantities are necessary to be determined.

When it came to development the first gas turbine, Frank Whittle faced immense challenges; we will mention a few here:

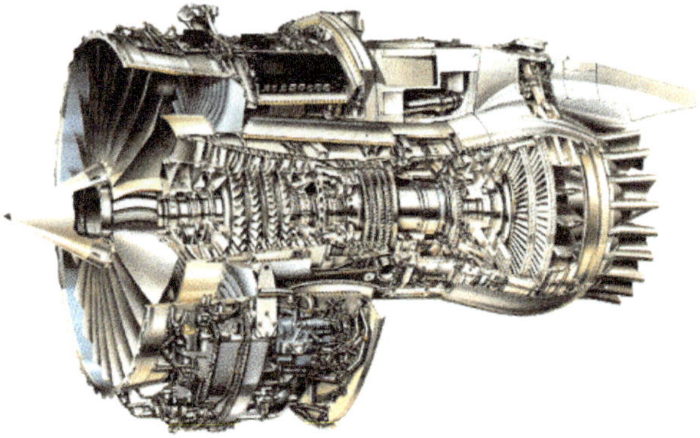

Fig. 8.8 Rolls Royce RB211-524 turbofan engine

- January 16, 1930 filed for a patent based on the principle of using a gas turbine for propulsion.
- October to December 1936 conducted several combustion experiments and detailed design was ready by end of 1936.
- First engine W.U. with a single combustor was tested on April 12, 1937; when the speed was raised from 2000 to 2500 RPM, the engine suddenly ran away; the problems were attributed to fuel system. Suspended tests August 23, 1937.
- New combustion tests October 22, 1937; throughout the rest of 1937 combustion testing and the first rebuild of the engine, W.U.
- April 29, 1938 the engine ran for 1 hour 45 minutes before failure attaining a thrust of 480 pounds at 13,000 RPM. Failure attributed to rubbing of the turbine nozzle assembly with the turbine wheel at high speed, causing severe overheating and failure of the turbine blades (even to day almost seven decades later turbine blade vibration failures is a dreaded phenomena for the designer and maintenance man).
- Whittle abandoned the single combustor for a series of ten small combustion chambers; May 30, 1938 reconstructed the experimental engine for a second time.
- July 19, 1938, Whittle's team concluded that the feasibility of jet propulsion for aircraft has been experimentally established for the first time; at 73.3% design speed quantitative verification of the engine has been obtained.
- October 26, 1938 testing on the third rebuild of the experimental engine.
- March 1939 14000 rpm maximum speed achieved; a turbine blade failure caused a two month delay.
- June 17, 1939 testing resumed with a new impeller; it had 29 blades instead of 30 to avoid resonant coupling with the 10 blade diffuser system; June 26 16000 rpm was reached.
- On February 22, 1941, the third rebuild was destroyed by a turbine failure after a total running of 170 hours.

Table 8.1 Significant events in the history of rotating machines

Year	Significant Event
2000000 BC	Beginning of Paleolithic or Old Stone Age – Dark Ages
13000 BC	Beginning of Mesolithic or Middle Stone Age - Earliest movement achieved to pull and push under log rollers
8000 BC	Beginning of Neolithic or New Stone Age – Agricultural villages
6000 BC	End of Stone Age – Pottery and pottery wheel
3000 BC	Wheel for transportation; sail ships
250 BC	Archimedes
200 BC	Hero's Aelopile – First reaction turbine
31 BC	Vitruvius and water wheel
700 AD	Wind mills
1543 AD	Copernicus and Scientific Revolution
1680 AD	Newton; Otto von Guericke
1690 AD	Papin; Savery; Newcomen
1750 AD	Euler
1780 AD	James Watt and Industrial Revolution
1883 AD	Laval and Parsons
	Edison, Pearl Street Station
1940 AD	Von Ohain and Frank Whittle
1980 AD	1500 MW steam turbine, cryogenic pumps accelerating to 100000 RPM in 4–5 seconds

- May 1941 the first engine shipped to Gloster and on 15th E28/39 took the flight for the first time.

Subsequently in May 1942, Whittle felt that the Ministry of Aircraft Production felt uneasy of the project as a whole and was in response to previous industry over-optimism; frequency of turbine blade failures was becoming the latest technological barrier to overcome. Rotors with mounted rotating parts continue to be the most stressed mechanical elements of all machinery and receive maximum attention in their design.

As a recap, the most significant events in the history of rotating machines are presented in Table 8.1.

It was almost two million years of existence of humans that can be described as Stone Age; the use of stone tools and the ability to walk straight up has set us apart from animal kingdom. Around 15000 years ago, the first breakthrough came in the Stone Age where man has learnt a faster way of moving things by inventing the sledge and roller. Though we are still dependent on human labor, the first step has been taken and ushered us into the Middle Stone Age from Old Stone Age. It is about 10000 years ago, we have settled down to live in a village rather than being totally nomadic and ushering in the New Stone Age. It took another 2000 years to invent the first wheel – pottery wheel for manufacturing applications. With this we have now learnt the ability to shape our own tools and the Stone Age has ended.

It is about 5000 years ago, we have transformed the log and pottery wheel to a wheel in transportation; bullocks and horses are used and we learnt to use animals extensively and replacing human work. Thus we began a new revolution with animal power to satisfy the growing human needs. Relieved from human labor for existence, we found that extra time to think about our surroundings; understanding physics, chemistry and biology – for example we postulated that the universe is made of five elements.

We began to notice that animal power relieved us from drudgery but it is not fast enough to deliver the human dreams and vision for energy. 2200 years ago we began new experimentation with Archimedean knowledge base; Hero made initial attempts to harness steam energy through a turbine. Subsequently, hydro energy began to be exploited in the form of water wheels; wind energy was exploited in the form of wind mills. No major breakthroughs came as we were shackled by religion and remained earth centric in our thinking. The first jolt came in 1543 AD, merely 450 years ago when we discovered that we are at the universe center and our understanding on earth and its position in the universe is totally wrong. It took us another 120 years, to understand basic laws of motion of bodies, a prerequisite to properly postulate motion and design machinery in a scientific manner. We have just about three centuries of knowledge base in our existence of more than two million years. We learnt around this time that atmosphere around us exerts pressure that can be utilized to derive motion of a body instead of using animal power. A century later we discovered that steam pressure can be used directly to give us more and efficient power rather than vacuum principles; this helped us in developing reciprocating steam engines on a firm footing and exploit the energy for our industrial needs; a rapid expansion took place ushering in Industrial Revolution. Reciprocating machines are not ideal for generating powers at the level that we are able to conceive our imagination so far and the dream of Hero in developing a rotating machine could be achieved only in the late 19th century. Tremendous strides in technology were made in the 20th century to improve the rotating machinery to the present day status in Power, Transportation, Oil and Gas, Space industries.

Looking at the history of rotating machinery, we might think that we now know a lot about them; it is probably right to say that we do not know still a lot about them and imagine what is in store for us in the next 100 years. We will now study the evolution of rotor dynamic analysis methods in 20th century in the next coming chapters.

References

1. *A Century of Progress. The General Electric Story 1876–1978* (1981) Hall of History, Schenectady, New York.
2. Conner, M. (2001) *Hans von Ohain: Elegance in Flight*, American Institute for Aeronautics and Astronautics, Inc., Reston, Virginia.
3. Parsons, C.A. (1911) *The Steam Turbine, The Rede Lecture*, Cambridge University Press.
4. Safran Group (2001) GE90-115B Exceeds 120,000 Pounds of Thrust during Test, http://www.le-webmag.com/article.php3?id_article=26&lang=en#.

5. Smil, V. (2005) *Creating the Twentieth Century: Technical Innovations of 1867–1914 and Their Lasting Impact*, Oxford University Press.
6. St. Peter, J. (1999) *The History of Aircraft Gas Turbine Engine Development in the United States? A Tradition of Excellence*, ASME Publication.

Chapter 9
Fundamentals of Elasticity

Fundamentals of Theory of Elasticity or physics of deformable bodies were established during the scientific revolution. Engineers would call these deformable bodies as structures. Hooke's law was discovered in 1660. Robert Hooke (1635–1703) was an English physicist. His important law of elasticity, known as Hooke's law (1660), states that the stretching of a solid is proportional to the force applied to it. He published his law in 1678 [4].

Cauchy generalized Hooke's law to three dimensional elastic bodies and stated that the six components of stress are linearly related to the six components of strain. Augustin Louis Cauchy (1789–1857), a French mathematician and military engineer, published his work in his memoire in 1822, see Cauchy (1822). Cauchy introduced the notion of stress at a point determined by the tractions per unit area across all plane elements through the point.

The behavior of deformable bodies is governed by 15 coupled partial differential equations, 3 equations of equilibrium, 6 strain displacement relations and 6 compatibility relations. With the help of these 15 partial differential equations, one can theoretically obtain the 15 unknowns, 3 displacements, 6 strains and 6 stresses that define the state of the deformable body. These equations are first derived by Navier [8]. Claude Louis Marie Henri Navier (1785–1836) was born in Dijon, France; introduced the basic principles of engineering science to a field that previously had been almost completely empirical. Cauchy in 1822 laid the basic foundations of theory of elasticity. The Navier–Cauchy equations are

Equations of equilibrium (3)

$$
\begin{aligned}
\frac{\partial \sigma_{xx}}{\partial x} + \frac{\partial \sigma_{yx}}{\partial y} + \frac{\partial \sigma_{zx}}{\partial z} + b_x = 0 \\
\frac{\partial \sigma_{xy}}{\partial x} + \frac{\partial \sigma_{yy}}{\partial y} + \frac{\partial \sigma_{zy}}{\partial z} + b_y = 0 \\
\frac{\partial \sigma_{xz}}{\partial x} + \frac{\partial \sigma_{yz}}{\partial y} + \frac{\partial \sigma_{zz}}{\partial z} + b_z = 0
\end{aligned}
\tag{9.1}
$$

Strain Displacement relations (6)

$$\varepsilon_{xx} = \frac{\partial u}{\partial x}, \quad \varepsilon_{yz} = \frac{1}{2}\left(\frac{\partial w}{\partial y} + \frac{\partial v}{\partial z}\right) = \varepsilon_{zy}$$

$$\varepsilon_{yy} = \frac{\partial v}{\partial y}, \quad \varepsilon_{zx} = \frac{1}{2}\left(\frac{\partial u}{\partial z} + \frac{\partial w}{\partial x}\right) = \varepsilon_{xz}$$

$$\varepsilon_{zz} = \frac{\partial w}{\partial z}, \quad \varepsilon_{xy} = \frac{1}{2}\left(\frac{\partial v}{\partial x} + \frac{\partial u}{\partial y}\right) = \varepsilon_{yx} \tag{9.2}$$

Compatibility relations (6)

$$\frac{\partial^2 \varepsilon_{xx}}{\partial y^2} + \frac{\partial^2 \varepsilon_{yy}}{\partial x^2} = 2\frac{\partial^2 \varepsilon_{xy}}{\partial x \partial y}, \quad \frac{\partial^2 \varepsilon_{xx}}{\partial y \partial z} = \frac{\partial}{\partial x}\left(-\frac{\partial \varepsilon_{yz}}{\partial x} + \frac{\partial \varepsilon_{zx}}{\partial y} + \frac{\partial \varepsilon_{xy}}{\partial z}\right)$$

$$\frac{\partial^2 \varepsilon_{yy}}{\partial z^2} + \frac{\partial^2 \varepsilon_{zz}}{\partial y^2} = 2\frac{\partial^2 \varepsilon_{yz}}{\partial y \partial z}, \quad \frac{\partial^2 \varepsilon_{yy}}{\partial z \partial x} = \frac{\partial}{\partial y}\left(\frac{\partial \varepsilon_{yz}}{\partial x} - \frac{\partial \varepsilon_{zx}}{\partial y} + \frac{\partial \varepsilon_{xy}}{\partial z}\right)$$

$$\frac{\partial^2 \varepsilon_{zz}}{\partial x^2} + \frac{\partial^2 \varepsilon_{xx}}{\partial z^2} = 2\frac{\partial^2 \varepsilon_{zx}}{\partial z \partial x}, \quad \frac{\partial^2 \varepsilon_{zz}}{\partial x \partial y} = \frac{\partial}{\partial z}\left(\frac{\partial \varepsilon_{yz}}{\partial x} + \frac{\partial \varepsilon_{zx}}{\partial y} - \frac{\partial \varepsilon_{xy}}{\partial z}\right) \tag{9.3}$$

This is quite a complex problem and even today we do not have a closed form solution of these equations even for simple structures. There are several textbooks that describe these equations, some of which include Dym and Shames [2], Sokolnikoff [10], Timoshenko and Goodier [11].

Therefore, elasticians have evolved approximate methods through energy principles to solve theory of elasticity problems. Energy principles are fundamental to the solution of all elasticity problems including those where the displacement field is time dependent (vibration problems). Initially these problems happened to be by and large beams, then plates and solid structures with intricate geometry. Again there are several texts that describe the energy methods in an elegant manner, some of them are: Fox [3], Lanczos [5], Langhaar [6], Love [7], Prescott [9], Washizu [12], Weinstock [13].

References

1. Cauchy, A.L. (1822) Memoir, communicated to Paris Academy.
2. Dym, C.L. and Shames I.H. (1973) *Solid Mechanics, A Variational Approach*, McGraw-Hill Book Co.
3. Fox, C. (1950) *An Introduction to the Calculus of Variations*, Oxford University Press.
4. Hooke, R. (1678) *De Potentia restitutiva*, London.
5. Lanczos, C. (1949) *The Variational Principle of Mechanics*, University of Toronto.
6. Langhaar, H.L. (1962) *Energy Methods Methods in Applied Mechanics*, John Wiley & Sons.
7. Love, A.E.H. (1944) *Mathematical Theory of Elasticity*, Dover.
8. Navier, Lois (1821) De l'équilibre et du mouvement des corps solides élastiques, Paper read to the Académie des Sciences, 14 May 1821.

9. Prescott, J. (1946) *Applied Elasticity*, Dover.
10. Sokolnikoff, I.S. (1956) *Mathematical Theory of Elasticity*, McGraw-Hill Book Co.
11. Timoshenko, S.P. and Goodier, J.N. (1951) *Theory of Elasticity*, McGraw-Hill Book Co.
12. Washizu, K. (1982) *Variational Methods in Elasticity and Plasticity*, Pergammon Press.
13. Weinstock, R. (1952) *Calculus of Variations with Applications to Physics and Engineering*, McGraw-Hill Book Co.

Chapter 10
Energy Methods

The fundamental principle in Physics is that the energy in the Universe is conserved; it can change in form but cannot be created or destroyed. Energy can be in various forms, important of these forms for vibration and rotor dynamics study is kinetic energy and potential energy (strain energy). In freely vibrating systems, these energies keep continuously change in these forms thus producing oscillatory motion for the mass.

Jakob (1655–1705) and Johann (Jean) Bernoulli (1667–1748) two brothers worked on a problem first discussed by Galileo (1564–1642) concerning finding the equation for the path followed by a particle from one point to another in the shortest time, if the particle is acted upon by gravity alone [14]. Johann proposed Cycloid and Jakob challenged this solution. This dispute led to the development of Calculus of Variations by Isaac Newton (1642–1727). This subject Calculus of Variations forms the basis of all vibration studies and optimization.

At this time, there were two contending views in mechanics regarding the conservation of energy, viz., Conservation of momentum and Conservation of the *vis viva*, or living force, a concept developed by Gottfried Wilhlem von Leibniz (1646–1716). The Leibnizian concept prevailed, eventually having its original term replaced by the modern kinetic energy, defined as the energy a body possesses by virtue of its motion. As a Leibnizian, Bernoulli ardently supported the *vis viva*, fully accepting the idea put forth by Leibniz's student Christian Wolff, who postulated the universal validity of the conservation of the living force. Agreeing with Wolff, Bernoulli identified the living force as one of the fundamental principles in mechanics in his essay "De vera notione virium vivarum" in 1735. Johann Bernoulli is perhaps the first to realize the importance of the principle of conservation [18].

Joseph Lagrange (1736–1813) transformed Newtonian mechanics into a branch of analysis, Lagrangian mechanics as it is now called, and exhibited the so-called mechanical "principles" as simple results of the variational calculus. It was developed by Swiss mathematician Leonhard Euler (1707–1783) and Lagrange developed Euler–Lagrange equation, or Lagrange's equation which is a differential equation whose solutions are the functions for which a given functional is stationary.

William Rowan Hamilton (1805–1865) formulated the principle of stationary action, called Hamilton's principle. It states that the dynamics of a physical system is determined by a variational problem for a functional based on a single function, the Lagrangian, which contains all physical information concerning the system and the forces acting on it. The variational problem is equivalent to and allows for the derivation of the differential equations of motion of the physical system. In Lagrangian mechanics, because of Hamilton's principle of stationary action, the evolution of a physical system is described by the solutions to the Euler–Lagrange equation for the action of the system. In classical mechanics, this is equivalent to Newton's laws of motion, but it has the advantage that it takes the same form in any system of generalized coordinates, and it is better suited to generalizations; thus most preferred in the Theory of Vibrations.

10.1 Euler–Lagrange Equations

Let us begin with Variational Calculus by considering a functional that would represent a beam problem:

$$I = \int_{x_1}^{x_2} F(x, y, y', y'')dx \tag{10.1}$$

where x is the independent variable in the range x_1 to x_2 with y as an admissible path (solution) and its derivatives with x. For y to be an extremizing path, we consider varied path defined by

$$\breve{y} = y + \varepsilon\eta \tag{10.2}$$

where ε is a small parameter and η is a differentiable function satisfying

$$\eta(x_1) = \eta(x_2) = 0 \tag{10.3}$$

The functional in (10.1) over the varied path is now written as

$$\begin{aligned} \breve{I} &= \int_{x_1}^{x_2} F(x, \breve{y}, \breve{y}', \breve{y}'')dx \\ &= \int_{x_1}^{x_2} F(x, y + \varepsilon\eta, y' + \varepsilon\eta', y'' + \varepsilon\eta'')dx \end{aligned} \tag{10.4}$$

We adopt the same strategy as in Calculus and expand the above functional as

$$\breve{I} = (\breve{I})_{\varepsilon=0} + \left(\frac{\partial \breve{I}}{\partial \varepsilon}\right)_{\varepsilon=0} \varepsilon + \left(\frac{\partial^2 \breve{I}}{\partial \varepsilon^2}\right)_{\varepsilon=0} \frac{\varepsilon^2}{2!} + \cdots \tag{10.5}$$

$$\breve{I} - (\breve{I})_{\varepsilon=0} = \breve{I} - I = \left(\frac{\partial \breve{I}}{\partial \varepsilon}\right)_{\varepsilon=0} \varepsilon + \left(\frac{\partial^2 \breve{I}}{\partial \varepsilon^2}\right)_{\varepsilon=0} \frac{\varepsilon^2}{2!} + \cdots \tag{10.5a}$$

We readily recognize that

$$\left(\frac{\partial \breve{I}}{\partial \varepsilon}\right)_{\varepsilon=0} = 0 \tag{10.6}$$

For y in (10.2) to be an extremizing path for the functional in (10.1), we have

$$\begin{aligned}
\left(\frac{\partial \breve{I}}{\partial \varepsilon}\right)_{\varepsilon=0} &= \left(\frac{d}{d\varepsilon}\int_{x_1}^{x_2} F\left(x, \breve{y}, \breve{y}', \breve{y}''\right) dx\right)_{\varepsilon=0} \\
&= \left\{\int_{x_1}^{x_2}\left(\frac{\partial F}{\partial \breve{y}}\eta + \frac{\partial F}{\partial \breve{y}'}\eta' + \frac{\partial F}{\partial \breve{y}''}\eta''\right) dx\right\}_{\eta=0} \\
&= \int_{x_1}^{x_2}\left(\frac{\partial F}{\partial y}\eta + \frac{\partial F}{\partial y'}\eta' + \frac{\partial F}{\partial y''}\eta''\right) dx = 0
\end{aligned} \tag{10.7}$$

Integrating by parts

$$\int_{x_1}^{x_2} \frac{\partial F}{\partial y'}\eta' dx = \left(\frac{\partial F}{\partial y'}\eta\right)_{x_1}^{x_2} - \int_{x_1}^{x_2} \frac{d}{dx}\left(\frac{\partial F}{\partial y'}\right)\eta dx$$

$$\begin{aligned}
\int_{x_1}^{x_2} \frac{\partial F}{\partial y''}\eta'' dx &= \left(\frac{\partial F}{\partial y''}\eta'\right)_{x_1}^{x_2} - \int_{x_1}^{x_2} \frac{d}{dx}\left(\frac{\partial F}{\partial y''}\right)\eta' dx \\
&= \left(\frac{\partial F}{\partial y''}\eta'\right)_{x_1}^{x_2} - \frac{d}{dx}\left(\frac{\partial F}{\partial y''}\eta\right)_{x_1}^{x_2} + \int_{x_1}^{x_2} \frac{d^2}{dx^2}\left(\frac{\partial F}{\partial y''}\right)\eta dx
\end{aligned} \tag{10.7a}$$

Substituting in (10.7)

$$\begin{aligned}
&\int_{x_1}^{x_2}\left[\frac{d^2}{dx^2}\left(\frac{\partial F}{\partial y''}\right) - \frac{d}{dx}\left(\frac{\partial F}{\partial y'}\right) + \frac{\partial F}{\partial y}\right]\eta dx \\
&\quad + \left(\frac{\partial F}{\partial y''}\eta'\right)_{x_1}^{x_2} + \left(\left\{-\frac{d}{dx}\left(\frac{\partial F}{\partial y''}\right) + \frac{\partial F}{\partial y'}\right\}\eta\right)_{x_1}^{x_2} = 0
\end{aligned} \tag{10.8}$$

To satisfy equation (10.8) here are two possibilities:

1. η and $\eta' = 0$ at x_1 and x_2. then we get Euler–Lagrange equation

$$\int_{x_1}^{x_2}\left[\frac{d^2}{dx^2}\left(\frac{\partial F}{\partial y''}\right) - \frac{d}{dx}\left(\frac{\partial F}{\partial y'}\right) + \frac{\partial F}{\partial y}\right]\eta dx = 0$$

$$\frac{d^2}{dx^2}\left(\frac{\partial F}{\partial y''}\right) - \frac{d}{dx}\left(\frac{\partial F}{\partial y'}\right) + \frac{\partial F}{\partial y} = 0 \tag{10.9}$$

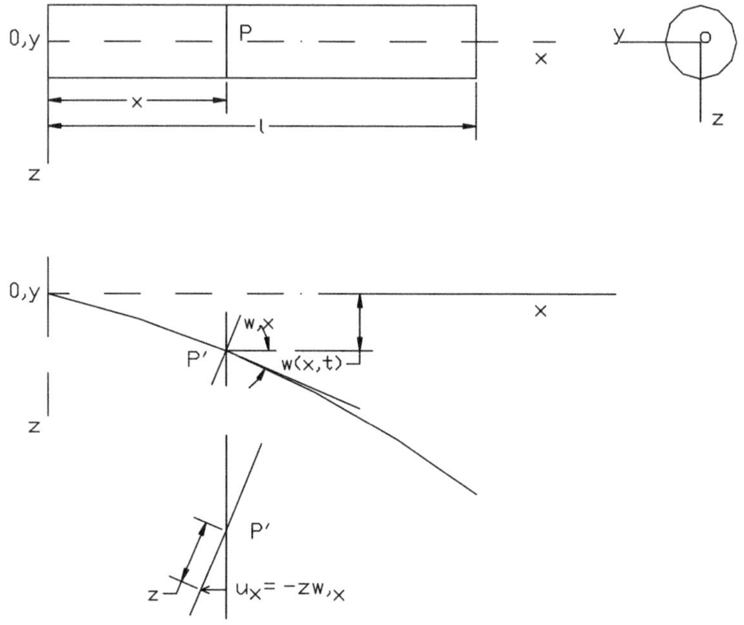

Fig. 10.1 Cantilever beam idealized according to Euler–Bernoulli

2. η and η' are prescribed at x_1 and x_2. Then we get the boundary conditions

$$\left(\frac{\partial F}{\partial y''}\right)_{x_1}^{x_2} = 0$$

$$\left(-\frac{d}{dx}\left(\frac{\partial F}{\partial y''}\right) + \frac{\partial F}{\partial y'}\right)_{x_1}^{x_2} = 0 \tag{10.10}$$

We will illustrate how Euler–Lagrange equations are set up for simplest possible structure, a one dimensional idealization of a cantilever beam as proposed by Euler and Bernoulli in 1750. Figure 10.1 shows this idealization.

The beam is of length l with a cross-sectional area A. We recognize only the elastic axis about which the beam bends with a cross-section P at distance x having lateral deflection w remaining plane before and after but rotating through the angle given by the slope $w_{,x}$. Notice that the deflection is a function of space coordinate x. For a vibrating beam, it will also be a function of time t. The cross-sectional properties required enter through the derivation. The rotation of the plane cross-section gives rise to another displacement in x direction at distance z from P'. Then in accordance to the Euler–Bernoulli assumption, the displacement field is approximated as

$$u_x = zw_{,x}$$
$$u_y = 0$$
$$u_z = w(x,t) \tag{10.11}$$

In the above assumption, we have violated the Theory of Elasticity and simplified the structure. The greatness of Euler and Bernoulli lies in converting the complex problem of solving 15 coupled partial differential equations to a simple problem and yet obtain a solution which is good enough for engineering applications. We can refer to this approach as Strength of Materials that allowed us to make structural designs.

We can now use the strain displacement relations (9.2)

$$\varepsilon_{xx} = \frac{\partial u}{\partial x}, \quad \varepsilon_{yz} = \frac{1}{2}\left(\frac{\partial w}{\partial y} + \frac{\partial v}{\partial z}\right) = \varepsilon_{zy}$$
$$\varepsilon_{yy} = \frac{\partial v}{\partial y}, \quad \varepsilon_{zx} = \frac{1}{2}\left(\frac{\partial u}{\partial z} + \frac{\partial w}{\partial x}\right) = \varepsilon_{xz}$$
$$\varepsilon_{zz} = \frac{\partial w}{\partial z}, \quad \varepsilon_{xy} = \frac{1}{2}\left(\frac{\partial v}{\partial x} + \frac{\partial u}{\partial y}\right) = \varepsilon_{yx} \tag{9.2}$$

and obtain a simplified field

$$\varepsilon_{xx} = z\frac{\partial^2 w}{\partial x^2} \tag{10.12}$$

Notice that only one normal strain along the x axis exists; the rest are made zero by not taking into account cognizance of Poisson's ratio.

The stress field is now obtained from

$$\sigma_{ij} = \lambda\delta_{ij}\varepsilon_{ii} + 2G\varepsilon_{ij} \tag{10.13}$$

Likewise we assume that there is only one normal stress and no other stresses including any shear stress. Since Poisson's ratio is not accounted for, $\lambda + 2G \to E$, therefore

$$\sigma_{xx} = E\varepsilon_{xx} = -Ez\frac{\partial^2 w}{\partial x^2} \tag{10.14}$$

Let the beam cross-section be uniform rectangular with breadth b and thickness t. The strain energy in the beam according to the stress-strain field above is

$$U = \frac{1}{2}\iint\int_v \sigma_{ij}\varepsilon_{ij}dv = \frac{1}{2}b\int_0^L\int_{-\frac{1}{2}t}^{\frac{1}{2}t} \sigma_{xx}\varepsilon_{xx}dzdx \tag{10.15}$$

Work W of loading $q(x)$ assumed to act on the Centroidal axis is

$$W = \frac{1}{2}\int_0^L q(x)w(x)dx \tag{10.16}$$

Total potential energy π is

$$\pi = U - W = \frac{1}{2}b\int_0^L \int_{-\frac{1}{2}t}^{\frac{1}{2}t} \sigma_{xx}\varepsilon_{xx}dzdx - \frac{1}{2}\int_0^L q(x)\,w(x)\,dx$$

$$= \frac{1}{2}b\int_0^L \int_{-\frac{1}{2}t}^{\frac{1}{2}t} \left(Ez^2\right)\left(\frac{d^2w}{dx^2}\right)^2 dzdx - \frac{1}{2}\int_0^L q(x)\,w(x)\,dx$$

$$= \int_0^L \left[\frac{1}{2}EI\left(\frac{d^2w}{dx^2}\right)^2 - qw\right]dx = \int_0^L Fdx$$

$$\text{where } F \text{ is a functional } = \frac{1}{2}EI\left(\frac{d^2w}{dx^2}\right)^2 - qw \tag{10.17}$$

In the above I is the second moment of area of the cross-section about the yy axis. The Euler–Lagrange equation derived from the principle of virtual work is

$$-\frac{d^2}{dx^2}\left(\frac{\partial F}{\partial w''}\right) + \frac{d}{dx}\left(\frac{\partial F}{\partial w'}\right) - \left(\frac{\partial F}{\partial w}\right) = 0 \tag{10.18}$$

where

$$F = \frac{1}{2}EI\left(\frac{d^2w}{dx^2}\right)^2 - qw$$

and $'$ represents the derivative with respect to the independent variable x.

The derivatives of the above functional are

$$\frac{\partial F}{\partial w''} = EI\left(\frac{d^2w}{dx^2}\right)$$

$$\frac{\partial F}{\partial w'} = 0$$

$$\frac{\partial F}{\partial w} = -q$$

Substituting in (10.18) we have the governing differential equation

$$\frac{d^2}{dx^2}\left[EI\left(\frac{d^2w}{dx^2}\right)\right] = q \tag{10.19}$$

The boundary conditions from (10.10) are given by

$$\left(\frac{\partial F}{\partial w''}\right)_{x_1}^{x_2} = \left\{EI\left(\frac{d^2w}{dx^2}\right)\right\}_{x_1}^{x_2} = 0$$

$$\left(-\frac{d}{dx}\left(\frac{\partial F}{\partial w''}\right) + \frac{\partial F}{\partial w'}\right)_{x_1}^{x_2} = \left[-\frac{d}{dx}\left\{EI\left(\frac{d^2w}{dx^2}\right)\right\}\right]_{x_1}^{x_2} = 0 \quad (10.20)$$

As we know, the first condition above refers to the bending moment of the beam and the second condition gives the shear force.

The variational calculus formulation leads us to the governing differential equation and boundary conditions. In several structures it is difficult sometimes to visualize all boundary conditions using Newtonian approach, whereas the energy approach fixes up the system completely. That is a major advantage.

The next step is to obtain a solution of the differential equation above, which is the simplest structure. It is feasible in very special case even for these simplest structures, e.g., a uniform cross-section beam. But that's not what engineers want; we need solutions for realistic structures. The beauty of energy methods is it leads to solution procedures as well. We will look at these methods and see how they evolved historically.

10.2 Lagrange Method

In the energy method approach, the solution of the equation is always assumed that satisfies at least the kinematic boundary conditions. For cantilever boundary conditions, we can assume the solution in polynomial form with $X = x/L$ that satisfies the boundary conditions in (10.20):

$$w(X) = a(6X^2 - 4X^3 + X^4)$$
$$\frac{d^2w}{dX^2} = \frac{a}{L^2}(12 - 24X + 12X^2) \quad (10.21)$$

The total potential energy of the beam is

$$\begin{aligned}\pi &= \frac{1}{2}\int_0^L EI\left(\frac{d^2w}{dx^2}\right)^2 dx - \int_0^L q(x)\,w(x)\,dx \\ &= \frac{1}{2}\frac{EI}{L^3}\int_0^1\left(\frac{d^2w}{dX^2}\right)^2 dX - L\int_0^1 q(X)a(6X^2 - 4X^3 + X^4)dX \\ &= \frac{1}{2}\frac{EI}{L^3}\int_0^1\left[a(12 - 24X + 12X^2)\right]^2 dX \\ &\quad - L\int_0^1 q(X)\,a(6X^2 - 4X^3 + X^4)dX \end{aligned} \quad (10.22)$$

Minimizing total potential energy, i.e., $\partial \pi / \partial a = 0$ and carrying out the definite integrals, we get

$$\frac{144}{5} \frac{EI}{L^3} a = \frac{6}{5} qL$$

$$\therefore a = \frac{1}{24} \frac{L^4}{EI} q$$

Hence the deflection in the beam from (10.21) is

$$w(x) = \frac{qL^4}{24EI} (6X^2 - 4X^3 + X^4)$$

Maximum deflection occurs at $x = L$, i.e., $X = 1$ given by

$$w_{x=1} = \frac{1}{8} \frac{qL^4}{EI}$$

This agrees with the exact solution.

10.3 Rayleigh's Energy Approach

Rayleigh gave an approximate method based on the energy conservation principle in 1877 [32]. For a conservative system vibrating in simple harmonic motion, the maximum potential energy that occurs when the system is at maximum displacement is equal to the maximum kinetic energy when the system passes through its mean equilibrium position with maximum velocity. Here again, the deflection shape (mode shape) is assumed.

The potential energy in bending of a beam is given by

$$U = \int_0^L \left[\frac{1}{2} EI \left(\frac{d^2 w}{dx^2} \right)^2 \right] dx \tag{10.23}$$

Let m be the mass per unit length, then the kinetic energy of the beam is

$$T = \int_0^L \left[\frac{1}{2} m \left(\frac{dw}{dt} \right)^2 \right] dx \tag{10.24}$$

Also let the beam vibrate in simple harmonic motion

$$w(x, t) = w(x) \sin \omega t$$

$$\frac{\partial w}{\partial t} = \omega w(x) \cos \omega t \tag{10.25}$$

The maximum kinetic energy in the beam is

$$T_{\max} = \frac{1}{2}\omega^2 \int_0^L m[w(x)]^2 dx \tag{10.26}$$

The maximum potential energy is

$$U_{\max} = \int_0^L \left[\frac{1}{2} EI \left(\frac{d^2 w}{dx^2}\right)^2\right] dx \tag{10.27}$$

Equating,

$$\omega^2 = \frac{\int_0^L \left[EI \left(\frac{d^2 w}{dx^2}\right)^2\right] dx}{\int_0^L m\,[w\,(x)]^2\, dx} \tag{10.28}$$

Rayleigh was looking for a method of calculating fundamental natural frequencies without adopting Euler and Lagrange approaches so that industry can adopt a tabular method. This was the first time industry looked into routine numerical operations so that the designers could be trained to turn out designs faster with simple checking rather than highly mathematical approaches. Thus Rayleigh may be given the credit for making highly complex mathematics into industry user friendly methods.

Stodola [38] and Kearton [21], among others, illustrated these methods at the beginning of the 20th century. We illustrate here an example given by Scanlan and Rosenbaum [37] for a cantilever beam.

Here we choose the deflection curve to be

$$y = y_0 \left(1 - \cos\frac{\pi x}{2l}\right) \tag{10.29}$$

where y_0 is the tip deflection. Then from (10.28),

$$\omega^2 = \frac{E\left(\frac{\pi}{2L}\right)^2 \int_0^L \left[I\left(\cos\frac{\pi x}{2L}\right)^2\right] dx}{\int_0^L m\left(1 - \cos\frac{\pi x}{2L}\right)^2 dx} \tag{10.30}$$

$$\omega = \frac{7.5\pi}{L^2}\sqrt{E}\sqrt{\frac{\int_0^L \left[I\left(\cos\frac{\pi x}{2L}\right)^2\right] dx}{\int_0^L m\left(1 - \cos\frac{\pi x}{2L}\right)^2 dx}}\ \text{cpm} \tag{10.31}$$

Note that, invariably, gravitational units were adopted after the industrial revolution in English speaking countries and Foot, Pounds were used in the British system called FPS. We will retain the same units here. The length of cantilever is 240 in and Young's modulus is 10.3×10^6 lb/in^2. Equation (10.31) is converted to a tabular method that can be performed by regular staff without knowing the energy methods and solutions of differential equations. Table 10.1 gives these numerical steps.

Table 10.1 Numerical calculations for Rayleigh's energy approach

Interval	x/l midpoint	$m(x)$ $\frac{\text{lb}}{386}/\text{in}$	$I(x)$ in^4	$\left(\frac{\pi x}{2l}\right)\times\left(\frac{180}{\pi}\right)$ degrees	$\cos\frac{\pi x}{2l}$	$\cos^2\frac{\pi x}{2l}$	$1-\cos\frac{\pi x}{2l}$	$\left(1-\cos\frac{\pi x}{2l}\right)^2$	$I\cos^2\frac{\pi x}{2l}$	$m\left(1-\cos\frac{\pi x}{2l}\right)^2$
1	0.05	0.0389	456	4.5	0.9969	0.995	0.0031	0.00001	454	0
2	0.15	0.0375	408	13.5	0.9724	0.945	0.0276	0.00076	386	0.000029
3	0.25	0.0337	360	22.5	0.9239	0.853	0.0761	0.00578	307	0.000195
4	0.35	0.0252	312	31.5	0.8526	0.726	0.1474	0.0218	227	0.000549
5	0.45	0.0124	264	40.5	0.7604	0.578	0.2396	0.0574	153	0.000712
6	0.55	0.0068	216	49.5	0.6495	0.422	0.3505	0.1220	91	0.000830
7	0.65	0.0054	168	58.5	0.5225	0.273	0.4775	0.2880	46	0.001231
8	0.75	0.0046	120	67.5	0.3827	0.147	0.6173	0.3820	18	0.001757
9	0.85	0.0038	72	76.5	0.2335	0.0545	0.7665	0.5870	4	0.002231
10	0.95	0.0018	24	85.5	0.07846	0.00615	0.9215	0.8500	-	0.001530
Σ↓						4.99965		2.25475	1686	0.009064

$$\omega=\frac{7.5\pi}{l^2}\sqrt{10.3\times10^6}\sqrt{\frac{1686}{0.009064}}=566\text{ cpm}$$

10.4 Ritz Method

The Ritz method [35] can be summarized as follows. It uses Rayleigh's average energy or maximum energy principle. For harmonic motion with frequency p, we can average the kinetic energy T in a system as $\bar{T}$

$$\bar{T}=\int_0^{2\pi/p} T dt \tag{10.32}$$

Similarly the potential energy U is also averaged as $\bar{U}$

$$\bar{U}=\int_0^{2\pi/p} U dt \tag{10.33}$$

As in the Lagrange method, we assume a shape function for the deflection y of the beam; here a polynomial is chosen

$$y=\sum_{i=1}^{n} A_i f_i(Z) q_i(t) \tag{10.34}$$

The above averaging process allows us to get rid of one independent parameter t. Then the Lagrangian will be a function of the arbitrary parameters in the assumed mode shapes,

$$L=L(A_i) \tag{10.35}$$

In the Ritz method, we minimize the Lagrangian with respect to A_i

$$\frac{\partial L}{\partial A_i}=0 \tag{10.36}$$

The Ritz method begins with an averaging process as in Rayleigh's method; the procedure is also called the Rayleigh–Ritz method or the Ritz averaging method. Since the method is not derived after an extremization process in the variational principle, the question of natural boundary conditions is not part of this procedure. If one uses Euler–Lagrangian equations, it is necessary that we satisfy the natural boundary conditions that arise out of the variational process. However, we should always satisfy the kinematic conditions at the boundaries. This is a major advantage of the Rayleigh–Ritz method.

Let $f_i(Z)$ be

$$f_i\,(Z) = A_{i-1}Z^{i-1} + A_i Z^i + A_{i+1}Z^{i+1} + A_{i+2}Z^{i+2} + A_{i+3}Z^{i+3} \qquad (10.37)$$

For a cantilever, the boundary conditions are

$$\begin{aligned} y = 0, \quad y' = 0 \quad \text{at } Z = 0 \\ y'' = 0, \quad y''' = 0 \quad \text{at } Z = 1 \end{aligned} \qquad (10.38)$$

From the boundary condition at $Z = 0$, the shape function (10.35) reduces to

$$f_i(Z) = A_{i+1}Z^{i+1} + A_{i+2}Z^{i+2} + A_{i+3}Z^{i+3} \qquad (10.39)$$

Taking derivatives of (10.39) with respect to Z, we have

$$\begin{aligned} f_i''(Z) &= A_{i+1}i(i+1)Z^{i-1} + A_{i+2}(i+1)(i+2)Z^i + A_{i+3}(i+2)(i+3)Z^{i+1} \\ f_i'''(Z) &= A_{i+1}(i-1)i(i+1)Z^{i-2} + A_{i+2}i(i+1)(i+2)Z^{i-1} \\ &\quad + A_{i+3}(i+1)(i+2)(i+3)Z^i \end{aligned} \qquad (10.40)$$

Substituting the above in the second boundary condition at $Z = 1$, we get

$$\begin{aligned} &\begin{bmatrix} i(i+1) & (i+1)(i+2) \\ (i-1)i(i+1) & i(i+1)(i+2) \end{bmatrix} \begin{Bmatrix} A_{i+1} \\ A_{i+2} \end{Bmatrix} \\ &= -A_{i+3} \begin{Bmatrix} (i+2)(i+3) \\ (i+1)(i+2)(i+3) \end{Bmatrix} \end{aligned} \qquad (10.41)$$

Solving we get

$$\begin{aligned} A_{i+1} &= A_{i+3}\frac{(i+2)(i+3)}{i(i+1)} \\ A_{i+2} &= A_{i+3}\frac{-2(i+3)}{(i+1)} \end{aligned} \qquad (10.42)$$

Using the above in equation (10.39) and writing A_{i+3} as A_i, we can write the polynomial series for the assumed shape function as

$$f_i(Z) = \sum A_i \left[(i+2)(i+3)Z^{i+1} - 2i(i+3)Z^{i+2} + i(i+1)Z^{i+3}\right] \quad (10.43)$$

The above represents a generalized polynomial function with as many terms as one wishes to taken in the analysis. If we take only one term, as we did in the illustration by Lagrange's method, for free vibrations we can write

$$\begin{aligned} w\,(Z,t) &= \hat{w}\sin pt \\ \dot{w} &= \hat{w}p\cos pt \\ w'' &= \hat{w}''\sin pt \end{aligned} \quad (10.44)$$

If we use maximum energy values in our formulation, the kinetic energy is

$$\hat{T} = \frac{1}{2}ml\int_0^1 \hat{w}^2 p^2 dZ \quad (10.45)$$

Similarly, maximum potential energy is

$$\hat{U} = \frac{1}{2}\frac{EI}{l^3}\int_0^1 \hat{w}''^2 p^2 dZ \quad (10.46)$$

Now, the Lagrangian function is

$$\begin{aligned} L &= \hat{T} - \hat{U} \\ &= \frac{1}{2}ml\int_0^1 \hat{w}^2 p^2 dZ - \frac{1}{2}\frac{EI}{l^3}\int_0^1 \hat{w}''^2 p^2 dZ \end{aligned} \quad (10.47)$$

$$w\,(Z,t) = (6Z^2 - 4Z^3 + Z^4)q(t) \quad (10.21)$$

Until the advent of computers in the second half of the 20th century, only one term is used in the analysis. Otherwise the same analysis based on energy principles continues to be the cornerstone of modern numerical solutions for vibration problems. For the present, continuing with one term approximation of (10.21) the Lagrangian function is

$$\begin{aligned} L = &\frac{1}{2}ml\int_0^1 A^2\left(6Z^2 - 4Z^3 + Z^4\right)^2 p^2 dZ \\ &- \frac{1}{2}\frac{EI}{l^3}\int_0^1 A^2\left(12 - 24Z + 12Z^2\right)^2 dZ \end{aligned} \quad (10.48)$$

Minimizing with respect to A

$$ml\int_0^1 \left(6Z^2 - 4Z^3 + Z^4\right)^2 p^2 dZ = \frac{EI}{l^3}\int_0^1 \left(12 - 24Z + 12Z^2\right)^2 dZ \quad (10.49)$$

Evaluating the integrals

$$\frac{104}{45}mlp^2 = \frac{144}{5}\frac{EI}{l^3} \tag{10.50}$$

The above gives

$$p = 3.53\sqrt{\frac{EI}{ml^4}}\ \text{rad/sec} \tag{10.51}$$

as against the exact value

$$p_{\text{exact}} = 3.5156\sqrt{\frac{EI}{ml^4}}\ \text{rad/sec}$$

10.5 Lagrange Method for Vibration Problems

We revisit the Lagrangian approach that we considered for the static case and now consider a time-dependent problem. The Euler–Lagrangian equations for this case can be shown to be

$$\frac{d}{dt}\left(\frac{\partial T}{\partial \dot{q}}\right) - \frac{\partial T}{\partial q} + \frac{\partial U}{\partial q} = 0 \tag{10.52}$$

Considering n number of terms in (10.34), we can write the kinetic energy in the system as

$$T = \frac{1}{2}m\int_0^l \dot{q}_i^2 f_i^2 dz \tag{10.53}$$

Similarly the strain energy is

$$U = \frac{1}{2}EI\int_0^l q_i^2 f_i''^2 dz \tag{10.54}$$

Then

$$\left(\frac{\partial T}{\partial \dot{q}}\right) = m\int_0^l \dot{q}_i f_i^2 dz$$

$$\frac{d}{dt}\left(\frac{\partial T}{\partial \dot{q}}\right) = m\ddot{q}_i\int_0^l \sum f_i f_j dz$$

$$\frac{\partial U}{\partial q} = EIq_i\int_0^l f_i''^2 dz = EIq_i\int_0^l \sum f_i'' f_j'' dz \tag{10.55}$$

Substituting in (10.52) we have the following matrix equations:

$$[M]\{\ddot{q}\} + [K]\{q\} = 0 \tag{10.56}$$

where

Table 10.2 Natural frequencies by the Lagrange method with 5 terms, in Hz

Exact values	No. of terms	Lagrange
248.65	1	249.67
248.65	2	248.68
1558.36		1606.4
248.65	3	248.68
1558.36		1558.50
4363.89		4686.10
248.65	4	248.68
1558.36		1558.50
4363.89		4368.60
8551.67		9638.70
248.65	5	248.68
1558.36		1558.40
4363.89		4366.60
8551.67		8599.70
14135.01		16901.00

$$[M] = m \int_0^l \sum f_i f_j dz$$

$$[K] = EI \int_0^l \sum f_i'' f_j'' dz \tag{10.57}$$

Thus we see how Lagrangian equations lead to an eigenvalue problem in matrix form that was suitable for computer programming after the 1960s in the 20th century. Vyas and Rao [45] gave the following example:

Length of the blade $l = 10$ cm
Second moment of area $I = 0.00675\ \text{cm}^4$
Young's modulus $E = 2.11 \times 10^{11}$ Pa
Area of cross-section of the beam $A = 0.9\ \text{cm}^2$
Density $\rho = 0.008\ \text{kg/cm}^3$

The results are given in Table 10.2.

The first thing we notice is that all energy methods lead to upper bound solutions for natural frequency determination. This is because; the energy methods use an assumed solution for the shape function which is different from the exact mode shape. This means we need to make extra effort to bring the assumed shape to the correct shape or the strain energystrain energy will be higher than what the exact mode shape provides. A larger number of terms in the solution means the ability to correct the approximate shape and bring it closer to the exact shape; thus the solution converges from an upper bound. The solution can reach an almost exact value from its upper bound but can never result in a lower bound value. A mathematical proof was given by Den Hartog [11] (see also [37]).

10.6 Galerkin Method

Boris Grigoryevich Galerkin (1871–1945), a Russian scientist, published an article in 1915 [15], in which he put forward an idea of differential equations boundary problems approximate solution method. He applied his method to a large number of pivot and plate analysis problems. Some time before, he had developed a similar approach for the variational problems solution, which he interpreted as a variant of the Ritz method. He did not associate the method developed by him with direct solutions to variational problems, but considered it to be common for solving differential equations and interpreted it using the probable displacements principle. We can interpret his method from Section 10.1, where for extremum of the dynamical system, we have

$$\int_{x_1}^{x_2} \left[\frac{d^2}{dx^2}\left(\frac{\partial F}{\partial y''}\right) - \frac{d}{dx}\left(\frac{\partial F}{\partial y'}\right) + \frac{\partial F}{\partial y} \right] \eta dx = 0 \tag{10.9}$$

$$\left(\frac{\partial F}{\partial y''}\right)_{x_1}^{x_2} = 0$$

$$\left(-\frac{d}{dx}\left(\frac{\partial F}{\partial y''}\right) + \frac{\partial F}{\partial y'}\right)_{x_1}^{x_2} = 0 \tag{10.10}$$

Equations (10.10) are boundary conditions and (10.9) leads to differential equation of the dynamical system. We can determine the extremized path (the solution) by satisfying the boundary conditions and equation (10.9). The differential equation in the rectangular brackets in (10.9) cannot be solved except in a very few special cases. Essentially the Galerkin method is to make the integral in (10.9) zero. This is done by assuming $\eta(x)$ to be the assumed solution itself, here (10.21). Duncan [12, 13] used the Galerkin method in solving differential equations.

The procedure can be written as follows. The dynamical system is described as

$$\int_{t_1}^{t_2} \left[\int_{x_1}^{x_2} (Ly)\, \eta dx + \left\{(gy)\, \eta'\right\}_{x_1}^{x_2} + \{(hy)\, \eta\}_{x_1}^{x_2} \right] dt = 0 \tag{10.58}$$

where L, g and h are differential operators. To satisfy the above equation exactly so that y is an extremizing path, we make

$$Ly = 0 \tag{10.59}$$

and

$$\{gy = 0\}_{x_1}^{x_2} \quad \text{and} \quad \{hy = 0\}_{x_1}^{x_2} \tag{10.60}$$

When the solution for equation (10.59) is not possible, we will assume an admissible path $\check{y}$ and obtain the error ε in the differential equation

$$\varepsilon = L\check{y} \tag{10.61}$$

We can still satisfy equation (10.58) above, provided we make

$$\int_{x_1}^{x_2} \left[L\breve{y}\right] \eta dx = 0 \tag{10.62}$$

The difficulty is, what η should we choose? In the Galerkin method, we choose $\breve{y}$ itself for η i.e.,

$$\int_{x_1}^{x_2} \varepsilon \breve{y} dx = 0 \tag{10.63}$$

As an illustration, let us consider a freely vibrating cantilever beam with rectangular cross-section tapering from fixed end to free end defined by the following relations for breadth b and thickness t (suffix 0 refers to the quantities at the root $Z = 0$).

$$\frac{d^2}{dx^2}\left[EI\left(\frac{d^2w}{dx^2}\right)\right] = mp^2w \tag{10.64}$$

$$\begin{aligned} b &= b_0 \left(1 - \alpha X\right) \\ t &= t_0 \left(1 - \beta X\right) \\ I &= I_0 \left(1 - \alpha X\right) \left(1 - \beta X\right)^3 \end{aligned} \tag{10.65}$$

where m is mass per unit length and p is the natural frequency. Substituting (10.65) in (10.64)

$$\left[(1 - \alpha X)(1 - \beta X)^3 w''\right]'' = \frac{q}{EI_0} \tag{10.66}$$

As in the Lagrange method, let us take one term approximation for the deflection w

$$\begin{aligned} w\left(X\right) &= a(6X^2 - 4X^3 + X^4) \\ \frac{d^2w}{dX^2} &= \frac{a}{L^2}(12 - 24X + 12X^2) \end{aligned} \tag{10.21}$$

Substituting (10.21) in (10.66), we get the error ε in the differential equation

$$\begin{aligned} \varepsilon = &\left[(1 - \alpha X)\left(1 - \beta X\right)^3 \frac{a}{L^2}(12 - 24X + 12X^2)\right]'' \\ &- a\frac{m_0p^2}{EI_0}\left(1 - \alpha X\right)\left(1 - \beta X\right)\left(6X^2 - 4X^3 + X^4\right) \end{aligned} \tag{10.67}$$

i.e.,

$$\varepsilon = \frac{12a}{L^4}\left[\begin{array}{l} 2(1-\alpha X)(1-\beta X)^3 + 12\beta(1-X)(1-\alpha X)\left(1-\beta X^2\right) \\ +4\alpha(1-X)(1-\beta X)^3 + 6\beta^2(1-X)(1-\beta X)(1-2X+X^2) \\ +6\alpha\beta(1-\beta X)^2(1-2X+X^2) \end{array}\right]$$

$$-a\frac{m_0 p^2}{EI_0}(1-\alpha X)(1-\beta X)(6X^2-4X^3+X^4) \tag{10.68}$$

Galerkin's procedure then gives

$$\int_0^1 \left\{\begin{array}{l} \frac{12a}{L^4}\left[\begin{array}{l} 2(1-\alpha X)(1-\beta X)^3 + 12\beta(1-X)(1-\alpha X)(1-\beta X^2) \\ +4\alpha(1-\alpha X)(1-\beta X)^3 + 6\beta^2(1-\alpha X)(1-\beta X)(1-2X+X^2) \\ +6\alpha\beta(1-\beta X)^2(1-2X+X^2) \end{array}\right] - \\ a\frac{m_0 p^2}{EI_0}(1-\alpha X)(1-\beta X)(6X^2-4X^3+X^4) \end{array}\right\}$$

$$\times(6X^2-4X^3+X^4)dX = 0 \tag{10.69}$$

We identify the following integrals in the above and evaluate them first:

$$I_1 = \int_0^1 \left\{\frac{12}{L^4}\left[2(1-\alpha X)(1-\beta X)^3\right]\right\}(6X^2-4X^3+X^4)dX$$

$$= \frac{12}{L^4}\left\{\frac{12}{5} - \frac{26}{15}\alpha - \frac{26}{5}\beta + \frac{142}{35}\alpha\beta + \frac{142}{35}\beta^2 - \frac{93}{28}\alpha\beta^2 - \frac{31}{28}\beta^3 + \frac{59}{63}\alpha\beta^3\right\}$$

$$I_2 = \int_0^1 \left\{\frac{12}{L^4}\left[12\beta(1-X)(1-\alpha X)(1-\beta X^2)\right]\right\}(6X^2-4X^3+X^4)dX$$

$$= \frac{12}{L^4}\left\{4\beta - \frac{16}{7}\alpha\beta - \frac{32}{7}\beta^2 + \frac{103}{35}\alpha\beta^2 + \frac{103}{70}\beta^3 - \frac{43}{42}\alpha\beta^3\right\}$$

$$I_3 = \int_0^1 \left\{\frac{12}{L^4}\left[4\alpha(1-\alpha X)(1-\beta X)^3\right]\right\}(6X^2-4X^3+X^4)dX$$

$$= \frac{12}{L^4}\left\{\frac{4}{3}\alpha - \frac{16}{7}\alpha\beta + \frac{103}{35}\alpha\beta^2 - \frac{43}{126}\alpha\beta^3\right\}$$

$$I_4 = \int_0^1 \left\{\frac{12}{L^4}\left[6\beta^2(1-\alpha X)(1-\beta X)(1-2X+X^2)\right]\right\}(6X^2-4X^3+X^4)dX$$

$$= \frac{12}{L^4}\left\{\frac{6}{7}\beta^2 - \frac{57}{140}\alpha\beta^2 - \frac{57}{140}\beta^3 + \frac{47}{210}\alpha\beta^3\right\}$$

$$I_5 = \int_0^1 \left\{ \frac{12}{L^4} \left[6\alpha\beta \left(1 - \beta X\right)^2 \left(1 - 2X + X^2\right) \right] \right\} (6X^2 - 4X^3 + X^4) dX =$$

$$= \frac{12}{L^4} \left\{ \frac{6}{7}\alpha\beta - \frac{57}{70}\alpha\beta^2 + \frac{47}{210}\alpha\beta^3 \right\}$$

$$I_6 = \int_0^1 \frac{m_0 p^2}{EI_0} (1 - \alpha X)(1 - \beta X)(6X^2 - 4X^3 + X^4)^2 dX =$$

$$= \frac{m_0 p^2}{EI_0} \left[\frac{104}{45} - \frac{584}{315}\alpha - \frac{584}{315}\beta + \frac{5353}{3465}\alpha\beta \right] \tag{10.70}$$

Substituting the above integrals in equation (10.69) and rearranging we have

$$p^2 = \frac{12EI_0}{m_0 L^4} \times \frac{\frac{12}{5} - \frac{2}{5}\alpha - \frac{6}{5}\beta + \frac{12}{35}\alpha\beta + \frac{12}{35}\beta^2 - \frac{9}{70}\alpha\beta^2 - \frac{3}{70}\beta^3 + \frac{2}{105}\alpha\beta^3}{\frac{104}{45} - \frac{584}{315}\alpha - \frac{584}{315}\beta + \frac{5353}{3465}\alpha\beta} \tag{10.71}$$

The above solution was provided by Rao [28] using the Galerkin [15] method to determine the fundamental natural frequency of a tapered cantilever blade. This solution is a one-term approximation. This was a time when digital computers began improving in speed and memory. Subsequently, solutions with three or more term approximations were used to solve tapered, twisted aerofoil cantilever blades mounted on a rotating disk, thus helping the designers. These are all one-element models when compared with modern finite element methods which came up subsequently.

Because they are one-element models, only the number of terms in the assumed solution controls the matrix size. In most cases the number of terms in the series is truncated to five and so the computational speed was reasonable even in the early day slow computers. However, these methods cannot capture the intricate geometry, e.g., the notch or dovetail in a bladed disk. Modern finite element methods therefore have taken over modern day design practices.

10.7 Hamilton's Principle

For one-dimensional structures Hamilton's principle can be stated as [17]

$$\delta \int_{t_1}^{t_2} \int_0^L \left[U(q) - T(q, \dot{q}) - W \right] dx dt = 0 \tag{10.72}$$

where U is potential energy, T is kinetic energy and work W done by external forces F, which are functions of the independent variable x. q represents the generalized coordinate.

Taking the variations

$$\delta U = \frac{\partial U}{\partial q}\delta q$$

$$\delta T = \frac{\partial T}{\partial \dot{q}}\delta \dot{q} + \frac{\partial T}{\partial q}\delta q$$

$$\delta W = F\delta q$$

Equation (10.72) becomes

$$\int_{t_1}^{t_2}\int_0^L \left[\frac{\partial U}{\partial q}\delta q - \left(\frac{\partial T}{\partial \dot{q}}\delta \dot{q} + \frac{\partial T}{\partial q}\delta q\right) - F\delta q\right] dxdt = 0 \tag{10.73}$$

We can perform the first integration by parts with respect to time of the $\dot{q}$ term to give

$$\int_{t_1}^{t_2}\left(\frac{\partial T}{\partial \dot{q}}\delta \dot{q}\right) dt = \left[\frac{\partial T}{\partial \dot{q}}\delta q\right]_{t_1}^{t_2} - \int_{t_1}^{t_2}\frac{d}{dt}\left(\frac{\partial T}{\partial \dot{q}}\right)\delta q dt$$

$$= -\int_{t_1}^{t_2}\frac{d}{dt}\left(\frac{\partial T}{\partial \dot{q}}\right)\delta q dt \tag{10.74}$$

and rewrite equation (10.73) as follows:

$$\int_{t_1}^{t_2}\int_0^L \left[\frac{\partial U}{\partial q} + \frac{d}{dt}\left(\frac{\partial T}{\partial \dot{q}}\right) - \frac{\partial T}{\partial q} - F\right]\delta q dxdt = 0 \tag{10.75}$$

With the help of the above equation, we can derive several approximate methods for solving dynamical problems. The modern Finite Element methods all emanate from this equation. Hence, Hamilton's principle is the most general form for all vibrating systems.

One of the most vexing problems in development of high speed and high capacity rotating machinery is the development of rotors and mounted parts to withstand thermal and centrifugal loads combined with alternating loads. Dynamics obviously plays a major role as evidenced by Frank Whittle. Therefore considerable time has been invested in understanding the design of turbomachine blades. Even today, they happen to be the most flexible and challenging design exercises faced in industry.

Analytically, the development of energy principles and solution methods were developed in the 1950s. The most significant of them is William Carnegie [6–9] who used Hamilton's principle extensively and one such derivation by Vyas and Rao [45] is illustrated here to show the power of Hamilton's principle.

A blade of length l and cross-sectional area A is mounted on a rotating disk of radius R. The angular velocity is ω rad/s.

$\tilde{M}\tilde{N}$ is a fixed axis system. The blade itself is located at angle $\Gamma = \omega_0 t + 1/2\alpha t^2$ in a coordinate system fixed to the rotating disk. The x axis is located at the root of the blade, parallel to the axis of rotation, and the y axis in the plane of the disk. The z axis is along the length of the blade passing through the centroids of all cross-

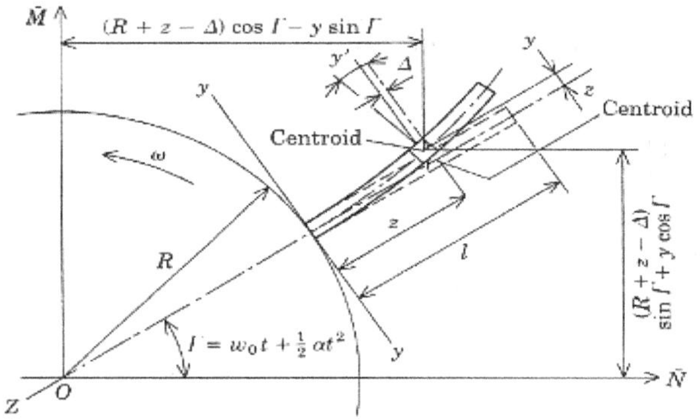

Fig. 10.2 Rotating blade mounted on a disk

sections. Pretwist and asymmetry are not included. The blade executes pure bending oscillations in the y–z plane. Δ is the inward displacement of the blade element dz as shown.

The displacements of an element at a distance z from the blade root in the $\tilde{M}$ and $\tilde{N}$ directions with respect to the disk center are

$$\begin{aligned} m &= (R + z - \Delta)\sin\Gamma + y\cos\Gamma \\ n &= (R + z - \Delta)\cos\Gamma - y\sin\Gamma \end{aligned} \tag{10.76}$$

The corresponding velocities are

$$\begin{aligned} \dot{m} &= \left\{\dot{y} + \dot{\Gamma}\,(R + z - \Delta)\right\}\cos\Gamma - \left(\dot{\Delta} + \dot{\Gamma}y\right)\sin\Gamma \\ \dot{n} &= \left\{\dot{y} + \dot{\Gamma}\,(R + z - \Delta)\right\}\sin\Gamma - \left(\dot{\Delta} + \dot{\Gamma}y\right)\cos\Gamma \end{aligned} \tag{10.77}$$

The kinetic energy due to translation and rotation of the blade element are

$$T_t = \int_0^l \frac{1}{2}\rho A(\dot{m}^2 + \dot{n}^2)dz \tag{10.78}$$

$$T_r = \int_0^l \frac{1}{2}\rho I_{xx}(\dot{\Gamma}^2 + \dot{\phi}_b^2)dz \tag{10.79}$$

where ϕ_b is the slope due to bending.

The instantaneous angular velocity is

$$\begin{aligned} \Gamma &= \omega_0 t + \frac{1}{2}\alpha t^2 \\ \dot{\Gamma} &= \omega_0 + \alpha t \end{aligned} \tag{10.80}$$

Now the total kinetic energy can be written as

$$T = \int_0^l \frac{1}{2}\rho A(\dot{y}^2 + \dot{\Delta}^2)dz + \int_0^l \frac{1}{2}\rho A\,(\omega_0 + \alpha t)^2 \left\{(R + z - \Delta)^2 + y^2\right\}dz$$
$$+ \int_0^l \rho A\,(\omega_0 + \alpha t)\left\{\dot{y}\,(R + z - \Delta) + y\dot{\Delta}\right\}dz$$
$$+ \int_0^l \frac{1}{2}\rho I_{xx}\left\{(\omega_0 + \alpha t) + \dot{\phi}_b^2\right\}^2 dz \qquad (10.81)$$

In equation (10.81), the first term on the right-hand side accounts for blade inertia, the second term accounts for the effect of centrifugal force, the third term accounts for the Coriolis force, while the fourth term refers to rotary inertia. The inward displacement of an inextensional blade is given by

$$\Delta = 1 - \int_0^z \cos\phi_b(\xi)d\xi.$$

Noting that $y' = \sin\phi_b$ one has $\Delta = 1 - \int_0^z \sqrt{1 - [y'(\xi)]^2}d\xi$. Expanding the square root term and truncating the series yields

$$\Delta = \frac{1}{2}\int_0^z y'^2 dz \quad \text{and} \quad \dot{\Delta} = \frac{1}{2}\frac{\partial}{\partial t}\int_0^z y'^2 dz \qquad (10.82)$$

Hence equation (10.81) becomes

$$T = \int_0^l \frac{1}{2}\rho A \begin{bmatrix} \dot{y}^2 + \left\{\frac{1}{2}\frac{\partial}{\partial t}\int_0^z y'^2 dz\right\}^2 \\ + (\omega_0 + \alpha t)^2 \left\{\left(R + z - \frac{1}{2}\int_0^z y'^2 dz\right)^2 + y^2\right\} \\ +2\,(\omega_0 + \alpha t)\left\{\dot{y}\left(R + z - \frac{1}{2}\int_0^z y'^2 dz\right) + \frac{1}{2}y\frac{\partial}{\partial t}\int_0^z y'^2 dz\right\} \end{bmatrix} dz$$
$$+ \int_0^l \frac{1}{2}\rho I_{xx}\left\{(\omega_0 + \alpha t) + \dot{\phi}_b^2\right\}^2 dz \qquad (10.83)$$

The potential energy taking into account shear deformation is

$$V = \int_0^l \left\{\frac{AG}{K}(y' - \phi_b)^2 + \frac{1}{2}EI_{xx}\phi'^2_b\right\} dz \qquad (10.84)$$

Now according to Hamilton's principle $\int_{t_1}^{t_2}(T - V)dt$ taken between any arbitrary intervals of time (t_1, t_2) is stationary for the dynamical system. Therefore

$$\int_{t_1}^{t_2} L dt = \int_{t_1}^{t_2} \int_0^l \frac{1}{2} \rho A \left[\dot{y}^2 + \left\{ \frac{1}{2} \frac{\partial}{\partial t} \int_0^z y'^2 dz \right\}^2 \right] dz dt +$$
$$+ \int_{t_1}^{t_2} \int_0^l \frac{1}{2} \rho A (\omega_0 + \alpha t)^2 \left\{ \left(R + z - \frac{1}{2} \int_0^z y'^2 dz \right)^2 + y^2 \right\} dz dt$$
$$+ \int_{t_1}^{t_2} \int_0^l \rho A (\omega_0 + \alpha t) \left\{ \dot{y} \left(R + z - \frac{1}{2} \int_0^z y'^2 dz \right) + \frac{1}{2} y \frac{\partial}{\partial t} \int_0^z y'^2 dz \right\} dz dt$$
$$+ \int_{t_1}^{t_2} \int_0^l \frac{1}{2} \rho I_{xx} \{ (\omega_0 + \alpha t) + \dot{\phi}_b^2 \}^2 dz dt$$
$$- \int_{t_1}^{t_2} \int_0^l \left\{ \frac{AG}{K} (y' - \phi_b)^2 + \frac{1}{2} E I_{xx} \phi'^2_b \right\} dz dt \tag{10.85}$$

With the help of the following relation derived by Carnegie [9]

$$\frac{\partial}{\partial t} \int_0^z y'^2 dz = \frac{\partial}{\partial t} y'^2 \int_z^l dz$$

the first term on the right-hand side of equation (10.85) can be written as

$$\int_{t_1}^{t_2} \int_0^l \frac{1}{2} \rho A \left[\dot{y}^2 + \left\{ \frac{1}{2} \frac{\partial}{\partial t} \int_0^z y'^2 dz \right\}^2 \right] dz dt$$
$$= \int_{t_1}^{t_2} \int_0^l \frac{1}{2} \rho A \left[\dot{y}^2 + \left\{ \frac{1}{2} \frac{\partial}{\partial t} y'^2 \int_z^l dz \right\}^2 \right] dz dt \tag{10.86}$$

This can be expressed as

$$\int_{t_1}^{t_2} \int_0^l \frac{1}{2} \rho A \left[\dot{y}^2 + \left\{ \frac{1}{2} \frac{\partial}{\partial t} \int_0^z y'^2 dz \right\}^2 \right] dz dt$$
$$= \int_{t_1}^{t_2} \int_0^l \frac{1}{2} \rho A [\dot{y}^2 + R_1^2 (y' \dot{y}')^2] dz dt \tag{10.87}$$

where $R_1 = l - z$. With higher powers of Δ neglected, the second term on the right-hand side of (10.85) can be written as

$$\int_{t_1}^{t_2} \int_0^l \frac{1}{2} \rho A (\omega_0 + \alpha t)^2 \left\{ \left(R + z - \frac{1}{2} \int_0^z y'^2 dz \right)^2 + y^2 \right\} dz dt$$
$$= \int_{t_1}^{t_2} \int_0^l \frac{1}{2} \rho A (\omega_0 + \alpha t)^2 \left\{ y^2 (R + z)^2 - (R + z) \int_0^z y'^2 dz \right\} dz dt \tag{10.88}$$

The third term is

$$\int_{t_1}^{t_2}\int_0^l \rho A\left(\omega_0+\alpha t\right)\left\{\dot{y}\left(R+z-\frac{1}{2}\int_0^z y'^2 dz\right)+\frac{1}{2}y\frac{\partial}{\partial t}\int_0^z y'^2 dz\right\}dzdt$$

$$=\int_{t_1}^{t_2}\int_0^l \rho A\left(\omega_0+\alpha t\right)\dot{y}\left(R+z\right)dzdt$$

$$+\int_{t_1}^{t_2}\int_0^l \rho A\left(\omega_0+\alpha t\right)\left\{-\frac{1}{2}\dot{y}\int_0^z y'^2 dz+\frac{1}{2}y\frac{\partial}{\partial t}\int_0^z y'^2 dz\right\}dzdt \quad (10.89)$$

We can now write

$$\int_{t_1}^{t_2} Ldt=\int_{t_1}^{t_2}\int_0^l \frac{1}{2}\rho A\left[\dot{y}^2+R_1^2\left(y'\dot{y}'\right)^2\right]dzdt+$$

$$+\int_{t_1}^{t_2}\int_0^l \frac{1}{2}\rho A\left(\omega_0+\alpha t\right)^2\left\{y^2\left(R+z\right)^2-\left(R+z\right)\int_0^z y'^2 dz\right\}dzdt$$

$$+\int_{t_1}^{t_2}\int_0^l \rho A\left(\omega_0+\alpha t\right)\dot{y}\left(R+z\right)dzdt$$

$$+\int_{t_1}^{t_2}\int_0^l \rho A\left(\omega_0+\alpha t\right)\left\{-\frac{1}{2}\dot{y}\int_0^z y'^2 dz+\frac{1}{2}y\frac{\partial}{\partial t}\int_0^z y'^2 dz\right\}dzdt$$

$$+\int_{t_1}^{t_2}\int_0^l \frac{1}{2}\rho I_{xx}\left\{\left(\omega_0+\alpha t\right)+\dot{\phi}_b^2\right\}^2 dzdt$$

$$-\int_{t_1}^{t_2}\int_0^l \left\{\frac{AG}{K}\left(y'-\phi_b\right)^2+\frac{1}{2}EI_{xx}\phi'^2_b\right\}dzdt \quad (10.90)$$

Carrying out extremization on the first term in (10.90) yields

$$\delta\int_{t_1}^{t_2}\int_0^l \frac{1}{2}\rho A\left[\dot{y}^2+R_1^2(y'\dot{y}')^2\right]dzdt$$

$$=\int_{t_1}^{t_2}\int_0^l \rho A\left[\dot{y}\delta\dot{y}+R_1^2(y'\dot{y}')(y'\delta\dot{y}'+\dot{y}'\delta y')\right]dzdt$$

Integrating by parts, neglecting higher order terms and noting that variations are zero at $t=t_1$ and $t=t_2$, one obtains

$$\delta\int_{t_1}^{t_2}\int_0^l \frac{1}{2}\rho A\left[\dot{y}^2+R_1^2(y'\dot{y}')^2\right]dzdt=-\int_{t_1}^{t_2}\int_0^l \rho A\ddot{y}\delta y dzdt \quad (10.91)$$

Similarly the second term on right-hand side of (10.90) yields

$$\delta \int_{t_1}^{t_2}\int_0^l \frac{1}{2}\rho A(\omega_0+\alpha t)^2\left\{y^2(R+z)^2-(R+z)\int_0^z y'^2dz\right\}dzdt$$
$$=\int_{t_1}^{t_2}\int_0^l \frac{1}{2}\rho A(\omega_0+\alpha t)^2\left\{y+R_2y''-(R+z)\,y'\right\}\delta ydzdt \qquad (10.92)$$

where

$$R_2=\int_z^l(R+z)dz=R(l-z)+\frac{1}{2}(l^2-z^2)$$

The third term on the right-hand side of equation (10.90) yields

$$\delta\int_{t_1}^{t_2}\int_0^l \rho A(\omega_0+\alpha t)\dot{y}(R+z)dzdt=\int_{t_1}^{t_2}\int_0^l \rho A\alpha(R+z)\delta ydzdt \qquad (10.93)$$

For the variation of the fourth term in equation (10.90), let

$$I=\int_{t_1}^{t_2}\int_0^l \rho A(\omega_0+\alpha t)\left\{-\frac{1}{2}\dot{y}\int_0^z y'^2dz+\frac{1}{2}y\frac{\partial}{\partial t}\int_0^z y'^2dz\right\}dzdt \qquad (10.94)$$

Upon replacing one of the z variables by θ,

$$I=\int_{t_1}^{t_2}\int_0^l \rho A(\omega_0+\alpha t)\{-\frac{1}{2}\dot{y}\int_0^\theta y'^2d\theta+\frac{1}{2}y\frac{\partial}{\partial t}\int_0^\theta y'^2d\theta\}dzdt \qquad (10.95)$$

For the variation of $\delta(y)$ of y it follows that

$$I+\delta I=\int_{t_1}^{t_2}\int_0^l \frac{1}{2}\rho A(\omega_0+\alpha t)\left\{\begin{array}{l}-(\dot{y}+\delta\dot{y})\int_0^\theta\left(y'+\delta y'\right)^2d\theta+\\ (y+\delta y)\frac{\partial}{\partial t}\int_0^\theta\left(y'+\delta y'\right)^2d\theta\end{array}\right\}dzdt \qquad (A1)$$

From the above two equations, we can write

$$\begin{aligned}\delta I=&\int_{t_1}^{t_2}\int_0^l \rho A(\omega_0+\alpha t)\,y\int_0^\theta\left(\dot{y}'\delta y'+y'\delta\dot{y}'\right)d\theta dzdt\\&+\int_{t_1}^{t_2}\int_0^l \rho A(\omega_0+\alpha t)\,y\int_0^\theta y'\dot{y}'d\theta dzdt\\&-\int_{t_1}^{t_2}\int_0^l \rho A(\omega_0+\alpha t)\,\dot{y}\int_0^\theta y'\delta y'd\theta dzdt\\&-\int_{t_1}^{t_2}\int_0^l \frac{1}{2}\rho A(\omega_0+\alpha t)\,\delta\dot{y}\int_0^\theta y'^2d\theta dzdt\end{aligned} \qquad (A2)$$

The first term in the above equation is

$$\int_{t_1}^{t_2}\int_0^l \rho A\left(\omega_0+\alpha t\right) y \int_0^\theta \left(\dot{y}'\delta y'+y'\delta\dot{y}'\right) d\theta dz dt$$
$$=\int_{t_1}^{t_2}\int_0^l \rho A\left(\omega_0+\alpha t\right) y \frac{\partial}{\partial t}\int_0^\theta y'\delta y' d\theta dz dt$$

Integrating by parts and noting that the variations are zero at $t = t_1$ and t_2, we obtain

$$\int_{t_1}^{t_2}\int_0^l \rho A\left(\omega_0+\alpha t\right) y \int_0^\theta \left(\dot{y}'\delta y'+y'\delta\dot{y}'\right) d\theta dz dt$$
$$=-\int_{t_1}^{t_2}\int_0^l \rho A\left\{\alpha y+\left(\omega_0+\alpha t\right)\dot{y}\right\}\left\{\left[y'\delta y\right]_0^\theta-\int_0^\theta y''\delta y d\theta\right\} dz dt$$

Replacing θ by z and since $\delta y = 0$ at $z = 0$, one has

$$\int_{t_1}^{t_2}\int_0^l \rho A\left(\omega_0+\alpha t\right) y \int_0^\theta \left(\dot{y}'\delta y'+y'\delta\dot{y}'\right) d\theta dz dt$$
$$=-\int_{t_1}^{t_2}\int_0^l \rho A\left\{\alpha y+\left(\omega_0+\alpha t\right)\dot{y}\right\}\left\{y'\delta y-\int_0^z y''\delta y dz\right\} dz dt$$

Effecting a Carnegie [9] transformation

$$\left\{\alpha y+\left(\omega_0+\alpha t\right)\dot{y}\right\}\int_0^z y''\delta y dz = y''\delta y\int_z^l \left\{\alpha y+\left(\omega_0+\alpha t\right)\dot{y}\right\} dz$$

one can write

$$\int_{t_1}^{t_2}\int_0^l \rho A\left(\omega_0+\alpha t\right) y \int_0^\theta \left(\dot{y}'\delta y'+y'\delta\dot{y}'\right) d\theta dz dt$$
$$=-\int_{t_1}^{t_2}\int_0^l \Bigg[\rho A\left\{\alpha y+\left(\omega_0+\alpha t\right)\dot{y}\right\} y'\delta y$$
$$-y''\delta y\int_z^l \left\{\alpha y+\left(\omega_0+\alpha t\right)\dot{y}\right\} dz\Bigg] dz dt \qquad \text{(A3)}$$

Similarly, the third term on the right-hand side of equation (A2) is

$$-\int_{t_1}^{t_2}\int_0^l \rho A\left(\omega_0+\alpha t\right)\dot{y}\int_0^\theta y'\delta y' d\theta dz dt$$
$$=-\int_{t_1}^{t_2}\int_0^l \rho A\left(\omega_0+\alpha t\right)\left\{\dot{y}y'-y''\int_z^l \dot{y} dz\right\}\delta y dz dt \qquad \text{(A4)}$$

and the fourth term is

$$-\int_{t_1}^{t_2}\int_0^l \frac{1}{2}\rho A\left(\omega_0+\alpha t\right)\delta\dot{y}\int_0^\theta y'^2 d\theta dz dt$$

$$=\int_{t_1}^{t_2}\int_0^l \frac{1}{2}\rho A\left\{\alpha\int_0^z y'^2 dz+\left(\omega_0+\alpha t\right)\frac{\partial}{\partial t}\int_0^z y'^2 dz\right\}\delta y dz dt \qquad \text{(A5)}$$

Now equation (A2) is

$$\delta I=-\int_{t_1}^{t_2}\int_0^l \rho A$$

$$\begin{bmatrix} \left\{\alpha y+\left(\omega_0+\alpha t\right)\dot{y}\right\}y'-y''\int_z^l\left\{\alpha y+\left(\omega_0+\alpha t\right)\dot{y}\right\}dz \\ +\left(\omega_0+\alpha t\right)\left\{\dot{y}y'-\int_0^z y'\dot{y}'dz-y''\int_z^l \dot{y}dz-\frac{1}{2}\frac{\partial}{\partial t}\int_0^z y'^2 dz\right\} \\ -\frac{1}{2}\alpha\int_0^z y'^2 dz \end{bmatrix}\delta y dz dt \qquad \text{(A6)}$$

Therefore, the variation of the fourth term in equation (10.90) is

$$\delta\int_{t_1}^{t_2}\int_0^l \rho A\left(\omega_0+\alpha t\right)\left\{-\frac{1}{2}\dot{y}\int_0^z y'^2 dz+\frac{1}{2}y\frac{\partial}{\partial t}\int_0^z y'^2 dz\right\}dz dt$$

$$=-\int_{t_1}^{t_2}\int_0^l \rho A \qquad (10.96)$$

$$\begin{bmatrix} \left\{\alpha y+\left(\omega_0+\alpha t\right)\dot{y}\right\}y'-y''\int_z^l\left\{\alpha y+\left(\omega_0+\alpha t\right)\dot{y}\right\}dz \\ +\left(\omega_0+\alpha t\right)\left\{\dot{y}y'-\int_0^z y'\dot{y}'dz-y''\int_z^l \dot{y}dz-\frac{1}{2}\frac{\partial}{\partial t}\int_0^z y'^2 dz\right\} \\ -\frac{1}{2}\alpha\int_0^z y'^2 dz \end{bmatrix}\delta y dz dt$$

The variation operation on the fifth and sixth terms in equation (10.90) gives

$$\delta\int_{t_1}^{t_2}\int_0^l \frac{1}{2}\rho I_{xx}\{\left(\omega_0+\alpha t\right)+\dot{\phi}_b^2\}^2 dz dt=-\int_{t_1}^{t_2}\int_0^l \rho I_{xx}\left(\alpha+\ddot{\phi}_b\right)\delta\phi_b dz dt \qquad (10.97)$$

$$\delta\int_{t_1}^{t_2}\int_0^l\left\{\frac{AG}{K}\left(y'-\phi_b\right)^2+\frac{1}{2}EI_{xx}\phi_b'^2\right\}dz dt$$

$$=\int_{t_1}^{t_2}\begin{bmatrix} \left[-\frac{AG}{K}\left(y'-\phi_b\right)\delta y\right]_0^l-\left[-EI_{xx}\phi_b'\delta\phi_b\right]_0^l \\ +\int_0^l\left\{\begin{matrix}\frac{AG}{K}\left(y'-\phi_b\right)\delta\phi_b+ \\ \frac{AG}{K}\left(y''-\phi_b'\right)\delta y+EI_{xx}\phi_b''\delta\phi_b\end{matrix}\right\}dz \end{bmatrix}dt \qquad (10.98)$$

Now with the help of Hamilton's principle we can write the governing equations of motion and the boundary conditions

$$\rho A\ddot{y} - \rho A\,(\omega_0 + \alpha t)^2\left\{y + R_2 y'' - (R+z)\,y'\right\}$$
$$+ 2\rho A(\omega_0 + \alpha t)\left\{\dot{y}y' - y''\int_z^l \dot{y}dz - \int_0^z y'\dot{y}'dz\right\}$$
$$+ \rho A\alpha\left\{(R+z) + yy' - \frac{1}{2}\int_0^z y'^2 dz - y''\int_z^l y dz\right\} - \frac{AG}{K}(y'' - \phi_b') = 0 \tag{10.99}$$

$$EI_{xx}\phi_b'' + \frac{AG}{K}(y' - \phi_b) - \rho I_{xx}(\alpha + \ddot{\phi}_b) = 0 \tag{10.100}$$

At $z = 0$

$$y = 0,\ \phi_b = 0$$

At $z = l$

$$\frac{AG}{K}(y' - \phi_b) = 0,\ EI_{xx}\phi_b' = 0 \tag{10.101}$$

We can eliminate ϕ_b from (10.99) and (10.100) and write one governing equation as follows.

$$EI_{xx}y'''' + \rho A\begin{bmatrix} \ddot{y} - (\omega_0 + \alpha t)^2\left\{y + R_2 y'' - (R+z)\,y'\right\} + \\ +2\,(\omega_0 + \alpha t)\left\{\dot{y}y' - y''\int_z^l \dot{y}dz - \int_0^z y'\dot{y}'dz\right\} \\ +\alpha\left\{(R+z) + yy' - \frac{1}{2}\int_0^z y'^2 dz - y''\int_z^l y dz\right\} \end{bmatrix}$$
$$-\left(\rho I_{xx} + \frac{k\rho EI_{xx}}{G}\right)\ddot{y}'' + I_{xx}k\rho^2\ddot{y} = 0 \tag{10.102}$$

The above can be written as

$$EI_{xx}y'''' + \rho A\begin{bmatrix} \ddot{y} - (\omega_0 + \alpha t)^2\left\{y + R_2 y'' - (R+z)\,y'\right\} + \\ +2\,(\omega_0 + \alpha t)\left\{\dot{y}y' - y''\int_z^l \dot{y}dz - \int_0^z y'\dot{y}'dz\right\} \\ +\alpha\left\{yy' - \frac{1}{2}\int_0^z y'^2 dz - y''\int_z^l y dz\right\} \end{bmatrix}$$
$$-\left(\rho I_{xx} + \frac{k\rho EI_{xx}}{G}\right)\ddot{y}'' + I_{xx}k\rho^2\ddot{y} = -\rho A\alpha\,(R+z) \tag{10.103}$$

This form of equation of motion reveals the presence of a pseudo-static force term $-\rho A\alpha(R+z)$ arising due to the accelerating conditions. The origin of the pseudo-static force term can be readily traced back to the Coriolis effects.

For a disk rotating with constant angular speed ($\alpha = 0$), and with shear deformation ignored ($k = 0$), equation (10.103) reduces to

$$EI_{xx}y'''' + \rho A\begin{bmatrix} \ddot{y} - \omega_0^2\left\{y + R_2 y'' - (R+z)\,y'\right\} + \\ +2\omega_0\left\{\dot{y}y' - y''\int_z^l \dot{y}dz - \int_0^z y'\dot{y}'dz\right\} \end{bmatrix} - (\rho I_{xx})\,\ddot{y}'' = 0$$

The above is the same as derived by Carnegie [9].

The solution for the above equation (10.103) is also obtained from energy methods. Neglecting shear (for long beams as in helicopter blades), let us rewrite equation (10.103) and identify each of the terms as in equation (10.104).

As can be seen, the equation is highly nonlinear with Coriolis and acceleration terms besides stress stiffening and spin softening terms. It cannot be easily solved or put in finite element formulation.

Translation and Rotational Inertias → $\ddot{y} + \frac{I_{xx}}{A}\ddot{y}''$

Beam Stiffness → $\frac{EI_{xx}}{\rho A} y''''$

Stiffness due to rotation: Hardening → $(R+Z)y'$; Softening → $-y - R_2 y''$

Nonlinear Coriolis Forces → $2(\omega_0+\alpha t)\{\ldots\}$

Nonlinear terms due to acceleration → $\alpha\{\ldots\}$

Force due to acceleration independent of y → $\alpha(R+Z)$

$$\ddot{y} + \frac{I_{xx}}{A}\ddot{y}'' + \frac{EI_{xx}}{\rho A} y'''' + (\omega_0 + \alpha t)^2 \left\{(R+Z)y' - y - R_2 y''\right\} + \\ + 2(\omega_0 + \alpha t)\left\{\dot{y}y' - y''\int_0^l \dot{y}\,dz - \int_0^z y'\dot{y}'\,dz\right\} + \\ + \alpha\left\{yy' - \frac{1}{2}\int_0^z y'^2\,dz - y''\int_0^l y\,dz\right\} + \alpha(R+Z) = 0 \tag{10.104}$$

The Galerkin method is powerful and can provide a solution for this equation [30]. There is no excitation or oscillatory external force; there is only a pseudo-static force. Obviously when the rotor begins to turn, transient vibrations set in; the response will be at natural frequencies of the system. We can assume that the response will be predominant at the fundamental frequency of the stationary rotor. Thus we can assume a one-term solution as

$$\begin{aligned} y &= f(Z)\,\theta(t) \\ f(Z) &= 6Z^2 - 4Z^3 + Z^4 \end{aligned} \tag{10.105}$$

The error in the differential equation due to the above single-term approximation is given by

$$\begin{aligned} \varepsilon = &\left[-\left(12\frac{\rho I_{xx}}{l^2}\right)\left(1 - 2Z + Z^2\right) + \left(6Z^2 + 4Z^3 + Z^4\right)\rho A\right]\ddot{\theta} \\ &+ \left[\frac{8\rho A(\omega_0 + \alpha t)}{l}\begin{pmatrix} -\frac{18}{5} + \frac{36}{5}Z - \frac{18}{5}Z^2 + 12Z^3 - 27Z^4 \\ + \frac{108}{5}Z^5 - \frac{36}{5}Z^6 + \frac{36}{25}Z^7 \end{pmatrix}\right]\theta\dot{\theta} \\ &+ \left[-\rho A(\omega_0 + \alpha t)^2 \begin{pmatrix} 6 - 12Z - 6Z^2 + 20Z^3 - 9Z^4 \\ + 4\bar{R}\left(3 - 12Z - 12Z^2 - 4Z^3\right) \end{pmatrix} + 24\frac{EI_{xx}}{l^4}\right]\theta \\ &+ \left[4\frac{\rho A\alpha}{l}\begin{pmatrix} -\frac{18}{5} + \frac{36}{5}Z - \frac{18}{5}Z^2 + 18Z^3 - 36Z^4 \\ + \frac{138}{5}Z^5 - \frac{46}{5}Z^6 + \frac{46}{25}Z^7 \end{pmatrix}\right]\theta^2 + \rho A\alpha(R+Z) \end{aligned} \tag{10.106}$$

Following the Galerkin method, we get

$$
\begin{aligned}
&\int_0^l \left[-12\frac{\rho I_{xx}}{l^2}\left(1-2Z+Z^2\right)+\left(6Z^2+4Z^3+Z^4\right)\rho A\right]\ddot{\theta}(6Z^2-4Z^3+Z^4)dZ\\
&+\int_0^l \left[\frac{8\rho A\left(\omega_0+\alpha t\right)}{l}\begin{pmatrix}-\frac{18}{5}+\frac{36}{5}Z-\frac{18}{5}Z^2+12Z^3\\ -27Z^4+\frac{108}{5}Z^5-\frac{36}{5}Z^6+\frac{36}{25}Z^7\end{pmatrix}\right]\theta\dot{\theta}(6Z^2-4Z^3+Z^4)dZ\\
&+\int_0^l \left[\begin{matrix}-\rho A\left(\omega_0+\alpha t\right)^2\begin{pmatrix}6-12Z-6Z^2+20Z^3-9Z^4\\ +4\bar{R}(3-12Z-12Z^2-4Z^3)\end{pmatrix}\\ +24\frac{EI_{xx}}{l^4}\end{matrix}\right]\theta(6Z^2-4Z^3+Z^4)dZ\\
&+\int_0^l \left[4\frac{\rho A\alpha}{l}\begin{pmatrix}-\frac{18}{5}+\frac{36}{5}Z-\frac{18}{5}Z^2+18Z^3\\ -36Z^4+\frac{138}{5}Z^5-\frac{46}{5}Z^6+\frac{46}{25}Z^7\end{pmatrix}\right]\theta^2(6Z^2-4Z^3+Z^4)dZ\\
&+\int_0^l \rho A\alpha(R+Z)(6Z^2-4Z^3+Z^4)dZ=0
\end{aligned}
\tag{10.107}
$$

After performing the integral above and non-dimensionalizing we have

$$
\begin{aligned}
a &= \frac{\frac{9}{52}+\frac{81}{52}\bar{R}}{1-\frac{135}{182}\frac{I_{xx}}{Al^2}}\\
b &= \frac{\frac{27}{52}\bar{R}+\frac{3}{8}}{1-\frac{135}{182}\frac{I_{xx}}{Al^2}}\\
c &= \frac{\frac{165}{104}}{1-\frac{135}{182}\frac{I_{xx}}{Al^2}}\\
p^2 &= \frac{\frac{162}{13}\frac{I_{xx}}{Al^4}}{1-\frac{135}{182}\frac{I_{xx}}{Al^2}}\\
\bar{R} &= \frac{R}{l}\\
\phi &= \frac{\theta}{l}
\end{aligned}
\tag{10.108}
$$

$$
\ddot{\phi}+\left[a(\omega_0+\alpha t)^2+p^2\right]\phi+\frac{c\alpha}{l}\phi^2=-\frac{b}{\alpha}
\tag{10.109}
$$

Making further adjustments

$$r_0 = \frac{\omega_0}{p}$$
$$r_\alpha = \frac{\alpha}{p^2}$$
$$\tau = pt \tag{10.110}$$

we get the following second-order differential equation

$$\frac{d^2\phi}{d\tau^2} + \left[a(r_0 + r_\alpha \tau)^2 + 1\right]\phi + r_\alpha c\phi^2 = -r_\alpha b \tag{10.111}$$

This equation is split into two first-order differential equations as given below.

$$\frac{d\phi}{d\tau} = \overline{\phi}$$
$$\frac{d\overline{\phi}}{d\tau} = -\left\{\left[a(r_0 + r_\alpha \tau)^2 + 1\right]\phi + r_\alpha c\phi^2 + r_\alpha b\right\} \tag{10.112}$$

These equations are solved in Matlab and the results obtained are given here.

The shock response of the accelerating blade is shown in Figure 10.3. We notice that the shock response due to acceleration is more intense for higher rates of acceleration and takes a longer time to die out. This kind of analysis is still not possible through commercial codes. The basic energy methods remain important in understanding the dynamic behavior of several nonlinear structures.

10.8 Complementary Virtual Work

Various forms of virtual work principle have been credited to Johann (Jean) Bernoulli (1667–1748) and Daniel Bernoulli (1700–1782). We can choose one of the two forms:

1. Impose equilibrium on real stresses and forces, by using consistent virtual displacements and strains in the virtual work equation.
2. Impose consistent displacements and strains, by using equilibriated virtual stresses and forces in the virtual work equation.

These two general scenarios give rise to two often stated variational principles. They are valid irrespective of material behavior.

In Section 10.1, we derived the principle of virtual displacements in variational notations for supported bodies. The virtual displacements and strains are specified as variations of the real displacements and strains using variational notation; these virtual displacements are made zero on the part of the surface that has prescribed displacements, and thus the work done by the reactions is zero. There remain only external surface forces on the part that do work. The virtual work equation then becomes the principle of virtual displacements.

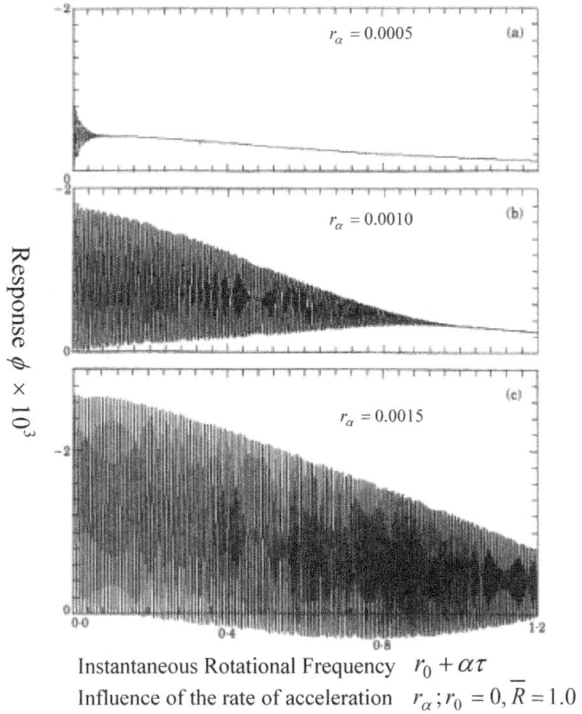

Fig. 10.3 Shock response of a long accelerating blade with Coriolis forces

In the principle of virtual forces, we specify virtual forces and stresses as variations of the real forces and stresses. Also virtual forces are made zero on the part of the surface that has prescribed forces and thus only surface forces (where displacements are prescribed) would do work. The virtual work equation becomes the principle of virtual forces. It is also called the principle of complementary virtual work.

In the virtual work principle, we determine the displacements and strains quite accurately, thus it is most useful in determining the natural frequencies. However, the stresses determined may be in error, because of the rapid deterioration in the accuracy of an approximate solution for the stress field. For engineers, the stress field τ_{ij} is equally important to achieve a good structural design. In the complementary virtual work principle, we vary the stress field and the body forces B_i keeping the displacements fixed. We will of course restrict ourselves to the variations of the stress field and the external forces which satisfy the equations of equilibrium and the boundary conditions, i.e., using tensor notation,

$$\{\delta\tau_{ij}\}_{,j} + \delta B_i = 0 \tag{10.113}$$

with

$$\{\delta\tau_{ij}\}\nu_j + \delta T_i^{(\nu)} \tag{10.114}$$

on the boundary S_2 and

$$\delta T_i^{(\nu)} = 0 \tag{10.115}$$

on the boundary S_1.

The complementary virtual work is defined as

$$\delta W^* = \iiint_V u_i \delta B_i dv + \iint_S u_i \delta T_i^{(\nu)} dA \tag{10.116}$$

The second term in the above is converted into stresses by using Cauchy's formula

$$\iint_S u_i \delta T_i^{(\nu)} dA = \iint_S u_i (\delta\tau_{ij}) \nu_j dA \tag{10.117a}$$

Next we can use the divergence theorem to express work of traction forces over the entire body volume

$$\iint_S u_i (\delta\tau_{ij}) \nu_j dA = \iiint_V [u_i(\delta\tau_{ij})]_{,j} dv \tag{10.117b}$$

$$\iiint_V u_i \delta B_i dv + \iint_S u_i \delta T_i^{(\nu)} dA$$
$$= \iiint_V u_i \left\{\delta B_i + \left(\delta\tau_{ij}\right)_{,j} dv\right\} + \iiint_V u_{i,j} \delta\tau_{ij} dv \tag{10.118}$$

In view of (10.113), the above reduces to

$$\iiint_V u_i \delta B_i dv + \iint_S u_i \delta T_i^{(\nu)} dA = \iiint_V u_{i,j} \delta\tau_{ij} dv \tag{10.119}$$

Using $u_{i,j} = \varepsilon_{ij} + \omega_{ij}$ and noting that the rotation tensor ω_{ij} is skew-symmetric, we have

$$\iiint_V u_i \delta B_i dv + \iint_S u_i \delta T_i^{(\nu)} dA = \iiint_V \varepsilon_{,j} \delta\tau_{ij} dv \tag{10.120}$$

Defining the following complementary strain energy density function [29], $U_0^* = \int_0^\tau \varepsilon d\tau$ so that $\varepsilon_{ij} = \partial U_0^* / \partial\tau_{ij}$, equation (10.120) becomes

$$\iiint_V u_i \delta B_i dv + \iint_S u_i \delta T_i^{(\nu)} dA = \iiint_V \frac{\partial U_0^*}{\partial\tau_{ij}} \delta\tau_{ij} dv \tag{10.121}$$

The right-hand side in the above equation is δU^*. Let us define a potential function

$$V^* = \iiint_V u_i B_i dv + \iint_S u_i T_i^{(\nu)} dA$$

so that

$$\delta V^* = \iiint_V u_i \delta B_i dv + \iint_S u_i \delta T_i^{(\nu)} dA$$

Therefore, equation (10.121) becomes

$$\delta(U^* + V^*) = 0 \tag{10.122}$$

In any approximation technique employing the above principle, we will get a good approximation for the stress field. However, the displacement field could be in error. It is however necessary that our displacement field is not in error to obtain good approximations for the natural frequencies of the system. Therefore the method of complementary virtual work is not very suitable for dynamic analysis. To get over the complication, where we want good approximations for both the displacement and stress fields, we find that Reissner's principle is most appropriate to determine the natural frequencies as also forced vibration displacement and stress response. This principle is discussed in the next section.

10.9 Hellinger–Reissner Variational Principle

The principles of virtual work and Hamilton's principle are minimum principles; the Hellinger–Reissner [19, 33] principle is only a stationary principle in the sense that the nature of the extremum of this functional is not known. Reissner's functional is

$$I_R = \iiint_V \left\{\tau_{ij}\varepsilon_{ij} - U_0^*\left(\tau_{ij}\right)\right\} dv - \iiint_V B_i u_i dv - \iint_S T_i^{(\nu)} u_i dA \tag{10.123}$$

In the above, the body forces and traction forces are both prescribed and we will vary the stresses, strains and displacements, subject to

$$\begin{aligned} \delta\varepsilon_{ij} &= \frac{1}{2}(\delta u_{i,j} + \delta u_{j,i}) \\ \tau_{ij} &= \tau_{ji} \end{aligned} \tag{10.124}$$

Now, taking variation of (10.123),

$$\begin{aligned} \delta I_R = &\iiint_V \left\{\delta\tau_{ij}\varepsilon_{ij} + \tau_{ij}\delta\varepsilon_{ij} - \frac{\partial U_0^*}{\partial \tau_{ij}}\delta\tau_{ij}\right\} dv \\ &- \iiint_V B_i \delta u_i dv - \iint_{S1} T_i^{(\nu)} \delta u_i dA \end{aligned} \tag{10.125}$$

Using (10.124), the second term in (10.125) can be written as

$$\iiint_V \tau_{ij}\delta\varepsilon_{ij}dv = \iiint_V \tau_{ij}\delta u_{i,j}dv$$

$$= \iiint_V \left(\tau_{ij}\delta u_i\right)_{,j} dv - \iiint_V \tau_{ij,j}\delta u_i dv \qquad (10.126)$$

We can further use the divergence theorem to write the above as

$$\iiint_V \tau_{ij}\delta\varepsilon_{ij}dv = \iint_S \tau_{ij}\nu_j\delta u_i dA - \iiint_V \tau_{ij,j}\delta u_i dv \qquad (10.127)$$

Equation (10.125) can now be written as

$$\delta I_R = \iiint_V \delta\tau_{ij}\varepsilon_{ij}dv + \iint_S \tau_{ij}\nu_j\delta u_i dA - \iiint_V \tau_{ij,j}\delta u_i dv$$

$$- \iiint_V \frac{\partial U_0^*}{\partial \tau_{ij}}\delta\tau_{ij}dv - \iiint_V B_i\delta u_i dv - \iint_{S1} T_i^{(\nu)}\delta u_i dA \qquad (10.128)$$

Regrouping

$$\delta I_R = \iiint_V \left\{\left(\varepsilon_{ij} - \frac{\partial U_0^*}{\partial \tau_{ij}}\right)\delta\tau_{ij} - \left(\tau_{ij,j} + B_i\right)\delta u_i\right\} dv$$

$$+ \iint_{S1} \left(\tau_{ij}\nu_j - T_i^{(\nu)}\right)\delta u_i dA + \iint_{S2} \tau_{ij}\nu_j\delta u_i dA \qquad (10.129)$$

For $\delta I_R = 0$ in V, equation (10.129) gives

$$\varepsilon_{ij} = \frac{\partial U_0^*}{\partial \tau_{ij}} \qquad (10.130)$$

$$\tau_{ij,j} + B_i = 0 \qquad (10.131)$$

and on the surface

$$\tau_{ij}\nu_j = T_i^{(\nu)} \text{ for } S_1 \qquad (10.132)$$

$$\delta u_i = 0 \text{ on } S_2 \qquad (10.133)$$

The above result is a special case of the Hellinger–Reissner variational principle and referred to simply as the Reissner principle. It gives stress-strain law (10.130) and equations of equilibrium (10.131) at the same time satisfying stress and displacement boundary conditions. The strain displacement relations (10.124) are already included in the process. Therefore we expect Reissner's principle to give better approximations for both stress and displacement fields unlike the virtual work, a principle which either gives a good approximation for displacements or stresses as the case may be.

We will show here how a rotating turbine blade vibration problem is addressed by Reissner's principle [39]. Figure 10.4 shows the airfoil rotating blade of length L with an asymmetry r_x. O is the center of flexure of the asymmetric aerofoil at

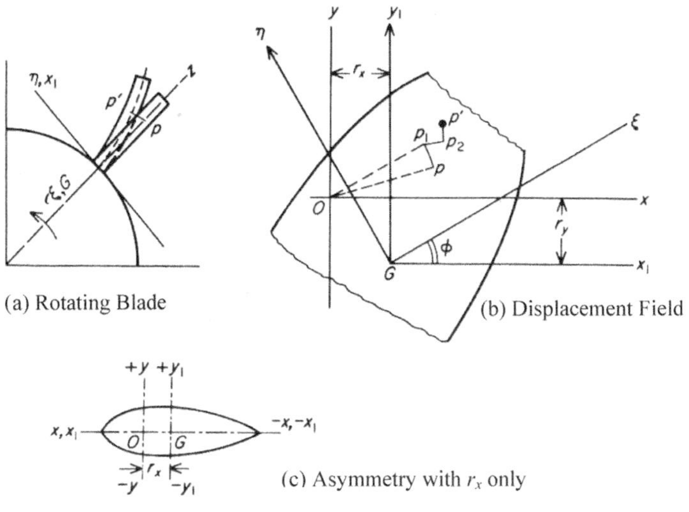

Fig. 10.4 Rotating turbine blade

distance z from the blade root; O is displaced by r_x and r_y from the centroid G in x_1y_1 axis system with x_1 in the plane of the disk. The O_{xyz} axis system is located at the blade root and the $Gx_1y_1z_1$ system is parallel to this as shown in Figure 10.4. The blade is mounted at a stagger angle ϕ on a disk of radius R rotating at ω rad/sec. The η axis is parallel to the plane of disk and ξ perpendicular to the plane.

The displacement field u_x, u_y, u_z of a particle p located at $\underset{\cdot}{x}\,\underset{\cdot}{y}\,z$ or $\underset{\cdot}{x_1}\,\underset{\cdot}{y_1}\,z$ is described by torsional deflection θ to p_1, x bending to p_2, and y bending to p'.

$$u_x = x - \underset{\cdot}{y}\,\theta = x_1 - \underset{\cdot}{y_1}\,\theta$$

$$u_y = y + \underset{\cdot}{x}\,\theta = y_1 + \underset{\cdot}{x_1}\,\theta$$

$$u_z = -\underset{\cdot}{x_1}\left(\phi_1 + \overline{r_y\theta}'\right) - y_1\left(\phi_2 + \overline{r_x\theta}'\right) + \phi_c\theta' \tag{10.134}$$

where ϕ_1 and ϕ_2 are bending slopes xz- and yz-planes. The strain and stress fields are calculated from

$$\varepsilon_{ij} = \frac{1}{2}(u_{i,j} + u_{j,i})$$

$$\sigma_{ij} = \lambda\delta_{ij}\varepsilon_{ii} + 2G\varepsilon_{ij} \tag{10.135}$$

The shear stresses in the above can be corrected to account for shear coefficient K. The following bending moments, shear forces, torsional stiffness, torsional constant and twisting moments including warping ϕ_c can be defined.

$$
\begin{aligned}
M_x &= \int_A \tau_{zz}\, y_1\, dA \\
M_y &= \int_A \tau_{zz}\, x_1\, dA \\
V_x &= \int_A \tau_{zx} dA \\
V_y &= \int_A \tau_{zy} dA \\
C &= KG \int_A \{(\phi_{c,x} - y_1)^2 + (\phi_{c,y} - x_1)^2\} dA \\
C_1 &= E \int_A (\phi_c + x_1\, y - y_1\, x)^2 dA \\
T_\theta &= C\theta' - C_1\theta'''
\end{aligned}
\tag{10.136}
$$

Now using (10.123) we can set up a Reissner functional as follows. It is assumed that the energy associated with C_1 is negligible in comparison with that due to C and that $x_1 x_1$ and $y_1 y_1$ are principal Centroidal axes of inertia so that I_{x1y1} vanishes and we have

$$
I_R = -\int_0^L \begin{bmatrix} M_x\phi_2' + M_y\phi_1' - V_x\left(x' - \phi_1\right) - V_y\left(y' - \phi_2\right) + T_\theta\theta' + \\ + \frac{M_x^2 I_{y1y1} + M_y^2 I_{x1x1}}{2EI_{x1x1}I_{y1y1}} + \frac{T_\theta^2}{2C} + \frac{V_x^2 + V_y^2}{2KGA} \end{bmatrix} dZ
\tag{10.137}
$$

The kinetic energy including centrifugal effects for zero stagger angle of the blade is

$$
T = \frac{1}{2}\int_0^L \left\{ \begin{array}{l} m\left(\dot{x} + r_y\dot{\theta}\right)^2 + m\left(\dot{y} + r_x\dot{\theta}\right)^2 + m_p\dot{\theta}^2 + \\ + m_{y1y1}\left(\dot{\phi}_1 + r_y\dot{\theta}'\right)^2 + m_{x1x1}\left(\dot{\phi}_2 + r_x\dot{\theta}'\right)^2 - \\ - m\omega^2\left[x_1'^2 + y_1'^2\right]\int_Z^L (R + Z)\, dZ + m\omega^2\left(x + r_y\theta\right)^2 \end{array} \right\} dZ
\tag{10.138}
$$

where m is mass per unit length, m_p is $\rho I_{pp}\ldots$

If one assumes asymmetry in one plane only as in Figure 10.4c, $r_y = 0$, we have coupled bending torsion vibrations in $(y - \theta)$ and uncoupled flexural vibrations in the $x - z$ plane. For coupled motion $(y - \theta)$, we have the Reissner functional

$$
L_R = T - I_R = \int_0^L \left\{ \begin{array}{l} \frac{m}{2}\left(\dot{y} + r_x\dot{\theta}\right)^2 + m_p\dot{\theta}^2 + m_{x1x1}\left(\dot{\phi}_2 + r_x\dot{\theta}'\right)^2 - \\ - m\omega^2\left[\left(y' - r_x\theta'\right)^2\right]\int_Z^L (R + Z)\, dZ + \\ + M_x\phi_2' - V_y\left(y' - \phi_2\right) - T_\theta\theta' + \frac{M_x^2}{2EI_{x1x1}} + \\ + \frac{T_\theta^2}{2C} + \frac{V_y^2}{2KGA} \end{array} \right\} dZ
\tag{10.139}
$$

where

$$
\begin{aligned}
M_x &= -EI_{x1x1}\phi_2' \\
V_y &= KGA\left(y' - \phi_2\right) \\
T_\theta &= C\theta' \\
M_x' &= V_y \\
V_y' &= 0 \\
T_\theta' &= 0
\end{aligned}
$$

The above Reissner functional (10.139) is seen to be a functional of the form

$$
L_R = f\left(M_x, V_y, T_\theta, y, y', \dot{y}, \phi_2, \phi_2', \dot{\phi}_2, \theta, \theta', \dot{\theta}, \dot{\theta}', Z, t\right) \tag{10.140}
$$

The time-averaged Reissner functional for harmonic motion with frequency p is

$$
\bar{L}_R = \int_0^{2\pi/p} (T - I_R)\, dt = \frac{\pi L}{p} \int_0^1
\times \left\{
\begin{array}{l}
p^2 \lambda E I_{x1x1}[(y^2 + 2L\bar{r}_x\theta y) + \bar{r}_{cf}^2 L^2\theta^2 + \frac{\bar{r}^2}{L^2}(\phi_2 + L\bar{r}_x\theta')^2] - \\
-\frac{\alpha^2 E I_{x1x1}}{2L^2}(y' + L\bar{r}_x\theta')^2[\bar{R}(1-z) + 0.5(1-z^2)] + \\
+ M_x\phi_2' - V_y(y' - \phi_2) - T_\theta\theta' + \frac{\bar{r}^2 L^2}{2EI_{x1x1}}\frac{V_y^2}{\bar{k}^2} + \frac{M_x^2}{2EI_{x1x1}} + \frac{T_\theta^2}{2\mu^2 EI_{x1x1}}
\end{array}
\right\} dz \tag{10.141}
$$

where

$$
\bar{R} = \frac{R}{L}, \quad \bar{r} = \frac{r}{L} \ldots, \quad r_{cf} = \left(\frac{I_p}{A} + r_x^2\right)^{1/2}, \quad \lambda = \frac{EI_{x1x1}}{mL^4}
$$

The terms containing α^2 represent the effects due to rotation, those associated with $\bar{r}^2$ account for the effects of rotary inertia, and those having either the shear slope $(y' - \phi_2)$ or the non-dimensional shear parameter $\bar{k} = (KG/E)^{1/2}$ indicate the effects of shear deformation. The following shape functions are assumed

$$
\begin{aligned}
y &= \sum_i (A_i z^i + A_{i+1} z^{i+1}) \\
\phi_2 &= \sum_i (B_i z^i + B_{i+1} z^{i+1}) \\
\theta &= \sum_i (C_i z^i + C_{i+1} z^{i+1})
\end{aligned} \tag{10.142}
$$

The arbitrary constants, A, B, C are eliminated using the boundary conditions

$$
\begin{aligned}
&y = \phi_2 = \theta = 0 \text{ at } z = 0 \\
&V_y = KGA(y' - \phi_2 t) = 0 \text{ at } Z = 0,\ L \text{ or } y \text{ prescribed} \\
&M_x = -EI_{x1x1}\phi_2' = 0 \text{ at } Z = 0,\ L \text{ or } \phi_2 \text{ prescribed} \\
&T_\theta = C\theta' = 0 \text{ at } Z = 0,\ L \text{ or } \theta \text{ prescribed}
\end{aligned}
\tag{10.143}
$$

i.e.,

$$
\begin{aligned}
&y = \phi_2 = \theta = 0 \text{ at } z = 0 \\
&(y' - \phi_2) = \phi_2' = \theta' = 0 \text{ at } z = 1
\end{aligned}
\tag{10.43a}
$$

Similarly the shape functions for the bending moment, shear force and twisting moment are assumed to be

$$
\begin{aligned}
M_x &= \sum_i \left\{E_i\,(1-z)^i + E_{i+1}\,(1-z)^{i+1}\right\} \\
V_y &= \sum_i \left\{D_i\,(1-z)^i + D_{i+1}\,(1-z)^{i+1}\right\} \\
T_\theta &= \sum_i \left\{F_i\,(1-z)^i + F_{i+1}\,(1-z)^{i+1}\right\}
\end{aligned}
\tag{10.144}
$$

that satisfy the boundary conditions $M_x = V_y = T_\theta = 0$ at $z = 1$. The arbitrary constants, D, E and F are eliminated by applying the boundary conditions

$$
V_y' = 0, \quad M_x' = V_y, \quad T_\theta' = 0 \text{ at } z = 0 \tag{10.145}
$$

The shape functions thus determined are substituted in (10.141) and Ritz minimization is applied according to

$$
\begin{aligned}
&\frac{\partial \bar{L}_R}{\partial A_i} = 0 \\
&\frac{\partial \bar{L}_R}{\partial B_i} = 0 \\
&\ldots \\
&\frac{\partial \bar{L}_R}{\partial F_i} = 0
\end{aligned}
\tag{10.146}
$$

This leads to an eigen-value problem

$$
[A] + p^2[B] = 0 \tag{10.147}
$$

With a one term approximation, we have the eigen-problem with 6×6 matrix size and six eigen-values. These elements can be obtained in closed form by integration of polynomials or by numerical integration. This is similar to a one-element finite element solution and therefore takes least time. This is also more accurate as displacement and stress fields are simultaneously varied and hence both frequency and stress problems can be simultaneously solved accurately. However the energy formulations with one element cannot capture intricate geometries at discontinuities and this is a major draw-back in modern designs where conventional factor of safety is not practiced (see also [31]).

10.10 Hu–Washizu Principle

Hu [20] and Washizu [46] considered a more general variational principle in which displacement, strain and stress all vary independently.

The dynamic equations of equilibrium (ρ mass density) from (9.1) are

$$\begin{aligned}
&\frac{\partial \sigma_{xx}}{\partial x} + \frac{\partial \sigma_{yx}}{\partial y} + \frac{\partial \sigma_{zx}}{\partial z} + b_x - \rho \ddot{u}_x = 0 \\
&\frac{\partial \sigma_{xy}}{\partial x} + \frac{\partial \sigma_{yy}}{\partial y} + \frac{\partial \sigma_{zy}}{\partial z} + b_y - \rho \ddot{u}_y = 0 \\
&\frac{\partial \sigma_{xz}}{\partial x} + \frac{\partial \sigma_{yz}}{\partial y} + \frac{\partial \sigma_{zz}}{\partial z} + b_z - \rho \ddot{u}_z = 0
\end{aligned} \tag{10.148}$$

In tensor notation, the above three equations are

$$\sigma_{ij,j} + b_i - \rho \ddot{u}_i = 0 \tag{10.148a}$$

The strain displacement relations are given by (9.2)

$$\begin{aligned}
&\varepsilon_{xx} = \frac{\partial u_x}{\partial x}, \quad \varepsilon_{yz} = \frac{1}{2}\left(\frac{\partial u_z}{\partial y} + \frac{\partial u_y}{\partial z}\right) = \varepsilon_{zy} \\
&\varepsilon_{yy} = \frac{\partial u_y}{\partial y}, \quad \varepsilon_{zx} = \frac{1}{2}\left(\frac{\partial u_x}{\partial z} + \frac{\partial u_z}{\partial x}\right) = \varepsilon_{xz} \\
&\varepsilon_{zz} = \frac{\partial u_z}{\partial z}, \quad \varepsilon_{xy} = \frac{1}{2}\left(\frac{\partial u_y}{\partial x} + \frac{\partial u_x}{\partial y}\right) = \varepsilon_{yx}
\end{aligned} \tag{10.149}$$

i.e.,

$$\varepsilon_{ij} = \frac{1}{2}(u_{i,j} + u_{j,i}) \tag{10.149a}$$

The strain energy density for an isotropic elastic medium is given by

$$W = \frac{1}{2}\lambda e^2 + \mu(\varepsilon_{xx}^2 + \varepsilon_{yy}^2 + \varepsilon_{zz}^2) + 2\mu(\varepsilon_{xy}^2 + \varepsilon_{xz}^2 + \varepsilon_{yz}^2) \tag{10.150}$$

where $e = \varepsilon_{xx} + \varepsilon_{yy} + \varepsilon_{zz}$ and λ and μ are Lame's constants. The stress strain relations are then written as

$$\sigma_{ij} = \frac{\partial W}{\partial \varepsilon_{ij}} = W_{,\varepsilon_{ij}} \tag{10.151}$$

Hu–Washizu for a static problem formed a quantity from which, if its first variation is put equal to zero, we will get the equilibrium equations (10.148). the strain-displacement relations (10.149), the stress-strain relations (10.151) and the boundary conditions. For dynamic problems, the principle is extended by introducing velocity components, v_i in terms of a kinetic energy density is defined and, as in Kirchoff's [22] variational equation, integrating over time between fixed limits at which the displacement variations are assumed to disappear. This procedure introduces inertia forces into the equilibrium equations and provides the relationship between $\dot{u}_i$ and the kinetic energy. Hu–Washizu's variational functional is [2],

$$J = \int_{t_0}^{t_1} \left\{ \begin{array}{l} \int_V \Big[\rho v_i \dot{u}_i - T(v_i) - W(\varepsilon_{ij}) + \\ + \left\{ \varepsilon_{ij} - \frac{1}{2}(u_{i,j} + u_{j,i}) \right\} \sigma_{ij} + \bar{b}_i u_i \Big] dV \\ + \int_{S_p} \bar{g}_i u_i dS + \int_{S_U} g_i (u_i - \bar{u}_i) dS \end{array} \right\} dt \tag{10.152}$$

In the above T is the kinetic energy density $\frac{1}{2}\rho(v_x^2 + v_y^2 + v_z^2)$, g_i represents surface traction and an overbar indicates quantities which are prescribed, surface traction being prescribed over the region S_p of the boundary and surface displacements over S_u. Taking the first variation of (10.152)

$$\delta J = \int_{t_0}^{t_1} \left\{ \begin{array}{l} \int_V \left[\begin{array}{l} \rho \dot{u}_i \delta v_i + \rho v_i \delta \dot{u}_i - T_{,v_i} \delta v_i - W_{,\varepsilon_{ij}} \delta \varepsilon_{ij} + \\ \varepsilon_{ij} \delta \sigma_{ij} + \sigma_{ij} \delta \varepsilon_{ij} - \frac{1}{2} \sigma_{ij} \delta (u_{i,j} + u_{j,i}) - \\ -\frac{1}{2}(u_{i,j} + u_{j,i}) \delta \sigma_{ij} + \bar{b}_i \delta u_i \end{array} \right] dV \\ + \int_{S_p} \bar{g}_i \delta u_i dS + \int_{S_u} \{(u_i - \bar{u}_i) \delta g_i + g_i \delta u_i\} dS \end{array} \right\} dt \tag{10.153}$$

The second term in the volume integral above is transformed and simplified as follows:

$$\int_{t_0}^{t_1} \int_V \rho v_i \delta \dot{u}_i dV dt = \int_V \rho v_i \delta u_i dV]_{t_0}^{t_1} - \int_{t_0}^{t_1} \int_V \rho \dot{v}_i \delta u_i dV dt = - \int_{t_0}^{t_1} \int_V \rho \dot{v}_i \delta u_i dV dt \tag{10.154}$$

The seventh term in (10.153) is integrated by parts to give

$$\int_{t_0}^{t_1} \left\{ \int_V -\frac{1}{2} \sigma_{ij} \delta (u_{i,j} + u_{j,i}) dV \right\} dt = \int_{t_0}^{t_1} \left\{ \int_V \sigma_{ij,j} \delta u_i dV - \int_S g_i \delta u_i dS \right\} dt \tag{10.155}$$

Using (10.154) and (10.155) in (10.153) gives

$$\delta J = \int_{t_0}^{t_1} \left\{ \begin{array}{l} \int_V \left[\begin{array}{l} \rho \dot{u}_i \delta v_i - \rho \dot{v}_i \delta u_i - T_{,v_i} \delta v_i - W_{,\varepsilon_{ij}} \delta \varepsilon_{ij} + \\ \varepsilon_{ij} \delta \sigma_{ij} + \sigma_{ij} \delta \varepsilon_{ij} + \sigma_{ij,j} \delta u_i - \\ -\frac{1}{2}(u_{i,j} + u_{j,i}) \delta \sigma_{ij} + b_i \delta u_i \end{array} \right] dV \\ + \int_{S_p} \bar{g}_i \delta u_i dS + \int_{S_u} (u_i - \bar{u}_i) \delta g_i dS \end{array} \right\} dt \quad (10.156)$$

Setting the first variation to zero and rearranging gives

$$\int_V [\sigma_{ij,j} + b_i - \rho \dot{v}_i] \delta u_i dV + \int_V [\sigma_{ij} - W_{,\varepsilon_{ij}}] \delta \varepsilon_{ij} dV$$
$$+ \int_V \left[\varepsilon_{ij} - \frac{1}{2}(u_{i,j} + u_{j,i}) \right] \delta \sigma_{ij} dV + \int_V [\rho \dot{u}_i - T_{,v_i}] \delta v_i dV$$
$$+ \int_{S_p} [\bar{g}_i - g_i] \delta u_i dS + \int_{S_u} (u_i - \bar{u}_i) \delta g_i dS = 0 \quad (10.157)$$

Since the variations of u_i, v_i, ε_{ij} and σ_{ij} are independent, the above equations lead to the following:

$$\begin{aligned} &\sigma_{ij,j} + b_i - \rho \dot{v}_i = 0 \\ &\sigma_{ij} = W_{,\varepsilon_{ij}} \\ &\varepsilon_{ij} = \frac{1}{2}(u_{i,j} + u_{j,i}) \\ &\rho \dot{u}_i - T_{,v_i} \end{aligned} \quad (10.158)$$

The first equation in the above represents the dynamic equations of equilibrium in equation (10.148); the second equation is the stress-strain relation in (10.151); the third equation is strain-displacement relations in (10.149) and the last equation is simply $v_i = \dot{u}_i$. Thus, Hu–Washizu's variational principle leads to all equations of elasticity.

Various authors have derived different higher order theories, e.g., Rao and Rao [27]; Rayleigh [32] – Love's theory [25], Bishop's theory [5], Mindlin and Herrmann's theory [26] for longitudinal vibration of rods; Saint-Venant's theory [36], Timoshenko's [41] and Gere's theory [16], Reissner [34] and Lo and Goulard's theory [24], Barr [1] for torsional vibrations of rods; Timoshenko's theory [40], Volterra's theory [44] and Krishna Murthy's theory [23] for beams.

The interest in higher order theories gradually decreased considerably with the advent of finite element methods that can handle large size practical problems. The higher order effects are found to be of not much importance in practical engineering applications. The above higher order theories can be referenced as given and no derivations are presented. However, we will present one last theory in torsion of rods which tried to remove certain discrepancies in the equations.

10.11 Different Theories of Torsion of Rods

Different theories for torsional vibration of rods can be derived starting from an appropriate state, calculating the strain and kinetic energies and using variational principles. We will first give several of these theories before deriving a refined theory.

10.11.1 Coulomb (1784) Elementary Theory , see Timoshenko and Goodier [42] for Circular Rods

State Assumption:

$$\begin{aligned} u_x &= 0 \\ u_y &= -z\theta \\ u_z &= y\theta \end{aligned} \tag{10.159a}$$

Differential Equation:

$$\rho I_p \ddot{\theta} - G I_p \theta_{,xx} = 0 \tag{10.159b}$$

Boundary Conditions:

$$\left[G I_p \theta_{,x}\, \delta\theta\right]_0^l = 0 \tag{10.159c}$$

Warping Differential Equation: None

Here ρ is mass density, I_p is polar moment of inertia of cross-section.

10.11.2 St. Venant (1853) Theory, see Todhunter [43] and Timoshenko and Goodier [42] for Circular Rods

State Assumption:

$$\begin{aligned} u_x &= \phi\theta_{,x} \\ u_y &= -z\theta \\ u_z &= y\theta \end{aligned} \tag{10.160a}$$

Differential Equation: (Longitudinal Inertia ignored)

$$\begin{aligned} &\rho I_p \ddot{\theta} - C_1 \theta_{,xx} = 0 \\ &C_1 = G \int_A [(\phi_{,y} - z)^2 + (\phi_{,z} + y)^2] dA \end{aligned} \tag{10.160b}$$

Boundary Conditions:

$$[C_1 \theta_{,x}\, \delta\theta]_0^l = 0 \tag{10.160c}$$

Warping Differential Equation:

$$\nabla^2\phi = 0 \tag{10.160d}$$

where ϕ is warping function.

10.11.3 Love's (1944) Theory

State Assumption:

$$\begin{aligned} u_x &= \phi\theta_{,x} \\ u_y &= -z\theta \\ u_z &= y\theta \end{aligned} \tag{10.161a}$$

Differential Equation:

$$\begin{aligned} &\rho I_p\ddot{\theta} - (C_1\partial_x^2 + EI_{\phi\phi}\partial_x^4)\theta = 0 \\ &I_{\phi\phi} = \int_A \phi^2 dA \end{aligned} \tag{10.161b}$$

Boundary Conditions:

$$\left[(\rho I_{\phi\phi}\ddot{\theta}_{,x} + C_1\theta_{,x})\delta\theta\right]_0^l = 0 \tag{10.161c}$$

Warping Differential Equation:

$$\begin{aligned} &g\nabla^2\phi - r\phi = 0 \\ &g = \int_0^l G\theta_{,x}^2\, dx \\ &r = \int_0^l \rho\theta_{,xt}^2\, dx \end{aligned} \tag{10.161d}$$

10.11.4 Timoshenko (1945) – Gere's (1954) Theory

State Assumption:

$$\begin{aligned} u_x &= \phi\theta_{,x} \\ u_y &= -z\theta \\ u_z &= y\theta \\ \varepsilon_{xx} &= \phi\theta_{,xx} \\ \varepsilon_{yy} &= -\nu\varepsilon_{xx} \\ \varepsilon_{zz} &= -\nu\varepsilon_{xx} \end{aligned}$$

Differential Equation:

$$\rho I_p \ddot{\theta} - (C_1 \partial_x^2 + E I_{\phi\phi} \partial_x^4)\theta = 0 \tag{10.162b}$$

Boundary Conditions:

$$\begin{aligned} &\left[E I_{\phi\phi} \partial_x^2 \theta \delta\theta,_x \right]_0^l = 0 \\ &\left[(C_1 \partial_x - E I_{\phi\phi} \partial_x^3)\theta \delta\theta\right]_0^l = 0 \end{aligned} \tag{10.162c}$$

Warping Differential Equation:

$$\begin{aligned} &-g\nabla^2\phi + d\phi = 0 \\ &d = \int_0^l E\theta,_{xx}^2 \, dx \end{aligned} \tag{10.162d}$$

10.11.5 Reissner (1952) and Lo–Goulard's (1955) Theory

State Assumption:

$$\begin{aligned} u_x &= \phi\alpha \\ u_y &= -z\theta \\ u_z &= y\theta \\ \varepsilon_{xx} &= \phi\alpha,_x \\ \varepsilon_{yy} &= -\nu\varepsilon_{xx} \\ \varepsilon_{zz} &= -\nu\varepsilon_{xx} \end{aligned} \tag{10.163a}$$

Differential Equation:

$$\begin{bmatrix} \rho I_p & 0 \\ 0 & 0 \end{bmatrix} \begin{Bmatrix} \ddot{\theta} \\ \ddot{\alpha} \end{Bmatrix} + \begin{bmatrix} -G I_p \partial_x^2 & -G L \partial_x \\ G L \partial_x & (G K - E I_{\phi\phi} \partial_x^2) \end{bmatrix} \begin{Bmatrix} \theta \\ \alpha \end{Bmatrix} = \begin{Bmatrix} 0 \\ 0 \end{Bmatrix}$$

$$\begin{aligned} L &= \int \left(y\phi,_z + z\phi,_y\right) dA \\ K &= \int \left(\phi,_y^2 + \phi,_z^2\right) dA \end{aligned} \tag{10.163b}$$

Boundary Conditions:

$$\begin{aligned} &\left[(G I_p \theta,_x + G L \alpha)\delta\theta\right]_0^l = 0 \\ &\left[- E I_{\phi\phi} \alpha,_x \delta\alpha\right]_0^l = 0 \end{aligned}$$

Warping Differential Equation:

$$-g_R \nabla^2 \phi + d_R \phi = 0$$

$$d_R = \int_0^l E\varepsilon,_x^2 \, dx$$

10.11.6 Barr's (1962) Theory

State Assumption:

$$\begin{aligned} u_x &= \phi\alpha \\ u_y &= -z\theta \\ u_z &= y\theta \\ \varepsilon_{xx} &= \phi\alpha,_x \\ \varepsilon_{yy} &= -\nu\varepsilon_{xx} \\ \varepsilon_{zz} &= -\nu\varepsilon_{xx} \end{aligned} \tag{10.164a}$$

Differential Equation:

$$\begin{bmatrix} \rho I_p & 0 \\ 0 & \rho I_{\phi\phi} \end{bmatrix} \begin{Bmatrix} \ddot{\theta} \\ \ddot{\alpha} \end{Bmatrix} + \begin{bmatrix} -GI_p \partial_x^2 & -GL\partial_x \\ GL\partial_x & \left(GK - EI_{\phi\phi}\partial_x^2\right) \end{bmatrix} \begin{Bmatrix} \theta \\ \alpha \end{Bmatrix} = \begin{Bmatrix} 0 \\ 0 \end{Bmatrix} \tag{10.164b}$$

Boundary Conditions:

$$\begin{aligned} &\left[(GI_p\theta,_x + GL\alpha)\delta\theta\right]_0^l = 0 \\ &[-EI_{\phi\phi}\alpha,_x \, \delta\alpha]_0^l = 0 \end{aligned} \tag{10.164c}$$

Warping Differential Equation:

$$-g_R \nabla^2 \phi + d_R \phi = 0 \tag{10.164d}$$

10.11.7 Refined Theory by Rao (1974)

The assumed state in Timoshenko–Gere theory does not satisfy the strain-displacement relations according to Poisson equations. To remove this discrepancy, the following state is assumed:

$$\begin{aligned} u_x &= \phi\theta_{,x} \\ u_y &= -z\theta - \nu\theta_{,xx} \int \phi dy \\ u_z &= y\theta - \nu\theta_{,xx} \int \phi dz \end{aligned} \tag{10.165}$$

The velocity field is then

$$
\begin{aligned}
u_{x,t} &= \phi\theta_{,xt} \\
u_{y,t} &= -z\theta_{,t} - \nu\theta_{,xxt}\int \phi dy \\
u_{z,t} &= y\theta_{,t} - \nu\theta_{,xxt}\int \phi dz
\end{aligned} \tag{10.166}
$$

The strain field is given by

$$
\begin{aligned}
\varepsilon_{xx} &= \phi\theta_{,xx} \\
\varepsilon_{yy} &= -\nu\phi\theta_{,xx} \\
\varepsilon_{zz} &= -\nu\phi\theta_{,xx}
\end{aligned}
$$

$$
\begin{aligned}
\varepsilon_{xy} &= \frac{1}{2}\left[(\phi_{,y} - z)\theta_{,x} - \nu\theta_{,xxx}\int \phi dy\right] \\
\varepsilon_{yz} &= -\frac{\nu\theta_{,xx}}{2}\left[\left(\int \phi dy\right)_{,z} + \left(\int \phi dz\right)_{,y}\right] \\
\varepsilon_{zx} &= \frac{1}{2}\left[(\phi_{,z} + y)\theta_{,x} - \nu\theta_{,xxx}\int \phi dz\right]
\end{aligned} \tag{10.167}
$$

The corresponding stresses are

$$
\begin{aligned}
\tau_{xx} &= E\phi\theta_{,xx} \\
\tau_{yy} &= 0 \\
\tau_{zz} &= 0
\end{aligned}
$$

$$
\begin{aligned}
\tau_{xy} &= G\left[(\phi_{,y} - z)\theta_{,x} - \nu\theta_{,xxx}\int \phi dy\right] \\
\tau_{yz} &= -\nu\theta_{,xx}G\left[\left(\int \phi dy\right)_{,z} + \left(\int \phi dz\right)_{,y}\right] \\
\tau_{zx} &= G\left[(\phi_{,z} + y)\theta_{,x} - \nu\theta_{,xxx}\int \phi dz\right]
\end{aligned} \tag{10.168}
$$

The strain and kinetic energies of the system are obtained as

$$2U = \int_0^l EI_{\phi\phi}\theta_{,xx}^2 dx + \int_0^l C_1 G\theta_{,x}^2 dx + \int_0^l \nu^2 GC_2\theta_{,xxx}^2 dx$$

$$- \int_0^l 2\nu GC_3\theta_{,x}\theta_{,xxx} dx + \int_0^l \nu^2 GC_4\theta_{,xx}^2 dx$$

$$2T = \int_0^l \rho I_p\theta_{,t}^2 dx + \int_0^l C_2\rho\nu^2\theta_{,xxt}^2 dx + \int_0^l 2\rho\nu C_5\theta_{,xxt}\theta_{,t} dx$$

$$+ \int_0^l \rho I_{\phi\phi}\theta_{,xt}^2 dx \tag{10.169}$$

where

$$C_2 = \int_A \left[\left(\int \phi dy\right)^2 + \left(\int \phi dz\right)^2\right] dA$$

$$C_3 = \int_A \left[(\phi_{,y} - z)\int \phi dy + (\phi_{,z} + y)\int \phi dz\right] dA$$

$$C_4 = \int_A \left[\left(\int \phi dy\right)_{,z} + \left(\int \phi dz\right)_{,y}\right]^2 dA$$

$$C_5 = \int_A \left[z\int \phi dy - y\int \phi dz\right] dA \tag{10.170}$$

Following Hamilton's principle we can get the governing differential equation of motion and boundary conditions.

$$\nu^2 GC_2\theta_{,xxxxxx} - (EI_{\phi\phi} + \nu^2 GC_4 + 2\nu GC_3)\theta_{,xxxx} - C_2\rho\nu^2\theta_{,xxxxtt}$$

$$+ C_1 G\theta_{,xx} + (\rho I_{\phi\phi} - 2\rho\nu C_5)\theta_{,xxtt} - \rho I_p\theta_{,tt} = 0 \tag{10.171}$$

$$\left[\left\{\begin{matrix} -\left(EI_{\phi\phi} + \nu^2 GC_4\right)\theta_{,xxxx} + C_1 G\theta_{,x} + \nu^2 GC_2\theta_{,xxxxxx} - 2\nu GC_3\theta_{,xxx} \\ -C_2\rho\nu^2\theta_{,xxxxtt} - \rho\nu C_5\theta_{,xtt} + \rho I_{\phi\phi}\theta_{,xtt} \end{matrix}\right\}\delta\theta\right]_0^l$$

$$+ \left[\{(EI_{\phi\phi} + \nu^2 GC_4)\theta_{,xx} - \nu^2 GC_2\theta_{,xxxx}\right.$$

$$\left. + \nu GC_3\theta_{,xx} + C_2\rho\nu^2\theta_{,xxtt} + \rho\nu C_5\theta_{,tt}\}\delta\theta_{,x}\right]_0^l$$

$$+ \left[\{\nu^2 GC_2\theta_{,xxx} - \nu GC_3\theta_{,x}\}\delta\theta_{,xx}\right]_0^l = 0 \tag{10.172}$$

With $\nu = 0$ the above equations reduce to the Timoshenko–Gere theory.

The linear system of equations can be solved as follows. For free vibrations $\theta = \theta \sin pt$ and (10.171) becomes

$$\nu^2 GC_2\theta_{,xxxxxx} - (EI_{\phi\phi} + \nu^2 GC_4 + 2\nu GC_3 - C_2\rho\nu^2 p^2)\theta_{,xxxx} + (C_1 G - \rho I_{\phi\phi} p^2 + 2\rho\nu C_5 p^2)\theta_{,xx} + \rho I_p p^2\theta = 0 \quad (10.173)$$

The above equation is written as

$$B_6\theta_{,xxxxxx} - (B_4 - D_4 p^2)\theta_{,xxxx} + (B_2 - D_2 p^2)\theta_{,xx} + D_0 p^2\theta = 0 \quad (10.173a)$$

The solution for the above is $\theta = Me^{bx}$ and hence

$$B_6 b^6 - (B_4 - D_4 p^2)b^4 + (B_2 - D_2 p^2)b^2 + D_0 p^2 = 0 \quad (10.174)$$

There are three roots; let them be b_1^2, b_2^2, b_3^2 and they in turn give six roots for b viz., $b_{11}, b_{12}, b_{21}, b_{22}, b_{31}, b_{32}$. Therefore

$$\theta = M_{11}e^{b_{11}x} + M_{12}e^{b_{12}x} + M_{21}e^{b_{21}x} + M_{22}e^{b_{22}x} + M_{31}e^{b_{31}x} + M_{32}e^{b_{32}x} \quad (10.175)$$

For fixed-fixed boundary conditions, $\theta = \theta_{,x} = \theta_{,xx} = 0$ at $x = 0$ and l and hence we get

$$\begin{vmatrix} 1 & 1 & 1 & 1 & 1 & 1 \\ b_{11} & b_{12} & b_{21} & b_{22} & b_{31} & b_{32} \\ b_1^2 & b_1^2 & b_2^2 & b_2^2 & b_3^2 & b_3^2 \\ e^{b_{11}l} & e^{b_{12}l} & e^{b_{21}l} & e^{b_{22}l} & e^{b_{31}l} & e^{b_{32}l} \\ b_{11}e^{b_{11}l} & b_{12}e^{b_{12}l} & b_{21}e^{b_{21}l} & b_{22}e^{b_{22}l} & b_{31}e^{b_{31}l} & b_{32}e^{b_{32}l} \\ b_1^2 e^{b_{11}l} & b_1^2 e^{b_{12}l} & b_2^2 e^{b_{21}l} & b_2^2 e^{b_{22}l} & b_3^2 e^{b_{31}l} & b_3^2 e^{b_{32}=0l} \end{vmatrix} = 0 \quad (10.176)$$

While the higher order effect from St. Venant's theory of noncircular rods will be very small and an engineer may not place much value on this kind of effort, the fact remains that we need basic principles from the science revolution period to give us an understanding of the physics of the system. While we had no computers until 1960s, and further improvements in speed and memory were necessary to utilize advances in finite element methods and gradually leave behind the strength of materials and tabular/graphical methods of simplified systems, it should be emphasized here that all finite elements emanated from energy principles of the scientific revolution and the approximations thereof should be understood by engineers to rely on the results. The world has come around and the strength of materials approach developed up to and during the 1960s has now been left behind in favor of the finite elements analysis approach (see Chapter 13). Therefore the finite element analysis engineers should assess carefully the limitations, if any, on commercial codes and should never depend on them blindly simply because the results from the post processors are attractive in making presentations.

References

1. Barr, A.D.S. (1962) Torsional Waves in Uniform Rods of Non-Circular Section, *Journal of Mechanical Engineering Sciences*, vol. 4, p. 127.
2. Barr, A.D.S. (1966) Variational Methods for the Derivation of Approximate Equations of Motion for Elastic Beams, in *Proceedings of Southampton Symposium on Variational Methods in Vibration Problems*, vol. 1, p. 15.
3. Bernoulli, Daniel (1751) *De vibrationibus et sono laminarum elasticarum commentationes physico-geometricae*, Commentari Academiae Scientiarum Imperialis Petropolitanae. T. 13 ad annum 1741, pp. 43, 105.
4. Bernoulli, Johann (1724) *Discours sur les loix de la communication du mouvement*, chs. 1–3, Prize Essay, Paris.
5. Bishop, R.E.D. (1952) Longitudinal Waves in Beams, *Aero Quaterly*, vol. 2, p. 280.
6. Carnegie, W. (1959) Vibrations of Rotating Cantilever Blading: Theoretical Approaches to the Frequency Problem Based on Energy Methods, *Journal of Mechanical Engineering Sciences*, vol. 1, p. 235.
7. Carnegie, W. (1962) Vibrations of Pre-twisted Cantilever Blading: An Additional Effect due to Torsion, *Proc. Instn. of Mech. Engrs.*, vol. 176, p. 315.
8. Carnegie, W. (1963) A Note on the Use of Variational Method to derive the Equations of Motion of a Vibrating Beam of Uniform Cross-Section taking account of Shear Deflection and Rotary Inertia, *Bulletin of Mechanical Engineering Education*, vol. 2, p. 66.
9. Carnegie, W. (1967) The Application of Variational Method to Derive the Equations of Motion of Vibrating Cantilever Blading under Rotation, *Bulletin Mechanical Engineering Education*, vol. 6, p. 29.
10. Coulomb, C. (1784) Recherches Théoriques et Expérimentales sur la Force de Torsion et sur l'Élasticité des Fils de Metal, Histoire de l'Acadamie, Paris.
11. Den Hartog, J.P. (1940) *Mechanical Vibration*, McGraw-Hill Book Co.
12. Duncan, W.J. (1937) *Galerkin's Method in Mechanics and Differential Equations*, R&M No. 1798.
13. Duncan, W.J. (1938) *The Principle of Galerkin Method*, R&M No. 1848.
14. Fleckenstein, J.O. (1949) *Johann und Jacob Bernoulli*, Basel.
15. Galerkin, B.G. (1915) Series Solution of Some Problems of Elastic Equilibrium of Rods and Plates, *VestnikInzhenerovi Tekhnikov*, vol. 19, p. 897 [English translation: NTIS Rept. TT-63-18924].
16. Gere, J.M. (1954) Torsional Vibration of Beams of Thin Walled Open-Section, *Journal of Applied Mechanics*, vol. 21, p. 381.
17. Hamilton, W.R. (1834–1835) On a General Method in Dynamics, *Philosophical Transaction of the Royal Society Part I*, 1834 pp. 247–308; Part II, 1835, pp. 95–144.
18. Heimann, P.M. (2007) Leibniz and Johann Bernoulli's Theory of Motion, *Centaurus*, vol. 21, no. 1, p. 1.
19. Hellinger, E. (1914) Die Algemeinen Ansätze der Mechanik der Kontinus, *Encyclopädie die mathematische Wissenschaft*, vol. 4, p. 654.
20. Hu, H.C. (1955) On some Variational Principles in the Theory of Elasticity and the Theory of Plasticity, *Scientia Sinica*, vol. 4, p. 33.
21. Kearton, W.J. (1926) *Steam Turbine Theory and Practice*, Pitman.
22. Kirchoff, G. (1876) *Vorlesungen über mathematische Physik: Mechanik*, Leipzig.
23. Krishna Murty, A.V. (1970) Vibration of Short Beams, *AIAA J.*, vol. 8, p. 34.
24. Lo, H. and Goulard, M. (1955) Torsion with Warping Restraint from Hamilton's Principle, in *Proceedings 2nd Midwestern Conference on Solid Mechanics*, p. 68.
25. Love, A.E.H. (1944) *Mathematical Theory of Elasticity*, Dover.
26. Mindlin, R.D. and Herrmann, G.A. (1954) One Dimensional Theory of Compressional Waves in Elastic Rods, in *Proceedings 1st US National Congress on Applied Mechanics*, p. 233.
27. Rao, D.K. and Rao, J.S. (1974) On the Use of Hu–Washizu's Variational Principle in Deriving the State Equations of Rods and Beams, *Journal of Aeronautical Society of India*, p. 85.

28. Rao, J.S. (1965) The Fundamental Flexural Vibration of a Cantilever Beam of Rectangular Cross-Section with Uniform Taper, *Aero. Quaterly*, vol. 16, p. 139.
29. Rao, J.S. (1974) A Refined Theory of Tortional Vibration of Bars, in *Proceedings of Noise, Shock & Vibration Conference*, Monash University, Melbourne, p. 171.
30. Rao, J.S. and Vyas, N.S. (1994) Shock in Rotor Blades During Speed Changes, *Journal of Sound and Vibration*, vol. 176, no. 4, p. 531.
31. Rao, J.S. et al. (1981) Coupled Bending Torsion Vibrations of Rotating Blades of Asymmetric Aerofoil Cross-Section Allowing for Shear Deflection and Rotary Inertia by Reissner Method, *Journal of Sound and Vibration*, vol. 75, p. 17.
32. Rayleigh, J.W.S. (1877) *Theory of Sound*, MacMillan, London.
33. Reissner, E. (1950) On a Variational Theorem in Elasticity, *Journal of Mathematical Physics*, vol. 29, p. 89.
34. Reissner, E. (1952) On Nonuniform Torsion of Cylindrical Rods, *Journal of Mathematical Physics*, vol. 31, p. 214.
35. Ritz, W. (1911) *Gesammelte Werke*, Gauthier-Villars.
36. Saint-Venant, *Memoir on Torsion*, June 1853.
37. Scanlan, R.H. and Rosenbaum, R. (1951) *Introduction to the Study of Aircraft Vibration and Flutter*, MacMillan.
38. Stodola, A. (1910) *Dampf- und Gasturbinen*, Springer, Berlin. Transl. (1927) *Steam and Gas Turbines*, McGraw-Hill.
39. Subrahmanyam, K.B., Kulkarni, S.V. and Rao, J.S. (1981) *Journal of Sound and Vibration*, vol. 75, no. 1, p. 17.
40. Timoshenko, S.P. (1921) On the Correction for Shear of the Differential Equation for Transverse Vibrations of Prismatic Bars, *Philosophical Magazine*, vol. 41, p. 744.
41. Timoshenko, S.P. (1945) Theory of Bending, Torsion and Buckling of Thin Walled Open Cross-Section, *Journal of Franklin Institute*, vol. 239, pp. 201, 249, 343.
42. Timoshenko, S.P. and Goodier, J.N. (1951) *Theory of Elasticity*, McGraw-Hill Book Co.
43. Todhunter, I. (1960) *A History of the Theory of Elasticity and of the Strength of Materials: From Galilei to Lord Kelvin*, Dover.
44. Volterra, E. (1961) Method of Internal Constraints and Certain of Its Applications, *Journal of Engineering Mechanics, Proceedings ASCE*, vol. 87, no. EM4, p. 103.
45. Vyas, N.S. and Rao, J.S. (1992) Equations of Motion of a Blade Rotating with Variable Angular Velocity, *Journal of Sound and Vibration*, vol. 156, no. 2, p. 327.
46. Washizu, K. (1955) On the Bounds of Eigenvalues, *The Quarterly Journal of Mechanics and Applied Mathematics*, vol. 8, no. 3, p. 311.

Chapter 11
20th Century Graphical and Numerical Methods

With the rapid growth of rotating machinery from the beginning of the 20th century, there was a need to determine the natural frequencies in an industrial environment by rapid means that can be established to handle various design problems. Methods were devised to streamline and develop fool-proof methods which semi-skilled engineers could handle in the shortest possible time with minimum errors. These methods continued to play a significant role until recently; they are only now being phased out in industry in favor of finite element methods.

Rayleigh's energy approach, given in Section 10.3 in a tabular form, is in fact directed at achieving the above objective

11.1 Stodola–Viannello (Rayleigh's Maximum Energy) Method in Graphical Form

Classical beam theory was in full force for design purposes at the turn of the 20th century. The most common form of solving the relevant differential equation was to first draw the shear force diagram, then the bending moment diagram and finally the deflection diagram, converting integration into summation in a graphical manner.

Since the inertia loading of a freely vibrating beam is $(w/g)p^2y$, Stodola–Viannello [12] considered the deflection y itself as a proportional load that produces the deflection. Consistent with technologies in vogue at that time, they determined the static deflection of the beam in place of the assumed shape function in a cosine form or polynomial form. Let the beam carry several loads $W_1, W_2, \ldots W_n$, then the static deflections $y_1, y_2, \ldots, y_n$ at the locations where the loads are applied are determined. The total strain energy of the system under one extreme position is

$$\hat{U} = \frac{1}{2}\sum Wy \qquad (11.1)$$

Table 11.1 Critical speed using the Stodola–Viannello method

Section	W	y	Wy	Wy^2
1	9.8	0.000260	0.00255	0.000000663
2	2.0	0.000450	0.00090	0.000000405
3	43.1	0.001046	0.04510	0.000045000
4	7.0	0.001325	0.00927	0.000012280
5	157.0	0.001355	0.21300	0.000289000
6	154.0	0.001339	0.20600	0.000276000
7	147.0	0.001223	0.18000	0.000220200
8	14.9	0.001120	0.01670	0.000018700
9	26.2	0.000845	0.02210	0.000018680
10	2.0	0.000480	0.00096	0.000000461
11	9.8	0.000250	0.00245	0.000000612
		Totals	0.69903	0.000882001

If p is the fundamental natural frequency, with the assumed static deflection as the shape function, the maximum kinetic energy in the system is given by

$$\hat{T} = \frac{1}{2g} p^2 \sum W y^2 \tag{11.2}$$

Therefore, the first mode critical speed is

$$N_c = \frac{60}{2\pi} \sqrt{\frac{g \sum W y}{\sum W y^2}} \text{ rpm} \tag{11.3}$$

The loads applied can be the disc weights and the distributed weight of the shaft for the section considered. To illustrate the method, let us consider a turbine rotor as shown in Figure 11.1. The graphical integration is reproduced in original form as given by Kearton [6]. The weights calculated are indicated in the figure. The total load is 621 lb and the bearing reactions are 301 lb at the left bearing A and 320 lb at the right bearing B.

The shearing force diagram is marked F. A scale is used in the original to draw the shearing force diagram with $1'' = 100$ lb. The length scale is $1'' = 4''$. The bending moment diagram is next drawn by integrating the shear force. The scale used is $1'' = 1000$ lb-in. It is marked M in the figure. The corrected M/I diagram is next obtained as shown. The scale on the original diagram was $1'' = 100$ lb/in^3.

Similarly the Slope Curve is obtained by another integration marked as shown. The scale was $1'' = 0.00005$ rad. Next the deflection curve is obtained by another integration shown as Static Deflection. Maximum deflection found is $0.001364''$ at point F (slope zero).

Table 11.1 gives the calculation of the critical speed.

Hence,

$$N_c = \frac{60}{2\pi} \sqrt{\frac{32.2 \times 12 \times 0.69903}{0.000882001^2}} = 5285 \text{ rpm}$$

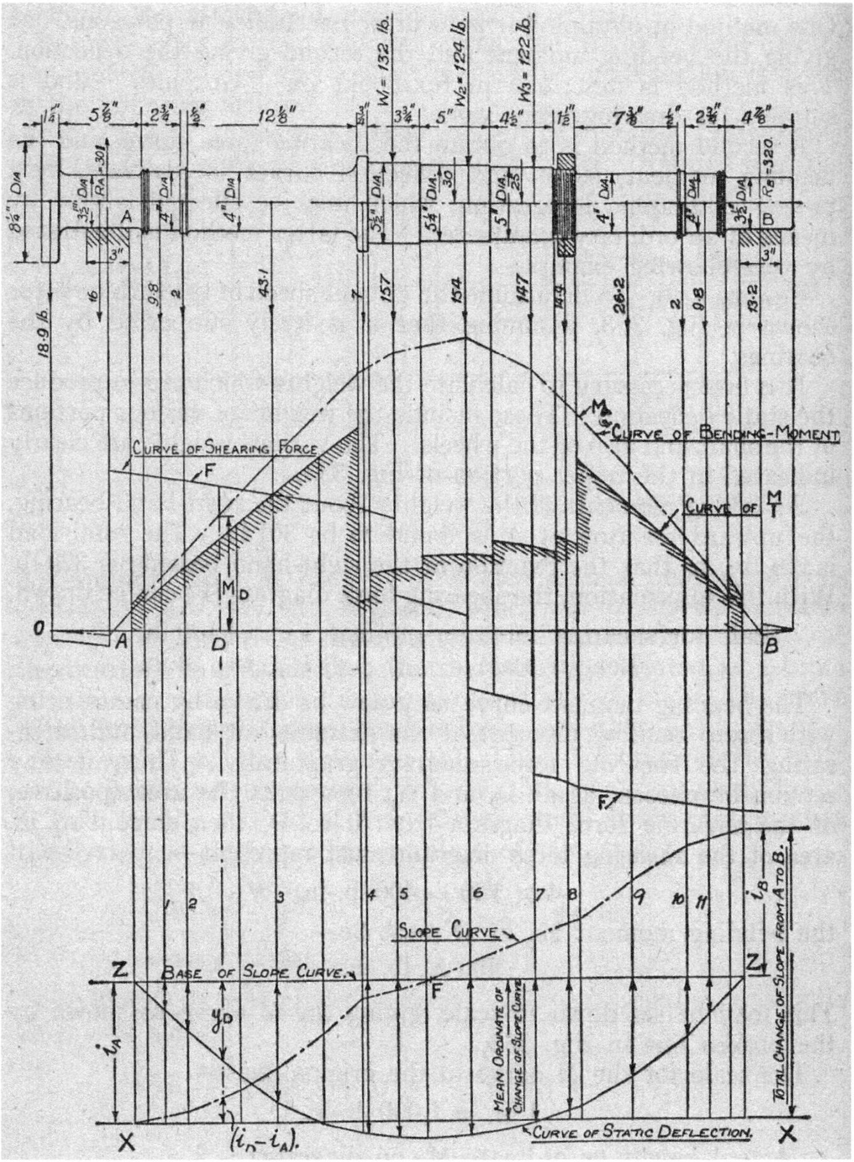

Fig. 11.1 Stodola–Viannello graphical method [12] for determining static deflection (from Kearton [6])

11.2 Stodola–Viannello Iterative Method in Tabular Form

In Section 10.3, we discussed a tabular form of Rayleigh's method with an assumed shape function for a cantilever beam as given in Table 10.1. The Stodola method

Table 11.2 Stodola–Viannello iterative method – Iteration 1

Station	$I(x)$ in^4	$m(x)$ $\frac{\text{lb}}{386}$/in	y_1	$m\,y_1$	S	M	M/I	slope	y_2
Factor →					Δx	$(\Delta x)^2$	$(\Delta x)^2$	$(\Delta x)^2/E$	$(\Delta x)^4/E$
10	0	0.0013	1.0	0.0013		0	0		0.008886
					0.0008			0.001272	
9	48	0.0026	0.8	0.0021		0.0008	0.000017		0.007614
					0.0029			0.001255	
8	96	0.0041	0.6	0.0025		0.0037	0.000039		0.006359
					0.0054			0.001216	
7	144	0.0048	0.5	0.0024		0.0091	0.000063		0.005143
					0.0078			0.001153	
6	192	0.0060	0.4	0.0024		0.0169	0.000088		0.003900
					0.0102			0.001065	
5	240	0.0088	0.3	0.0050		0.0271	0.000113		0.002925
					0.0152			0.000752	
4	288	0.0180	0.2	0.0038		0.0423	0.000147		0.001973
					0.0190			0.000805	
3	336	0.0304	0.15	0.0046		0.0613	0.000182		0.001168
					0.0236			0.000623	
2	384	0.0360	0.1	0.0036		0.0849	0.000221		0.000545
					0.0272			0.000402	
1	432	0.0384	0.05	0.0019		0.1121	0.000259		0.000143
					0.0291			0.000143	
0	480	0.0390	0	0		0.1421	0.000294		0

developed in graphical form takes more time and depends on the accuracy with which the drawing was made using drawing tools of the day. A tabular (numerical) method will be more accurate with analog multiplying devices again with the tools available nearly a century ago. We will discuss a tabular procedure to determine the critical speed by adopting the Stodola–Viannello method [11].

The inertia loading is proportional to mass, frequency squared and the deflection curve $y(x)$. As before we can choose an arbitrary deflection shape and if we assume the frequency ω to be equal to unity, then the loading is $my(x)$ and we can determine the deflection under this loading by successive integrations. The deflection that we obtain will not be same as the assumed shape unless the frequency is equal to unity and the assumed shape function is one of the modes (for the present fundamental mode). That is if $y(x)$ is exact mode shape and ω is exact frequency, the deflection curve under $m\omega^2 y(x)$ loading will be $y(x)$. Or if the beam is loaded with $my(x)$, the deflection will be $y(x)/\omega^2$. Therefore if the assumed shape is exact, the frequency can be obtained by comparing the deflections at any point on the assumed curve $y_1(x)$ and the resulting deflection curve $y_2(x)$. Then

$$\omega^2 = \frac{y_1(x)}{y_2(x)} \tag{11.4}$$

Since $y_1(x)$ is not exact and if we want more accurate result for the frequency, we can start with the resulting shape $y_2(x)$ and repeat the process all over again, thus developing an iterative algorithm to determine the frequency accurately. This probably is the beginning of an iterative approach that can be adopted effectively to determine the desired accurate solutions for linear as well as nonlinear solutions. This is illustrated in an example given in Section 10.3, Table 10.1 and in Tables 11.2 and 11.3.

Table 11.3 Stodola–Viannello iterative method – Iteration 2

Station	$I(x)$ in^4	$m(x)$ $\frac{\text{lb}}{386}$/in	y_2 normalized	$m\,y_2$	S	M	M/I	slope	Y_3	y_2/y_3
Factor →					Δx	$(\Delta x)^2$	$(\Delta x)^2$	$(\Delta x)^2/E$	$(\Delta x)^4/E$	
10	0	0.0013	1.0	0.0013		0	0		0.008954	111.68
					0.0008			0.001246		
9	48	0.0026	0.857	0.0022		0.0008	0.000017		0.007658	111.91
					0.0030			0.001279		
8	96	0.0041	0.716	0.0029		0.0038	0.000040		0.006379	112.24
					0.0059			0.001239		
7	144	0.0048	0.579	0.0028		0.0097	0.000067		0.005140	112.65
					0.0087			0.001172		
6	192	0.0060	0.440	0.0027		0.0184	0.000098		0.003968	113.18
					0.0114			0.001076		
5	240	0.0088	0.329	0.0029		0.0298	0.000124		0.002892	113.76
					0.0143			0.000952		
4	288	0.0180	0.222	0.0042		0.0441	0.000153		0.001940	114.43
					0.0185			0.000799		
3	336	0.0304	0.131	0.0040		0.0826	0.000186		0.001141	114.81
					0.0225			0.000613		
2	384	0.0360	0.061	0.0022		0.0851	0.000222		0.000528	115.53
					0.0247			0.000391		
1	432	0.0384	0.016	0.0006		0.1098	0.000254		0.000137	116.79
					0.0253			0.000151		
0	480	0.0390	0	0		0.1351	0.000281		0	Av.113.70

Table 11.2 gives the first iteration. The beam is divided into 10 stations 0 to 10 from the root to tip. The root is at station 9 and the tip at station 10 with 10 equal intervals λ between each successive station. The summation begins at free end station 10. The second moment of area and mass properties are given in the second and third columns. The assumed displacement vector at each station is given in column 4. The loading values my_1 are listed in column 5. Column 6 lists the shear starting from the free end station 10. Note the distance $\Delta x = 24''$ between successive stations and taken as a common factor in the shear column. The starting value for shear is taken as

$$S_{10} = \Delta x \frac{(3my_{10} + my_9)}{8}$$

For the next steps, shear is obtained by successive adding of the loads. The bending moment is found by successive addition of shear values. In the slope column, which is obtained by summing from the root out, the slope at the root station 0 is obtained similar to the shear at station 10 and after that M/I values are successively added from the root towards the tip till station 10. The deflections are next obtained starting from the root station 0 by successively adding the slope values to the tip. This deflection curve is then normalized with respect to the tip value set at unity. Table 11.3 gives the iteration starting from the new normalized shape. Finally, we take the average ratio of y_2/y_3 which is

$$\frac{y_2}{y_3} = \frac{E}{(\Delta x)^4} \times 113.79$$

$$= \frac{10.3 \times 10^6}{24^4} \times 113.79$$

Therefore

$$\omega^2 = \frac{y_2}{y_3} = \frac{10.3 \times 10^6}{24^4} \times 113.79 = 3530$$

giving $\omega = 59.41$ rad/sec or 567 cpm.

11.3 Dunkerley's Method

Dunkerley [2] conducted several experiments and arrived at a semi-empirical formula

$$\frac{1}{{N_c}^2} = \frac{1}{{N_a}^2} + \frac{1}{{N_1}^2} + \frac{1}{{N_2}^2} + \frac{1}{{N_3}^2} \cdots \tag{11.5}$$

where N_c is the critical speed of the complete shaft system; N_s is the critical speed of shaft alone; N_1 is the critical speed of the shaft supporting load W_1 only neglecting the influence of the shaft; N_2 is the critical speed of the shaft supporting load W_2 only neglecting the influence of the shaft; N_3 is the critical speed of the shaft supporting load W_3 only neglecting the influence of the shaft; ...

A major advantage of this empirical rule is in the determination of natural frequency of the shaft with one load each time and using equation (11.5) thus simplifying the procedure and saving time. Whereas all energy methods give upper bound solutions for the natural frequency, Dunkerley's rule gives a lower bound value. Therefore one can take the average of Dunkerley's result and the Rayleigh maximum energy principle result to get a more accurate estimate for the fundamental frequency.

11.4 Proof of the Dunkerley Formula by Blaess [1]

Blaess used the law of deformation work of a rotor with several masses $m_1, m_2, \ldots$ Let ω_k be the critical speed with all the masses simultaneously mounted on the shaft and $y_1, y_2, \ldots$ be the deflections of the actual elastic curve e' at the critical speed (see Figure 11.2). The inertia loading on the shaft then is given by $P_1 = m_1 y_1 \omega_k^2$, $P_2 = m_2 y_2 \omega_k^2 \ldots$ The deformation work φ_a of these forces is written as

$$\varphi_a = \frac{1}{2} \sum (P_1 y_1 + P_2 y_2 + \cdots) = \frac{1}{2} (\sum m_i y_i^2) \omega_k^2 \tag{11.6}$$

If only mass m_1 is mounted on the assumed weightless shaft, the deflection at the critical speed ω_1 now prevailing can also be made y_1 and then we have

$$\varphi_1 = \frac{1}{2} m_1 y_1 {\omega_1}^2 \tag{11.7}$$

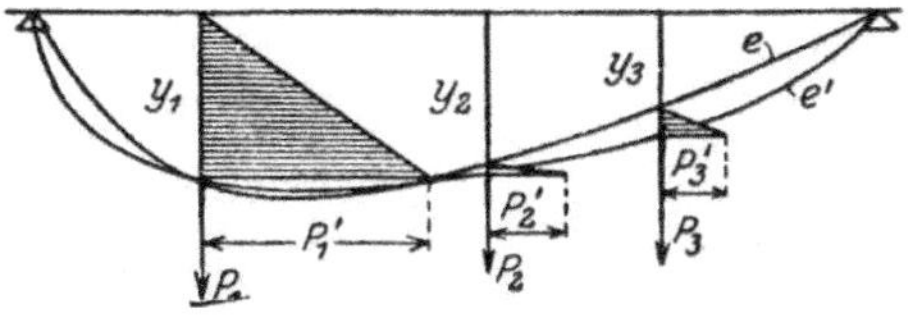

Fig. 11.2 Elastic curve e and actual elastic curve e'

Similarly $\varphi_2 = \frac{1}{2}m_2y_2\omega_2{}^2$, $\varphi_3 = \frac{1}{2}m_3y_3\omega_3{}^2$ etc. Using these results in equation (11.6) we get

$$\frac{\varphi_a}{\omega_k{}^2} = \frac{\varphi_1}{\omega_1{}^2} + \frac{\varphi_2}{\omega_2{}^2} + \frac{\varphi_3}{\omega_3{}^2} + \cdots \tag{11.8}$$

Figure 11.2 shows the exact elastic curve e at the critical speed ω_1. By laying off horizontally the force $P_1' = m_1y_1\omega_1^2$, we get in the hatched triangular area the elastic energy φ_1. To get the curve corresponding to ω_k, we can, starting at e, add at the points of application of m_2, m_3 forces increasing from zero until they become P_2, P_3, ... Since then, the point of application of m_1 will also move forward, the force applied there must decrease from P_1' to P_1, so that finally the load consists of the original forces $P_1, P_2, \ldots$, as it should. The works of the complementary forces are represented by the small cross-hatched triangular areas. These areas are much smaller than that of φ_1. We can therefore say that $\varphi_1 = \varphi_2 = \varphi_3 = \cdots = \varphi_a$. Hence

$$\frac{1}{\omega_k^2} \approx \frac{1}{\omega_1^2} + \frac{1}{\omega_2^2} + \frac{1}{\omega_3^2} + \cdots \tag{11.9}$$

which is Dunkerley's formula.

11.5 Hahn's Proof Using Matrix Algebra [3]

Hahn formulated his problem using the method of Influence Coefficients. Under the inertia loading $P_1 = m_1y_1\omega^2$, $P_2 = m_2y_2\omega^2$, ... the deflections $y_1, y_2 \ldots$ are written in terms of influence coefficients α_{ij} as follows:

$$\begin{aligned} y_1 &= \alpha_{11}P_1 + \alpha_{12}P_2 + \cdots + \alpha_{1n}P_n \\ y_2 &= \alpha_{21}P_1 + \alpha_{22}P_2 + \cdots + \alpha_{2n}P_n \\ &\ldots \\ y_n &= \alpha_{n1}P_1 + \alpha_{n2}P_2 + \cdots + \alpha_{nn}P_n \end{aligned} \tag{11.10}$$

Substituting for values of P,

$$(\alpha_{11}m_1\omega^2 - 1)y_1 + \alpha_{12}m_2\omega^2 y_2 + \cdots + \alpha_{1n}m_n\omega^2 y_n = 0$$
$$\alpha_{21}m_1\omega^2 y_1 + (\alpha_{22}m_2\omega^2 - 1)y_2 + \cdots + \alpha_{2n}m_n\omega^2 y_n = 0$$
$$\cdots$$
$$\alpha_{n1}m_1\omega^2 y_1 + +\alpha_{n2}m_2\omega^2 y_2 + \cdots + (\alpha_{nn}m_n\omega^2 - 1)y_n = 0 \tag{11.11}$$

For finite deflections to exist the following determinant should be zero:

$$D = \begin{vmatrix} (\alpha_{11}m_1\omega^2 - 1) & \alpha_{12}m_2\omega^2 & \cdots & \alpha_{1n}m_n\omega^2 y_n \\ \alpha_{21}m_1\omega^2 & (\alpha_{22}m_2\omega^2 - 1) & \cdots & \alpha_{2n}m_n\omega^2 \\ \cdots & \cdots & \cdots & \cdots \\ \alpha_{n1}m_1\omega^2 & \alpha_{n2}m_2\omega^2 & \cdots & (\alpha_{nn}m_n\omega^2 - 1) \end{vmatrix} = 0 \tag{11.12}$$

We can expand this determinant to a polynomial equation of degree $2n$ to give

$$D = D_n\omega^{2n} - D_{n-1}\omega^{2(n-1)} + \cdots + D_2(-1)^{n-2}\omega^4 + D_1(-1)^{n-1}\omega^2 + (-1)^n = 0 \tag{11.13}$$

The term D_1 in the above can be shown to be

$$D_1 = \alpha_{11}m_1 + \alpha_{22}m_2 + \cdots + \alpha_{nn}m_n \tag{11.14}$$

Let us suppose that this shaft without mass is loaded with only mass m_1, then (11.11) gives

$$(\alpha_{11}m_1\omega^2 - 1)y_1 = 0 \tag{11.15}$$

For $y_1 \neq 0$, we get the critical speed with mass m_1 only.

$$\omega_1{}^2 = \frac{1}{\alpha_{11}m_1} \tag{11.16}$$

In a similar manner

$$\omega_2{}^2 = \frac{1}{\alpha_{22}m_2};\ \omega_3{}^2 = \frac{1}{\alpha_{33}m_3};\ \cdots \tag{11.17}$$

Using the above two equations, equation (11.14) reduces to

$$D_1 = \frac{1}{\omega_1{}^2} + \frac{1}{\omega_2{}^2} + \cdots + \frac{1}{\omega_n{}^2} \tag{11.18}$$

For the fundamental frequency, neglecting higher harmonics, equation (11.13) gives

$$D_1 = \frac{1}{\omega^2} \tag{11.19}$$

Equation (11.18) then gives Dunkerley's result.

$$\frac{1}{\omega^2} = \frac{1}{\omega_1{}^2} + \frac{1}{\omega_2{}^2} + \cdots + \frac{1}{\omega_n{}^2} \tag{11.20}$$

The greatest advantage of Dunkerley's method is in using a simple single mass and shaft model for each of the masses in obtaining the overall result of the shaft carrying several masses. Therefore it became a popular industrial practice in the early days of rotating machinery development.

11.6 Holzer Method for Torsional Vibration

One of the main problems faced by industry related to marine transmission units from the reciprocating steam engine drives with heavy pulsation torques through couplings and gear transmission units and to the long propeller shafts before the propellers get their power. Propeller shafts designed according to statics failed and when their stiffness was increased the failures occurred sooner. It was suspected around World War I that there is a need to calculate higher torsional modes of drive trains and prevent any resonances. The available energy methods and empirical rules could provide for a good estimate of the fundamental mode but not higher modes. Holzer in 1922 provided this numerical procedure in a tabular form to determine fundamental as well as higher modes, which is popular even today because of its simplicity.

The Holzer method [4, 5] is inverse to Stodola's method in the sense that a frequency is assumed to begin an iterative process. This method can be easily applied to torsional vibrations which have only two state quantities at every station, torque and deflection. It can be applied to bending problems as well; however we have to deal with four state quantities, deflection, slope, shear and bending moment and we have to wait for another two decades for this to happen.

A Railway Diesel-Electric Drive train is modeled as inertia and stiffness system for torsional vibrations given Table 11.4 [10]. Here a discrete system is obtained by assuming the mass moment of inertia of each part as indicated. The torsional stiffness between these inertias is determined by a test or by calculation and modeled as a discrete stiffness. We can determine the approximate value of the fundamental mode by observing the system inertias and minimum stiffness member; the node point with zero deflection can be expected between stations 10 and 11. We can sum up all the inertias upto station 11 – connecting this total inertia with the minimum stiffness member to the rest of the inertias clubbed together.

With this assumed frequency 86 rad/s as given in Table 11.4 and unit torsional amplitude at station 1, i.e., the first inertia for free vibration, we can determine $I\omega^2\theta = I_1\omega^2\theta_1 = 2735\ \mathrm{kgm^2/s^2}$. With the help of stiffness between stations 1 and 2, the difference in amplitude between stations 1 and 2 is given by

$$\frac{1}{K_1} I_1\omega^2\theta_1 = 0.0171$$

We next proceed to determine the torsional amplitude at station $i = 2$ by using

$$\theta_i = \theta_{i-1} - \frac{1}{K_i} \sum I\omega^2\theta = 0.9829$$

and continue the numerical computations in Table 11.3 until we find at station 12

$$\sum I\omega^2\theta = 317 \approx 0$$

consistent with numerical accuracy.

Holzer's method is a trial and error method and we can determine higher modes also in a similar manner by assuming a frequency and watching the mode shape that we get once the boundary condition is satisfied. It can be used to determine torsional vibration characteristics of shafts, blades, aircraft wings, etc. A similar, successful, attempt at bending vibrations was made 22 years later by Myklestad [7, 8].

11.7 The Myklestad Method [7,8]

As mentioned, Myklestad's method basically follows Holzer's method, previously applied for bending vibrations of beams, now applicable for aircraft wings and turbine blades. It was the time when World War II was over and basic scientific works began in earnest. World War II had seen a lot of action of aircraft and as mentioned, Ohain and Whittle began development of jet engines. Turbine blades are distributed mass systems and therefore energy methods could easily be applied for bending vibration dynamics, whereas aircraft wings are necessarily built up structures with Spars, Ribs, Stringers and Skin. They can be modeled as discrete structures but not as continuous systems. Therefore it became necessary to have a method by which uncoupled bending vibrations can be determined. Holzer's method already provided such a method for torsional vibrations. Myklestad provided the tabular method to handle four state quantities, viz., deflection, slope, shear and bending moments rather than torsional vibration which had only two state quantities.

A major advantage of Myklestad's method is its applicability in determining higher modes as in the case of Holzer's method; the energy methods though are theoretically capable of determining higher modes; we had to wait for the 1960s to have this capability with the advent of main frame computers, however primitive they now appear when compared to later years. With the advanced computational facilities, finite element methods gradually overtook these tabular and energy methods, except for some special cases of Coriolis forces or accelerating through a critical speed as we have discussed earlier.

Let us consider a cantilever beam with masses $m_1, m_2, m_3, \ldots\ m_n$ and m_{n+1} at distances $x_1, x_2, x_3, \ldots$ from the cantilever end. The length between masses m_1 and m_2 is denoted by l_1, between masses m_2 and m_3 is denoted by $l_2 \ldots$ Let the beam vibrate with frequency ω rad/s. The shear and bending moments in terms of inertia forces at station n are written as

Table 11.4 Diesel locomotive transmission system – assumed $\omega = 86$ rad/s

Sl. No.	I kgm^2	Description	K MNm/rad	$\theta_i = \theta_{i-1} - \frac{1}{K_i}\sum I\omega^2\theta$	$I\omega^2\theta$	$\sum I\omega^2\theta$	$\frac{1}{K_i}\sum I\omega^2\theta$
1	0.3698	Pulley		1.0000	2735	2735	
			0.1599				0.0171
2	0.2080	Coupling Outer part		0.9829	1512	4247	
			0.0275				0.1546
3	0.0618	Coupling Inner part		0.8283	379	4626	
			0.1707				0.0271
4	0.1413	Shaft and Flange		0.8012	1017	5643	
			1.0488				0..0052
5	3.8259	Damper		0.7960	22343	27986	
			17.4618				0.0016
6	0.9153	Cyl 4 and 8		0.7944	5377	33363	
			5.886				0.0057
7	0.8015	Cyl 3 and 7		0.7887	4676	38039	
			5.886				0.0065
8	0.8015	Cyl 2 and 6		0.7822	4637	42676	
			5.886				0.0073
9	0.9153	Cyl 1 and 5		0.7749	5244	47920	
			7.0632				0.0068
10	3.3452	Coupling		0.7681	19004	66924	
			0.0932				0.7181
11	0.5856	Suri Mach Primary		0.0500	218	67143	
			0.032				0.7205
12	13.476	Secondary		-0.6705	-66826	317	
						≈ 0	

$$S_n = \sum_{i-1}^{n} m_i y_i \omega^2$$

$$M_n = \sum_{i-1}^{n-1} m_i y_i \omega^2 (x_i - x_n) \tag{11.21}$$

Myklestad defined four different elastic constants for each beam segment l_n between masses m_n and m_{n+1}:

- d_{Fn} – linear deflection at station n relative to station $(n+1)$, the beam being assumed built in at station $(n+1)$ for a unit force applied at station n.
- v_{Fn} – angular deflection at station n relative to station $(n+1)$ for a unit force applied at station n.
- d_{Mn} – linear deflection at station n relative to station $(n+1)$ for a unit moment applied at station n.
- v_{Mn} – angular deflection at station n relative to station $(n+1)$ for a unit moment applied at station n.

Since each of the sections between masses is assumed to be a beam of uniform cross-section, we can use simple cantilever beam rules to determine all the above coefficients. The slopes α and deflections can be built from station 1 onwards (in a similar manner as in Holzer's method), using the following relations for sections between stations n and $n+1$:

$$\alpha_{n+1} - \alpha_n - S_n v_{Fn} - M_n v_{Mn}$$

$$y_{n+1} = y_n - l_n \alpha_{n+1} - S_n d_{Fn} - M_n d_{Mn} \tag{11.22}$$

Using (11.21) in the above we get

$$\alpha_{n+1} = \alpha_n - \omega^2 v_{Fn} \sum_{i-1}^{n} m_i y_i - \omega^2 v_{Mn} \sum_{i-1}^{n-1} m_i y_i (x_i - x_n)$$

$$y_{n+1} = y_n - l_n \alpha_{n+1} - \omega^2 d_{Fn} \sum_{i-1}^{n} m_i y_i - \omega^2 d_{Mn} \sum_{i-1}^{n-1} m_i y_i (x_i - x_n) \quad (11.23)$$

We can choose the free end deflection $y_1 = 1$ and assign the slope φ that should be determined later by using the boundary conditions. In terms of unknown φ the slope and deflection at station n are expressed as

$$\alpha_n = h_n \varphi - f_n$$

$$y_n = g_n - k_n \varphi \quad (11.24)$$

where h_n, f_n, g_n and k_n are coefficients which are independent of φ. Equations (11.24) give at starting station 1.

$$\varphi = h_1 \varphi - f_1 \Rightarrow h_1 = 1, \quad f_1 = 0$$

$$1 = g_1 - k_1 \varphi \Rightarrow g_1 = 1, \quad k_1 = 0 \quad (11.25)$$

Continuing further, Myklestad derived the following recursive relations for the coefficients h_n, f_n, g_n and k_n at station n:

$$h_n = h_{(n-1)} + v_{F(n-1)} K_{(n-1)} + v_{M(n-1)} K'_{(n-1)}$$

$$k_n = k_{(n-1)} + l_{(n-1)} h_n - d_{F(n-1)} K_{(n-1)} - d_{M(n-1)} K'_{(n-1)}$$

$$f_n = f_{(n-1)} + v_{F(n-1)} G_{(n-1)} + v_{M(n-1)} G'_{(n-1)}$$

$$g_n = g_{(n-1)} + l_{(n-1)} f_n - d_{F(n-1)} G_{(n-1)} - d_{M(n-1)} G'_{(n-1)} \quad (11.26)$$

where

$$K_n = \sum_{i=1}^{n} m_i \omega^2 k_i$$

$$K'_n = \sum_{i=1}^{n-1} l_i K_i$$

$$G_n = \sum_{i=1}^{n} m_i \omega^2 g_i$$

$$G'_n = \sum_{i=1}^{n-1} l_i G_i \quad (11.27)$$

Table 11.5 Elastic and geometric properties of a cantilever beam

n	$d_F \times 10^6$	$v_F \times 10^6$ $d_M \times 10^6$	$v_M \times 10^6$	m	x	l
1	21.3	0.875	0.0535	0.104	220	40
2	11.7	0.454	0.0248	0.181	180	40
3	8.0	0.308	0.0163	0.249	140	40
4	10.7	0.352	0.0148	0.648	100	50
5	2.9	0.132	0.0080	1.430	50	34
6	0.5	0.026	0.0033	1.550	16	16
7_{Root}				7.77	0	

Table 11.6 Myklestad's tabular procedure to evaluate k_n, K_n, K'_n, h_n, g_n, G_n, G'_n and f_n

n	$\frac{m\omega^2}{10^6}$	$k_n = k_{(n-1)} + l_{(n-1)}h_n - d_{F(n-1)}K_{(n-1)} - d_{M(n-1)}K'_{(n-1)}$	$K_n \times 10^{-6} = \sum_{i=1}^{n} \frac{m_i\omega^2}{10^6} k_i$	$K'_n \times 10^{-6} = \sum_{i=1}^{n} l_i K_i$	$h_n = h_{(n-1)} + v_{F(n-1)}K_{(n-1)} + v_{M(n-1)}K'_{(n-1)}$	$g_n = g_{(n-1)} + l_{(n-1)}f_n - d_{F(n-1)}G_{(n-1)} - d_{M(n-1)}G'_{(n-1)}$	$G_n \times 10^{-6} = \sum_{i=1}^{n} \frac{m_i\omega^2}{10^6} g$	$G'_n \times 10^{-6} = \sum_{i=1}^{n} l_i G_i$	$f_n = f_{(n-1)} + v_{F(n-1)}G_{(n-1)} + v_{M(n-1)}G'_{(n-1)}$
1	0.000374	0	0	0	1	1	0.000374	0	0
2	0.000652	40	0.02606	0	1	1.00511	0.001029	0.014960	0.000327
3	0.000896	80.168	0.09793	1.04256	1.0118	1.03288	0.001954	0.056120	0.001165
4	0.002330	121.420	0.38084	7.95964	1.0584	1.10724	0.004534	0.134280	0.002682
5	0.005150	178.917	1.30226	24.0016 4	1.2664	1.32471	0.011356	0.360980	0.006265
6	0.005580	227.405	2.57118	68.2784 8	1.6303	1.60630	0.020319	0.747084	0.010652
7	0.027970	255.088	9.70599	109.417	1.9214	1.79505	0.070527	1.072188	0.013646

$\frac{f_R}{h_R} = 0.007102$

At the root of the cantilever, the slope and deflection are

$$\alpha_R = h_R\varphi - f_R$$

$$y_R = g_R - k_R\varphi \tag{11.28}$$

If we make the slope α_R at the root zero in (11.28), we have the slope at the starting station given by $\varphi = f_R/h_R$. This gives the root deflection from (11.28) to be $y_R = g_R - k_R(f_R/h_R)$. This deflection for a cantilever should be zero and if in the numerical calculation it turns out to be zero, then the assumed value for ω is correct, else we make another guess for ω and repeat the process until a convergence is reached. Let us consider the data in Table 11.5 that gives the elastic and geometric properties of a cantilever beam.

Table 11.6 gives the Myklestad tabular procedure for an assumed value of $\omega =$ 60 rad/s. From the results, $\varphi = f_R/h_R = 0.007102$ and

$$y_R = g_R - k_R\varphi = 1.79505 - 0.007102 \times 255.088 = -0.0166$$

A frequency of 60 rad/s assumed is high. Decrease this to 55 rad/s and repeat; do an interpolation to find correct frequency, until the desired accuracy for $y_R \approx 0$ is achieved.

11.8 Prohl's Method [9]

After Myklestad published his paper, which mainly addressed aircraft wings, Prohl published his work addressing the rotating machine shaft calculations, which is fundamentally the same as Myklestad's method. We will briefly discuss here his procedure. For a fully balanced rotor whirling steadily at a critical speed ω with its central bent axis in a bowed shape, the governing differential equation is the same as that of a beam. Therefore we can write

$$EI\frac{d^2y}{dx^2} = M \tag{11.29}$$

$$\frac{d^2M}{dx^2} = \mu\omega^2 y \tag{11.30}$$

where μ is mass per unit length lb-sec^2/in^2. Let the shaft be divided into a series of appropriate sections, then the above equations are written as

$$\Delta\left(\frac{dy}{dx}\right) = \left(\frac{\Delta x}{EI}\right) M_{\text{avg}} \tag{11.31}$$

$$\Delta\left(\frac{dM}{dx}\right) = \left(\mu\omega^2\Delta x\right) y_{\text{avg}} \tag{11.32}$$

where Δx is the length of a given section and M_{avg} and y_{avg} are the average values of bending moment and deflection for that section. Prohl followed Holzer's approach for calculating the natural frequencies of torsional vibration. Here however there are four integrations involved instead of two and additional complications arise in dealing with the boundary conditions. The bending moment and shear force relation is

$$\frac{dM}{dx} = V \tag{11.33}$$

At each mass m there is a finite change in the shear force given by

$$\Delta V = m\omega^2 y \tag{11.34}$$

Assume the shear force V_0, bending moment M_0, slope θ_0 and the deflection y_0, at the left-hand end where the computations will begin. There will be a change in shear force at point 0 due to the inertia force of mass m_0, therefore the shearing force V_1, in section 1 is

$$V_1 = V_0 + m_0\omega^2 y_0 \tag{11.35}$$

The bending moment M_1 at point 1 is

$$M_1 = M_0 + V_1(\Delta x)_1 \tag{11.36}$$

The bending moment at distance x from the left-hand end of section 1 is

$$M = M_0 + \frac{M_1 - M_0}{(\Delta x)_1} x \tag{11.37}$$

The slope θ for the first section is obtained from

$$\begin{aligned}\theta &= \frac{1}{(EI)_1}\int_0^x M dx + c \\ &= \frac{1}{(EI)_1}\left[M_0 x + \frac{M_1 - M_0}{(\Delta x)_1}\frac{x^2}{2}\right] + \theta_0 \end{aligned}\tag{11.38}$$

where c is a constant of integration. The deflection y is similarly obtained as follows:

$$\begin{aligned}y &= \int_0^x \theta dx + c' \\ &= \frac{1}{(EI)_1}\left[M_0 \frac{x^2}{2} + \frac{M_1 - M_0}{(\Delta x)_1}\frac{x^3}{6}\right] + \theta_0 x + y_0 \end{aligned}\tag{11.39}$$

At the end section 1, i.e., point 1, the above values are ($\beta_1 = (\Delta x/EI)_1$)

$$\begin{aligned}\theta_1 &= \beta_1\left(\frac{M_0}{2} + \frac{M_1}{2}\right) + \theta_0 \\ y_1 &= \beta_1\left(\frac{M_0}{3} + \frac{M_1}{6}\right)(\Delta x)_1 + \theta_0(\Delta x)_1 + y_0 \end{aligned}\tag{11.40}$$

Now, the shear force in section 2 and bending moment at point 2 are

$$\begin{aligned}V_2 &= V_1 + m_1\omega^2 y_1 \\ M_2 &= M_1 + V_2(\Delta x)_2 \end{aligned}\tag{11.41}$$

Once the bending moment in shaft section 2 is defined, the slope θ_2 and deflection y_2 can be evaluated and this process can be repeated successively by the following general relations:

$$\begin{aligned}V_n &= V_{n-1} + m_{n-1}\omega^2 y_{n-1} \\ M_n &= M_{n-1} + V_n(\Delta x)_n \\ \theta_n &= \beta_n\left(\frac{M_{n-1}}{2} + \frac{M_n}{2}\right) + \theta_{n-1} \\ y_n &= \beta_n\left(\frac{M_{n-1}}{3} + \frac{M_n}{6}\right)(\Delta x)_n + \theta_{n-1}(\Delta x)_n + y_{n-1} \end{aligned}\tag{11.42}$$

Prohl in his paper extended the analysis (unlike Myklestad) to include the rotary inertia of the disks; this takes care of so-called gyroscopic effects as his application is meant for steam turbine rotors compared to Myklestad's approach for aircraft structures. Both these methods found wide applications in the structures and rotor dynamics community until the advent of the computer age.

References

1. Blaess, Z. (1914) VDI, p. 187.
2. Dunkerley, S. (1894) On the Whirling of Vibration of Shafts, *Phil. Trans. Roy. Soc., Series A*, vol. 185, p. 279.
3. Hahn (1918) Note sur la vitesse critique et la formule du Dunkerley, *Schweiz. Bauztg.*, vol. 72, p. 191.
4. Holzer, H. (1921) *Die Berechnung der Drehschwingungen*, Springer Verlag, Berlin.
5. Holzer, H. (1922) Tabular method for torsional vibration analysis of multiple-rotor shaft systems, *Machine Design*, May, p. 141.
6. Kearton, W.J. (1926) *Steam Turbine Theory and Practice*, Pitman.
7. Myklestad, N.O. (1944) *Vibration Analysis*, McGraw-Hill Book Co.
8. Myklestad, N.O. (1956) *Fundamentals of Vibration Analysis*, McGraw-Hill Book Co.
9. Prohl, M.A. (1945) A General Method of Calculating Critical Speeds of Flexible Rotors, *Journal of Applied Mechanics, Trans. ASME, Series E*, vol. 67, p. 142.
10. Rao, J.S. (1983) *Rotor Dynamics*, John Wiley & Sons, New Age International 1996.
11. Scanlan, R.H. and Rosenbaum, R. (1951) *Introduction to the Study of Aircraft Vibration and Flutter*, MacMillan.
12. Stodola, A. (1910) *Dampf- und Gasturbinen*, Springer, Berlin. Translation (1927) *Steam and Gas Turbines*, McGraw-Hill.

Chapter 12
Matrix Methods

The kinetic and potential energies in a free vibration problem are expressible as homogeneous quadratic forms in the velocities $\dot{q}_i$ and coordinates q_i respectively, leads to important conclusions to be drawn concerning normal coordinates. Consider a problem with three generalized coordinates

$$T = \frac{1}{2}\begin{pmatrix} a_{11}\dot{q}_1^2 + a_{12}\dot{q}_1\dot{q}_2 + a_{13}\dot{q}_1\dot{q}_3 \\ +a_{21}\dot{q}_2\dot{q}_1 + a_{22}\dot{q}_2^2 + a_{23}\dot{q}_2\dot{q}_3 \\ +a_{31}\dot{q}_3\dot{q}_1 + a_{32}\dot{q}_3\dot{q}_2 + a_{33}\dot{q}_3^2 \end{pmatrix}, \quad a_{ij} = a_{ji}; i = j = 1, 2, 3 \tag{12.1}$$

$$U = \frac{1}{2}\begin{pmatrix} b_{11}q_1^2 + b_{12}q_1q_2 + b_{13}q_1q_3 \\ +b_{21}q_2q_1 + b_{22}q_2^2 + b_{23}q_2q_3 \\ +b_{31}q_3q_1 + b_{32}q_3q_2 + b_{33}q_3^2 \end{pmatrix}, \quad b_{ij} = b_{ji}; i = j = 1, 2, 3 \tag{12.2}$$

The determinants

$$\begin{vmatrix} a_{11} & a_{12} & a_{13} \\ a_{21} & a_{22} & a_{23} \\ a_{31} & a_{32} & a_{33} \end{vmatrix} \tag{12.3}$$

and

$$\begin{vmatrix} b_{11} & b_{12} & b_{13} \\ b_{21} & b_{22} & b_{23} \\ b_{31} & b_{32} & b_{33} \end{vmatrix} \tag{12.4}$$

are the discriminants of T and U respectively. Using Lagrange equations

$$\frac{d}{dt}\left(\frac{\partial T}{\partial \dot{q}_r}\right) - \frac{\partial T}{\partial q_r} + \frac{\partial U}{\partial q_r} = 0 \tag{10.52}$$

we can obtain the equations of motion in matrix form as

$$\begin{bmatrix} a_{11} & a_{12} & a_{13} \\ a_{21} & a_{22} & a_{23} \\ a_{31} & a_{32} & a_{33} \end{bmatrix} \begin{Bmatrix} \ddot{q}_1 \\ \ddot{q}_2 \\ \ddot{q}_3 \end{Bmatrix} + \begin{bmatrix} b_{11} & b_{12} & b_{13} \\ b_{21} & b_{22} & b_{23} \\ b_{31} & b_{32} & b_{33} \end{bmatrix} \begin{Bmatrix} q_1 \\ q_2 \\ q_3 \end{Bmatrix} = 0 \tag{12.5}$$

The above equations can be written as

$$[A]\{\ddot{q}\} + [B]\{q\} = 0 \tag{12.6}$$

where $[A]$ is called inertia matrix and $[B]$ the stiffness matrix. Take inverse of matrix $[A]$ and premultiply equation (12.6) to get

$$\{\ddot{q}\} + [A]^{-1}[B]\{q\} = 0$$

or

$$\{\ddot{q}\} + [C]\{q\} = 0 \tag{12.7}$$

where $[C]$ is called the dynamic matrix. It is now desired to find a linear transformation

$$\begin{Bmatrix} q_1 \\ q_2 \\ q_3 \end{Bmatrix} = \begin{bmatrix} \alpha_{11} & \alpha_{12} & \alpha_{13} \\ \alpha_{21} & \alpha_{22} & \alpha_{23} \\ \alpha_{31} & \alpha_{32} & \alpha_{33} \end{bmatrix} \begin{Bmatrix} r_1 \\ r_2 \\ r_3 \end{Bmatrix} \tag{12.8}$$

such that when substituted into (12.7), we will obtain

$$\begin{Bmatrix} \ddot{r}_1 \\ \ddot{r}_2 \\ \ddot{r}_3 \end{Bmatrix} + \begin{bmatrix} \omega_1^2 & 0 & 0 \\ 0 & \omega_2^2 & 0 \\ 0 & 0 & \omega_3^2 \end{bmatrix} \begin{Bmatrix} r_1 \\ r_2 \\ r_3 \end{Bmatrix} = 0$$

$$\ddot{r}_i + \omega_i^2 r_i = 0, \quad i = 1, 2, 3 \tag{12.9}$$

From (12.7) and (12.8) we can write

$$[\alpha]\{\ddot{r}\} + [C][\alpha]\{r\} = 0$$

or

$$\{\ddot{r}\} + [\alpha]^{-1}[C][\alpha]\{r\} = 0 \tag{12.10}$$

Comparing (12.9) and (12.10), the desired matrix is such that

$$[\alpha]^{-1}[C][\alpha] = \begin{bmatrix} \omega_1^2 & 0 & 0 \\ 0 & \omega_2^2 & 0 \\ 0 & 0 & \omega_3^2 \end{bmatrix} \tag{12.11}$$

i.e., a matrix $[\alpha]$ is so chosen that by collinearity transformation, $[C]$ is transformed to a diagonal matrix. Sylvester (1852) has proved that the transformation to normal coordinates is always a real transformation and the roots all being real. This transformation is fundamental to modal analysis of Multidegree of freedom vibrating systems that is quite so often employed in determining the free or forced dynamic response of any vibrating system.

We can derive the now familiar orthogonality conditions in matrix form here. For harmonic motion, equation (12.6) for the modal column $\{q^{(i)}\}$ associated with frequency ω_i gives

$$\omega_i^2[A]\{q^{(i)}\} = [B]\{q^{(i)}\} \tag{12.12}$$

Similarly for the modal column $\{q^{(j)}\}$ associated with frequency ω_j

$$\omega_j^2[A]\{q^{(j)}\} = [B]\{q^{(j)}\} \tag{12.13}$$

Premultiply equation (12.12) by the transpose of $\{q^{(i)}\}$ and equation (12.13) by transpose of $\{q^{(i)}\}$ and, we get

$$\omega_i^2\{q^{(j)}\}^T[A]\{q^{(i)}\} = \{q^{(j)}\}^T[B]\{q^{(i)}\} \tag{12.14}$$

$$\omega_j^2\{q^{(i)}\}^T[A]\{q^{(j)}\} = \{q^{(i)}\}^T[B]\{q^{(j)}\} \tag{12.15}$$

Since the transpose of a product is the product of the transpose in reverse order and $[A]$ and $[B]$ are symmetric, equation (12.14) becomes

$$\omega_i^2\{q^{(i)}\}^T[A]\{q^{(j)}\} = \{q^{(j)}\}^T[B]\{q^{(i)}\} \tag{12.14a}$$

Subtracting (12.14a) from (12.15), we get

$$(\omega_j^2 - \omega_i^2)\{q^{(i)}\}^T[A]\{q^{(j)}\} = 0 \tag{12.16}$$

Since $\omega_j^2 \neq \omega_i^2$ it can be also seen that

$$\{q^{(i)}\}^T[A]\{q^{(j)}\} = \{q^{(i)}\}^T[B]\{q^{(j)}\} = 0 \tag{12.17}$$

This is the familiar orthogonality condition.

The matrix form of writing equations of motion and mathematical solution for eigen-values is attractive for engineers to determine their dynamic characteristics; however the computational needs of practical systems were not present prior to the dawn of the computer age. That was of course the reason for having developed several methods, graphical or tabular, in determining the most important frequency for rotors, a fundamental mode. As the machines became bigger in size and more flexible, the higher modes are necessary in the design; that's where methods like Holzer's helped in designing better drive trains. Most of the developments in this direction came from organizations, e.g., British Internal Combustion Engineering Research Institute, who published some very highly valuable data [8,21–23]. Reference may be made to excellent text books in this field by Biezeno and Grammel [1], Den Hartog [2] and Timoshenko [20].

Table 12.1 A closely coupled torsional system

Rotor No.	Inertia I		Stiffness k	
	Description	kgm^2	Description	MNm/rad
1	Damper Outer Part	3.2262		
			Sleeve Spring	0.33
2	Damper Inner Part	0.3628		
			Shaft upto Cyl. 8	5.9817
3	Cyl. No. 8	0.5482		
			Shaft Cyl 8 - 7	4.5794
4	Cyl. No. 7	0.5482		
			Shaft Cyl 7 - 6	4.5794
5	Cyl. No. 6	0.5482		
			Shaft Cyl 6 - 5	4.5794
6	Cyl. No. 5	0.5482		
			Shaft Cyl 5 - 4	4.5794
7	Cyl. No. 4	0.5482		
			Shaft Cyl 4 - 3	4.5794
8	Cyl. No. 3	0.5482		
			Shaft Cyl 3 - 2	4.5794
9	Cyl. No. 2	0.5482		
			Shaft Cyl 2 - 1	4.5794
10	Cyl. No. 1	0.5482		
			Shaft Cyl 1 to Flywheel	3.6871
11	Flywheel + Coupling Primary	15.3563		
			Vulcan Coupling	0.0882
12	Secondary Part of Coupling	0.5952		
			Generator Shaft	2.1966
13	Generator	16.0525		

12.1 Torsional Vibration Systems

Setting up equations of motion in torsional vibration is a straight forward matter using Newton's IInd law or D'Alembert's principle. We will illustrate this through a drive train example given by Rao and Rao [11]. The drive train data is given in Table 12.1.

Jacobsen and Ayre [7] discussed vibration problems of n masses with different ties or couplings. Closely coupled systems are those where the governing equation of a disk is affected by only the neighboring links, such as mass spring or inertia-stiffness systems. The example given in Table 12.1 has 13 masses (inertias) coupled by stiffnesses and the governing equations can be written as follows:

$$
\begin{aligned}
&I_1\ddot{\theta}_1 + k_1\theta_1 - k_1\theta_2 = 0\\
&I_2\ddot{\theta}_2 - k_1\theta_1 + (k_1 + k_2)\theta_2 - k_3\theta_3 = 0\\
&\dots\\
&I_{13}\ddot{\theta}_{13} + k_{12}\theta_{12} + k_{12}\theta_{13} = 0
\end{aligned}
\tag{12.18}
$$

In matrix notation it is

$$[M]\{\ddot{\theta}\} + [K]\{\theta\} = 0 \tag{12.19}$$

where the mass and stiffness matrices are

$$[M] = \begin{bmatrix} I_1 & & & & \\ & I_2 & & & \\ & & I_3 & & \\ & & & \cdots & \\ & & & & I_{13} \end{bmatrix}$$

$$[K] = \begin{bmatrix} k_1 & -k_1 & & & & \\ -k_1 & k_1 + k_2 & -k_2 & & & \\ & -k_2 & k_2 + k_3 & -k_3 & & \\ & & & \cdots & -k_{12} & \\ & & & -k_{12} & k_{12} & \end{bmatrix} \tag{12.20}$$

The free vibration solution of (12.19) is given by the eigen-value problem below

$$\begin{aligned} &\theta_i = \Theta_i \cos pt, \quad i = 1, 2, \ldots, 13 \\ &[[K] - p^2[M]]\,\{\Theta\} = 0 \\ &\left|[K] - p^2[M]\right| = 0 \end{aligned} \tag{12.21}$$

For the system given in Table 12.1, there are N $(= 13)$ eigen-values and 13 N associated modes represented by the modal matrix

$$[\Theta] = \begin{bmatrix} \Theta_{11} & \Theta_{12} & \Theta_{13} & \ldots & \Theta_{1N} \\ \Theta_{21} & \Theta_{22} & \Theta_{23} & \ldots & \Theta_{2N} \\ \Theta_{31} & \Theta_{32} & \Theta_{33} & \ldots & \Theta_{3N} \\ \ldots & & & & \\ \Theta_{N1} & \Theta_{N2} & \Theta_{N3} & \ldots & \Theta_{NN} \end{bmatrix} \tag{12.22}$$

The main problem in the first half of the 20th century was on dealing numerically with the eigen-values of equation (12.21) and determining the mode shapes in (12.22).

Matrices upto 4×4 could be dealt with in closed form; Jacobsen and Ayre [7] considered such a problem with both ends tied to the ground. The system then consisted of stiffnesses k_1 to k_5 with four masses m_1 to m_4 with k_1 and k_5 tied to the ground. In free vibration the system governing equations are given by

$$\begin{bmatrix} d_1 & -k_2 & 0 & 0 \\ -k_2 & d_2 & -k_3 & 0 \\ 0 & -k_3 & d_3 & -k_4 \\ 0 & 0 & -k_4 & d_4 \end{bmatrix} \begin{Bmatrix} X_1 \\ X_2 \\ X_3 \\ X_4 \end{Bmatrix} = 0 \tag{12.23}$$

where

$$\begin{aligned} d_1 &= k_1 + k_2 - m_1\omega^2 \\ d_2 &= k_2 + k_3 - m_2\omega^2 \\ d_3 &= k_3 + k_4 - m_3\omega^2 \\ d_4 &= k_4 + k_5 - m_4\omega^2 \end{aligned} \tag{12.24}$$

The natural frequency equation is obtained by setting the determinant in (12.23) equal to zero.

$$\Delta = \begin{vmatrix} d_1 & -k_2 & 0 & 0 \\ -k_2 & d_2 & -k_3 & 0 \\ 0 & -k_3 & d_3 & -k_4 \\ 0 & 0 & -k_4 & d_4 \end{vmatrix} = 0 \tag{12.25}$$

i.e.,

$$d_1 d_2 d_3 d_4 - \left(k_2^2 d_3 d_4 + k_3^2 d_4 d_1 + k_4^2 d_1 d_2\right) + k_2^2 k_4^2 = 0$$

The above is expressed in non-dimensional form for the frequency $z = \omega^2/(k/m)$ as

$$z^4 - C_3 z^3 + C_2 z^2 - C_1 z^1 + C_0 z^0 = 0 \tag{12.25a}$$

where

$$C_3 = \frac{m}{k}\left(\frac{k_1 + k_2}{m_1} + \frac{k_2 + k_3}{m_2} + \frac{k_3 + k_4}{m_3} + \frac{k_4 + k_5}{m_4}\right)$$

$$C_2 = \frac{m^2}{k^2}\left[\begin{array}{l} \frac{k_1k_2+k_2k_3+k_3k_4}{m_1m_2} + \frac{k_2k_3+k_3k_4+k_4k_2}{m_2m_3} + \frac{k_3k_4+k_4k_5+k_5k_3}{m_3m_4} \\ + \frac{(k_1+k_2)(k_3+k_4)}{m_1m_3} + \frac{(k_2+k_3)(k_4+k_5)}{m_2m_4} + \frac{(k_4+k_5)(k_1+k_2)}{m_4m_1} \end{array}\right]$$

$$C_1 = \frac{m^3}{k^3}\left[\begin{array}{l} \frac{k_1k_2k_3+k_2k_3k_4+k_3k_4k_1+k_4k_1k_2}{m_1m_2m_3} + \frac{k_2k_3k_4+k_3k_4k_5+k_4k_5k_2+k_5k_2k_3}{m_2m_3m_4} \\ + \frac{(k_1+k_2)(k_3k_4+k_4k_5+k_5k_3)}{m_3m_4m_1} + \frac{(k_4+k_5)(k_1k_2+k_2k_3+k_3k_4)}{m_4m_1m_2} \end{array}\right]$$

$$C_0 = \frac{m^4}{k^4}\left[\frac{k_1k_2k_3k_4 + k_2k_3k_4k_5 + k_3k_4k_5k_1 + k_4k_5k_1k_2 + k_5k_1k_2k_3}{m_1m_2m_3m_4}\right]$$

Now it is a matter of solving the roots of a polynomial equation in (12.25).

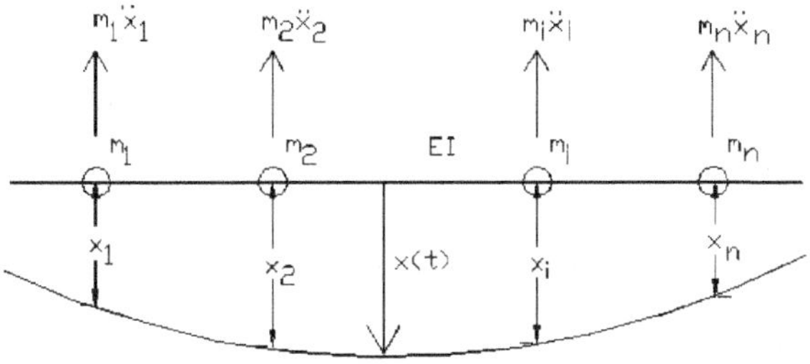

Fig. 12.1 Far-coupled system

12.2 Far-Coupled Systems

As mentioned before, Hahn [5] used the method of Influence Coefficients to prove Dunkerley's empirical result. In Figure 12.1 a far-coupled system (lateral vibrations of a beam) is shown.

Using equation (11.10), we can write, for the vibrating beam in Figure 12.1

$$\alpha_{11}m_1\ddot{x}_1 + \alpha_{12}m_2\ddot{x}_2 + \cdots + \alpha_{1n}m_n\ddot{x}_n + x_1 = 0$$
$$\alpha_{21}m_1\ddot{x}_1 + \alpha_{22}m_2\ddot{x}_2 + \cdots + \alpha_{2n}m_n\ddot{x}_n + x_2 = 0$$
$$\ldots$$
$$\alpha_{n1}m_1\ddot{x}_1 + \alpha_{n2}m_2\ddot{x}_2 + \cdots + \alpha_{nn}m_n\ddot{x}_n + x_n = 0 \tag{12.26}$$

We can define the influence coefficient matrix

$$[\alpha] = \begin{bmatrix} \alpha_{11} & \alpha_{12} & \cdots & \alpha_{1n} \\ \alpha_{21} & \alpha_{22} & \cdots & \alpha_{2n} \\ \cdots & \cdots & \cdots & \cdots \\ \alpha_{n1} & \alpha_{n2} & \cdots & \alpha_{nn} \end{bmatrix}$$

and obtain the following eigen-value problem:

$$[\alpha][M]\{\ddot{x}\} + [I]\{x\} = 0 \tag{12.27}$$

The above equation (12.27) is the same as (12.19), since

$$[M]\{\ddot{x}\} + [\alpha]^{-1}[I]\{x\} = 0$$
$$[M]\{\ddot{x}\} + [K]\{x\} = 0 \tag{12.28}$$

However, inversion of the influence coefficient matrix $[\alpha]$ is itself a problem in the pre-digital computer era. Jacobsen and Ayre [7] expanded the matrix equation (12.27) to obtain the polynomial frequency for a four mass far-coupled system as follows.

$$\begin{bmatrix} \alpha_{11}m_1 - \frac{1}{\omega^2} & \alpha_{12}m_2 & \alpha_{13}m_3 & \alpha_{14}m_4 \\ \alpha_{21}m_1 & \alpha_{22}m_2 - \frac{1}{\omega^2} & \alpha_{23}m_3 & \alpha_{24}m_4 \\ \alpha_{31}m_1 & \alpha_{32}m_2 & \alpha_{33}m_3 - \frac{1}{\omega^2} & \alpha_{34}m_4 \\ \alpha_{41}m_1 & \alpha_{42}m_2 & \alpha_{43}m_3 & \alpha_{44}m_4 - \frac{1}{\omega^2} \end{bmatrix} \begin{Bmatrix} X_1 \\ X_2 \\ X_3 \\ X_4 \end{Bmatrix} = 0 \tag{12.29}$$

The above is expanded to give the frequency equation

$$\left(\frac{1}{\omega^2}\right)^4 - A\left(\frac{1}{\omega^2}\right)^3 + B\left(\frac{1}{\omega^2}\right)^2 - C\left(\frac{1}{\omega^2}\right)^1 + D\left(\frac{1}{\omega^2}\right)^0 = 0 \tag{12.30}$$

where

$$A = \alpha_{11}m_1 + \alpha_{22}m_2 + \alpha_{33}m_3 + \alpha_{44}m_4$$

$$B = \begin{vmatrix} \alpha_{11}m_1 & \alpha_{12}m_2 \\ \alpha_{21}m_1 & \alpha_{22}m_2 \end{vmatrix} + \begin{vmatrix} \alpha_{11}m_1 & \alpha_{13}m_3 \\ \alpha_{31}m_1 & \alpha_{33}m_3 \end{vmatrix} + \begin{vmatrix} \alpha_{11}m_1 & \alpha_{14}m_4 \\ \alpha_{41}m_1 & \alpha_{44}m_4 \end{vmatrix}$$
$$+ \begin{vmatrix} \alpha_{22}m_2 & \alpha_{23}m_3 \\ \alpha_{32}m_2 & \alpha_{33}m_3 \end{vmatrix} + \begin{vmatrix} \alpha_{22}m_2 & \alpha_{24}m_4 \\ \alpha_{42}m_2 & \alpha_{44}m_4 \end{vmatrix} + \begin{vmatrix} \alpha_{33}m_3 & \alpha_{34}m_4 \\ \alpha_{43}m_3 & \alpha_{44}m_4 \end{vmatrix}$$

$$C = \begin{vmatrix} \alpha_{11}m_1 & \alpha_{12}m_2 & \alpha_{13}m_3 \\ \alpha_{21}m_1 & \alpha_{22}m_2 & \alpha_{23}m_3 \\ \alpha_{31}m_1 & \alpha_{32}m_2 & \alpha_{33}m_3 \end{vmatrix} + \begin{vmatrix} \alpha_{11}m_1 & \alpha_{12}m_2 & \alpha_{14}m_4 \\ \alpha_{21}m_1 & \alpha_{22}m_2 & \alpha_{24}m_4 \\ \alpha_{41}m_1 & \alpha_{42}m_2 & \alpha_{44}m_4 \end{vmatrix}$$
$$+ \begin{vmatrix} \alpha_{11}m_1 & \alpha_{13}m_3 & \alpha_{14}m_4 \\ \alpha_{31}m_1 & \alpha_{33}m_3 & \alpha_{34}m_4 \\ \alpha_{41}m_1 & \alpha_{43}m_3 & \alpha_{44}m_4 \end{vmatrix} + \begin{vmatrix} \alpha_{22}m_2 & \alpha_{23}m_3 & \alpha_{24}m_4 \\ \alpha_{32}m_2 & \alpha_{33}m_3 & \alpha_{34}m_4 \\ \alpha_{42}m_2 & \alpha_{43}m_3 & \alpha_{44}m_4 \end{vmatrix}$$

$$D = \begin{vmatrix} \alpha_{11}m_1 & \alpha_{12}m_2 & \alpha_{13}m_3 & \alpha_{14}m_4 \\ \alpha_{21}m_1 & \alpha_{22}m_2 & \alpha_{23}m_3 & \alpha_{24}m_4 \\ \alpha_{31}m_1 & \alpha_{32}m_2 & \alpha_{33}m_3 & \alpha_{34}m_4 \\ \alpha_{41}m_1 & \alpha_{42}m_2 & \alpha_{43}m_3 & \alpha_{44}m_4 \end{vmatrix}$$

Though it is a simple matter today to find the roots of polynomials in (12.25) and (12.30), it was difficult in the pre-digital computer era. So we had to find other means of calculating the required roots.

12.3 Gräffe's Method of Successive Approximations

Germinal Pierre Dandelin in 1826 and Karl Heinrich Gräffe in 1837 developed independently a method of finding roots of a polynomial equation [6,9]. Lobachevsky also discovered the principal idea of the method in 1834. Until the advent of the digital computer era, this method became useful in determining the natural frequencies of torsion (close-coupled) and bending (far-coupled) of one dimensional structures.

Restricting to polynomials having only real roots we consider here a fourth degree polynomial.

$$c_4 z^4 - c_3 z^3 + c_2 z^2 - c_1 z^1 + c_0 z^0 = 0 \tag{12.31}$$

Let the four real roots be r_1 r_2, r_3 and r_4, and then the equivalent of (12.31) is

$$(z - r_1)(z - r_2)(z - r_3)(z - r_4) = 0 \tag{12.32}$$

Expanding this equation we get

$$\begin{aligned} &z^4 - (r_1 + r_2 + r_3 + r_4)\, z^3 + (r_1 r_2 + r_1 r_3 + r_1 r_4 + r_2 r_3 + r_2 r_4 + r_3 r_4)\, z^2 \\ &\quad - (r_1 r_2 r_3 + r_1 r_2 r_4 + r_1 r_3 r_4 + r_2 r_3 r_4)\, z + r_1 r_2 r_3 r_4 = 0 \end{aligned} \tag{12.33}$$

Therefore

$$\begin{aligned} c_4 &= 1 \\ c_3 &= r_1 + r_2 + r_3 + r_4 \\ c_2 &= r_1 r_2 + r_1 r_3 + r_1 r_4 + r_2 r_3 + r_2 r_4 + r_3 r_4 \\ c_1 &= r_1 r_2 r_3 + r_1 r_2 r_4 + r_1 r_3 r_4 + r_2 r_3 r_4 \\ c_0 &= r_1 r_2 r_3 r_4 \end{aligned} \tag{12.34}$$

Let the roots be $r_4 > r_3 > r_2 > r_1$ we can obtain crude approximations for the smallest and highest roots from

$$\begin{aligned} r_1 &\approx \frac{c_0}{c_1} \\ r_4 &\approx \frac{c_3}{c_4} \end{aligned} \tag{12.35}$$

Now, let the roots be separated by higher margins then $r_4 \gg r_3 \gg r_2 \gg r_1$. This leads from (12.34) to

$$\begin{aligned} &c_4 = 1 \\ &c_3 \approx r_4 \text{ from which } r_4 \approx c_3 \\ &c_2 \approx r_3 r_4 \text{ from which } r_3 \approx \frac{c_2}{c_3} \\ &c_1 \approx r_2 r_3 r_4 \text{ from which } r_2 \approx \frac{c_1}{c_2} \\ &c_0 \approx r_1 r_2 r_3 r_4 \text{ from which } r_1 \approx \frac{c_0}{c_1} \end{aligned} \tag{12.36}$$

Therefore, if the differences between the four successive roots are very large or can be made very large by some means, the four roots are given by approximate expressions in (12.36).

First let us examine how the squaring of the four roots above will affect the coefficients of the new polynomial equation from whose new coefficients of the four approximate roots may be found according to (12.36). Let the polynomial have four purely imaginary roots of the same numerical values as the four roots of (12.31), then

$$(z + r_1)(z + r_2)(z + r_3)(z + r_4) = 0 \tag{12.37}$$

Multiplying (12.32) and (12.37)

$$(z^2 - r_1^2)(z^2 - r_2^2)(z^2 - r_3^2)(z^2 - r_4^2) = 0 \tag{12.38}$$

The above is a polynomial equation whose roots are squares of the original equation. Equation (12.38) can now be written as a new polynomial equation of the same form as (12.32)

$$(Z - R_1)(Z - R_2)(Z - R_3)(Z - R_4) = 0 \tag{12.39}$$

The polynomial equation corresponding to (12.37) is

$$c_4 z^4 + c_3 z^3 + c_2 z^2 + c_1 z^1 + c_0 z^0 = 0 \tag{12.40}$$

in which the coefficients have same values as in (12.33). The new polynomial in z^2 or in Z is obtained by multiplying (12.31) and (12.40)

$$c_4^2 Z^4 - (c_3^2 - 2c_4c_2)Z^3 + (c_2^2 - 2c_1c_3 + 2c_4c_0)Z^2 - (c_1^2 - 2c_2c_0)Z^1 + c_0^2 Z^0 = 0 \tag{12.41}$$

which can be written as

$$C_4 Z^4 - C_3 Z^3 + C_2 Z^2 - C_1 Z^1 + C_0 Z^0 = 0 \tag{12.42}$$

The above is equivalent to (12.39). We can adopt an iteration process now as given by Jacobsen and Ayre [7], given in Table 12.1.

12.4 Matrix Iteration Method

The solution of large size eigen-value problems was difficult to achieve in pre-digital computer era. Matrix methods however showed a good promise for vibration problems. Duncan and Collar [3] developed such a method. This is essentially an extension of the Stodola–Viannello method in Section 11.2.

Let us pick up equation (11.10) here

Table 12.2 Solution of $z^4 - 4.450z^3 + 5.9333z^2 - 2.3667z + 0.200 = 0$.

	c_4	r_4		c_3	r_3		c_2	r_2		c_1	r_1		c_0
1	1	(4.450)	-	4.450	(1.333)	+	5.9333	(0.3989)	-	2.3667	(0.0845)	+	0.200
2			-	19.802		+	35.205		-	5.601		+	0.04
3			+	11.867		-	21.063		+	2.373			
4						+	0.400						
5	1	(2.817)	-	7.935	(1.355)	+	14.542	(0.4711)	-	3.228	(0.1113)	+	0.040
6	1		-	62.967		+	211.467		-	10.420		+	$16\ 10^{-4}$
7			+	29.084		-	51.228		+	1.163			
8						+	0.080						
9	1	(2.413)	-	33.883	(1.475)	+	160.319	(0.4902)	-	9.257	(0.1147)	+	$16\ 10^{-4}$
10	1		-	1148.02		+	$257.06\ 10^{2}$		-	85.692		+	$256\ 10^{-8}$
11			+	320.64		-	$6.27\ 10^{2}$		+	0.257			
12						+	0						
13	1	(2.316)	-	827.38	(1.532)	+	$250.79\ 10^{2}$	(0.4915)	-	85.435	(0.1147)	+	$256\ 10^{-8}$
14	1		-	$68.46\ 10^{4}$		+	$62895\ 10^{4}$		-	$7.293\ 10^{3}$			$6.5\ 10^{-12}$
15			+	$4.02\ 10^{4}$		-	$14\ 10^{4}$		+	0			
16						+	0						
17	1	(2.305)	-	$63.44\ 10^{4}$	(1.540)	+	$62.881\ 10^{4}$	(0.4915)	-	$7.293\ 10^{3}$	(0.1147)	+	$6.5\ 10^{-12}$
18	1		-	$40.246\ 10^{10}$		+	$39.540\ 10^{16}$		-	$53.10\ 10^{6}$			$43\ 10^{-24}$
19			+	$0.126\ 10^{10}$		-	0		+	0			
20						+	0						
21	1	(2.305)	-	$40.120\ 10^{10}$	(1.539)	+	$39.540\ 10^{16}$	(0.4915)	-	$53.19\ 10^{8}$	(0.1147)	+	$43\ 10^{-24}$

The four roots are 0.1147, 0.4915, 1.539 and 2.305.

$$y_1 = \alpha_{11} P_1 + \alpha_{12} P_2 + \cdots + \alpha_{1n} P_n$$
$$y_2 = \alpha_{21} P_1 + \alpha_{22} P_2 + \cdots + \alpha_{2n} P_n$$
$$\cdots$$
$$y_n = \alpha_{n1} P_1 + \alpha_{n2} P_2 + \cdots + \alpha_{nn} P_n \tag{11.10}$$

For normal modes of vibration with frequency ω, the forces P_j are $m_j y_j \omega^2$ and equation (11.10) can be rewritten as

$$y_1 = \omega^2 \left(m_1 \alpha_{11} y_1 + m_2 \alpha_{12} y_2 + \cdots + m_n \alpha_{1n} y_n \right)$$
$$y_2 = \omega^2 \left(m_1 \alpha_{21} y_1 + m_2 \alpha_{22} y_2 + \cdots + m_n \alpha_{2n} y_n \right)$$
$$\cdots$$
$$y_n = \omega^2 \left(m_1 \alpha_{n1} y_1 + m_2 \alpha_{n2} y_2 + \cdots + m_n \alpha_{nn} y_n \right) \tag{12.43}$$

We recast the above as

$$\begin{Bmatrix} y_1 \\ y_2 \\ \cdots \\ y_n \end{Bmatrix} = \omega^2 \begin{bmatrix} \alpha_{11} & \alpha_{12} & \cdots & \alpha_{1n} \\ \alpha_{21} & \alpha_{22} & \cdots & \alpha_{2n} \\ \cdots & \cdots & \cdots & \cdots \\ \alpha_{n1} & \alpha_{n2} & \cdots & \alpha_{nn} \end{bmatrix} \begin{bmatrix} m_1 & 0 & 0 & 0 \\ 0 & m_2 & 0 & 0 \\ 0 & 0 & \cdots & 0 \\ 0 & 0 & 0 & m_n \end{bmatrix} \begin{Bmatrix} y_1 \\ y_2 \\ \cdots \\ y_n \end{Bmatrix} \tag{12.44}$$

i.e.

$$\{y\} = \omega^2 [\alpha][M]\{y\}$$
$$\{y\} = \omega^2 [U]\{y\} \tag{12.45}$$

where $[U]$ is a dynamic matrix.

We follow the Stodola–Viannello iteration method as illustrated through an example given by Scanlan and Rosenbaum [16] of a wing 240 in long fixed at the midpoint of the fuselage. The influence coefficient matrix and mass matrix obtained for the wing discretized at seven stations starting from 0″ to 240″ at the midpoints, 16, 50, 100, 140, 180 and 220″ of six sections of the wing, are

$$[\alpha] = 10^{-7}\begin{bmatrix} 2 & & & & & \\ 11 & 68 & & & \text{sym} & \\ 25 & 204 & 790 & & & \\ 36 & 322 & 1260 & 2255 & & \\ 47 & 445 & 1725 & 3255 & 5000 & \\ 57 & 561 & 2200 & 4270 & 6900 & 10000 \end{bmatrix} \text{in/lb}$$

$$[M] = \begin{bmatrix} 1.55 & & & & & \\ & 1.43 & & & & \\ & & 0.648 & & & \\ & & & 0.249 & & \\ & & & & 0.181 & \\ & & & & & 0.104 \end{bmatrix} \text{lb.sec.}^2/\text{in.}$$

Then $[U] = [\alpha][M]$ and

$$\frac{10^7}{\omega^2}\begin{Bmatrix} y_1 \\ y_2 \\ y_3 \\ y_4 \\ y_5 \\ y_6 \end{Bmatrix} = \begin{bmatrix} 3.10 & 15.73 & 16.20 & 8.96 & 8.51 & 5.93 \\ 17.50 & 97.24 & 132.19 & 80.18 & 80.55 & 58.34 \\ 38.75 & 291.72 & 511.92 & 313.74 & 312.23 & 228.80 \\ 55.80 & 460.46 & 816.48 & 561.50 & 589.16 & 444.08 \\ 72.85 & 636.35 & 1117.80 & 810.50 & 905.00 & 717.60 \\ 88.35 & 802.23 & 1425.60 & 1063.23 & 1248.90 & 1040.00 \end{bmatrix}\begin{Bmatrix} y_1 \\ y_2 \\ y_3 \\ y_4 \\ y_5 \\ y_6 \end{Bmatrix}$$

The iteration process is clear, assume the first mode shape in the right-hand vector, determine deflections on the left side and compare with the assumed values; repeat next iteration with the newly determined deflections as assumed ones and continue until convergency is achieved. This was the same way Stodola–Viannello proceeded in their graphical method. Here the accuracy will be better since drawings are avoided.

We can begin with a normalized mode shape with the last station having a unit deflection; $y_1 = 0.01$, $y_2 = 0.06$, $y_3 = 0.22$, $y_4 = 0.43$, $y_5 = 0.69$ and $y_6 = 1.0$. The first iteration result is

Table 12.3 Matrix iteration for I mode of a cantilever beam

	(0)	Obtained	Normalized (1)	Obtained	Normalized (2)	Obtained	Normalized (3)
y_1	0.01	20.2	0.007	22	0.0075	21	0.0073
y_2	0.06	183.5	0.067	200	0.0684	157	0.0680
y_3	0.22	793.1	0.291	774	0.2647	763	0.2636
y_4	0.43	1299.9	0.478	1409	0.4819	1392	0.4808
y_5	0.69	1975.4	0.726	2130	0.7285	2107	0.7278
y_6	1.00	2721.6	1.000	2924	1.0000	2895	1.0000
$\frac{10^7}{\omega^2}$			2721.6		2924		2895

$$\frac{10^7}{\omega^2}\begin{Bmatrix} y_1 \\ y_2 \\ y_3 \\ y_4 \\ y_5 \\ y_6 \end{Bmatrix}^{(1)} =$$

$$= \begin{bmatrix} 3.10 & 15.73 & 16.20 & 8.96 & 8.51 & 5.93 \\ 17.50 & 97.24 & 132.19 & 80.18 & 80.55 & 58.34 \\ 38.75 & 291.72 & 511.92 & 313.74 & 312.23 & 228.80 \\ 55.80 & 460.46 & 816.48 & 561.50 & 589.16 & 444.08 \\ 72.85 & 636.35 & 1117.80 & 810.50 & 905.00 & 717.60 \\ 88.35 & 802.23 & 1425.60 & 1063.23 & 1248.90 & 1040.00 \end{bmatrix} \begin{Bmatrix} 0.01 \\ 0.06 \\ 0.22 \\ 0.43 \\ 0.69 \\ 1.0 \end{Bmatrix}^{(0)}$$

$$= \begin{Bmatrix} 20.2 \\ 183.5 \\ 793.1 \\ 1299.9 \\ 1975.4 \\ 2721.6 \end{Bmatrix}$$

giving the first approximation for the frequency $\omega = \sqrt{10^7/2721.6} = 60.61$ rad/s.

Normalizing the 1st iteration deflection values we can repeat the above steps as the iteration; the results are given in Table 12.3.

The third iteration result for the frequency $\omega = \sqrt{10^7/2895} = 58.77$ rad/s.

12.5 Method of Priebs [10]

While solution of equation (12.19) is an easy matter today with the advent of digital computers, it was an uphill task to handle matrices of size of more than three or at most four stations. Until such times (late 1960s and early 1970s), searches were on to deal with the solution of vibration problems through matrix methods.

After having set up the governing equations of motion, the eigen-value problem is easily written down. The frequency equations can be in principle developed by matrix multiplication, e.g., (12.25) or (12.30), however when the number of stations are more than 4, a closed form way of obtaining the polynomial frequency equations is difficult. If we can obtain the polynomial frequency equation then we could use Gräffe numerical scheme as illustrated in Table 12.1. Priebs provided a method of setting up a polynomial frequency equation for a multidegree of freedom system for free-free systems. Rao [12] used this method to solve vibration problems of cantilevers in torsion.

The equation of motion of the ith rotor from (12.18) is given by

$$I_i\ddot{\theta}_i + k_i(\theta_i - \theta_{i-1}) + k_{i+1}(\theta_i - \theta_{i+1}) = 0 \tag{12.46}$$

Writing the torsional moment in ith shaft as

$$m_i = k_i(\theta_{i-1} - \theta_i) \tag{12.47}$$

Equation (12.46) becomes

$$I_i\ddot{\theta}_i + m_i - m_{i+1} = 0 \tag{12.48}$$

Writing $\theta_i = \Theta_i \cos\omega t$ and $m_i = M_i \cos\omega t$ and with $Z = \omega^2$, we get

$$\begin{aligned} M_{i+1} &= M_i + I_i\Theta_i Z \\ \Theta_i &= \Theta_{i-1} - \frac{M_i}{k_i} \end{aligned} \tag{12.49}$$

Starting from zeroth rotor at free end $M_0 = 0$ we can move to the cantilever end where $\Theta_n = 0$ as follows:

$$\begin{aligned} M_1 &= I_0\Theta_0 Z \\ \Theta_1 &= \Theta_0 - \frac{M_1}{k_1} \\ M_2 &= M_1 + I_1\Theta_1 Z \\ \Theta_2 &= \Theta_1 - \frac{M_2}{k_2} \\ &\ldots \\ M_i &= M_{i-1} + I_{i-1}\Theta_{0-1} Z \\ \Theta_i &= \Theta_{i-1} - \frac{M_i}{k_i} \\ &\ldots \\ M_n &= M_{n-1} + I_{n-1}\Theta_{n-1} Z \\ \Theta_n &= \Theta_{n-1} - \frac{M_n}{k_n} = 0 \end{aligned} \tag{12.50}$$

Writing down the expressions for the amplitudes of torsional moment and the amplitudes of the rotors as

$$M_1 = \theta_0(A_{11}Z)$$
$$M_2 = \theta_0(A_{21}Z - A_{22}Z^2)$$
$$\ldots$$
$$M_n = \theta_0(A_{n1}Z - A_{n2}Z^2 + \cdots (-1)^n A_{nn}Z^n)$$

and

$$\theta_1 = \theta_0 (1 - B_{11}Z)$$
$$\theta_2 = \theta_0 \left(1 - B_{21}Z + B_{22}Z^2\right)$$
$$\ldots$$
$$\theta_n = \theta_0 \left(1 - B_{n+1,1}Z + B_{n+1,2}Z^2 + \cdots (-1)^n B_{n+1,n}Z^n\right) \quad (12.51)$$

from equations (12.50) and (12.51) we can obtain

$$A_{11} = I_0$$
$$A_{21} = A_{11} + I_1; \quad A_{22} = I_1 B_{21}$$
$$A_{31} = A_{31} + I_2; \quad A_{32} = A_{22} + I_2 B_{31}; \quad A_{33} = I_2 B_{32}$$
$$A_{41} = A_{31} + I_3; \quad A_{42} = A_{32} + I_3 B_{41}; \quad A_{43} = A_{33} + I_3 B_{42}; \quad A_{44} = I_3 B_{43}$$
$$\ldots \quad (12.52)$$

$$B_{21} = \frac{A_{11}}{k_1}$$
$$B_{31} = B_{21} + \frac{A_{21}}{k_2}; \quad B_{32} = \frac{A_{22}}{k_2}$$
$$B_{41} = B_{31} + \frac{A_{31}}{k_3}; \quad B_{42} = B_{32} + \frac{A_{32}}{k_3}; \quad B_{43} = \frac{A_{33}}{k_3}$$
$$B_{51} = B_{41} + \frac{A_{41}}{k_4}; \quad B_{52} = B_{42} + \frac{A_{42}}{k_4}; \quad B_{53} = B_{43} + \frac{A_{43}}{k_4}; \quad B_{54} = \frac{A_{44}}{k_4}$$
$$\ldots \quad (12.53)$$

Choosing reference values I^* and k^*, we can express the inertias and stiffnesses as nondimensional numbers for easier numerical calculations. Then

$$I'_n = \frac{I_n}{I^*} \quad \text{and} \quad k'_n = \frac{k_n}{k^*}$$

$$A_{11} = I_0 = I'_0 I^* = A'_{11} I^*$$

$$\dots \tag{12.54}$$

and

$$B_{21} = \frac{A_{11}}{k_1} = \frac{A'_{11}}{k'_1}\frac{I^*}{k^*} = B'_{21}\frac{I^*}{k^*}$$

$$\dots \tag{12.55}$$

From (12.51) we can then write

$$\theta_n = \theta_0 \left[1 - B'_{n+1,1} Z \frac{I^*}{k^*} + B'_{n+1,2} Z^2 \left(\frac{I^*}{k^*}\right)^2 + \cdots (-1)^n B'_{n+1,n} Z^n \left(\frac{I^*}{k^*}\right)^n \right] \tag{12.56}$$

i.e.,

$$\theta_n = \theta_0 \left[1 - B'_{n+1,1}\zeta + B'_{n+1,2}\zeta^2 + \cdots (-1)^n B'_{n+1,n}\zeta^n \right] \tag{12.57}$$

where $\zeta = Z\frac{I^*}{k^*}$.

The polynomial frequency equation for cantilever is then

$$1 - B'_{n+1,1}\zeta + B'_{n+1,2}\zeta^2 + \cdots (-1)^n B'_{n+1,n}\zeta^n = 0 \tag{12.58}$$

Consider a cantilever with five inertias and stiffnesses as follows:

$I_0 = 1.36 \times 10^{-5}$ lb.in.sec^2	$k_1 = 5.4054 \times 10^4$ lb.in./rad
$I_1 = 1.36 \times 10^{-5}$ lb.in.sec^2	$k_2 = 5.4054 \times 10^4$ lb.in./rad
$I_2 = 1.36 \times 10^{-5}$ lb.in.sec^2	$k_3 = 5.4054 \times 10^4$ lb.in./rad
$I_3 = 1.36 \times 10^{-5}$ lb.in.sec^2	$k_4 = 5.4054 \times 10^4$ lb.in./rad
$I_4 = 1.36 \times 10^{-5}$ lb.in.sec^2	$k_5 = 5.4054 \times 10^4$ lb.in./rad

Choosing $I^* = 10^{-5}$ lb.in.sec^2 and $k^* = 10^4$ lb.in./rad, the coefficients in (12.58) are determined as shown in Table 12.2. The frequency equation obtained is

$$1 - 2.787051\zeta + 1.413019\zeta^2 - 0.255203\zeta^3 + 0.018925\zeta^4 - 0.000492\zeta^5 = 0$$

which is rewritten as

$$\zeta^5 - 38.47\zeta^4 + 518.71\zeta^3 - 2871.99\zeta^2 + 5664.74\zeta - 2032.52 = 0$$

This polynomial is solved by Gräffe's Method in Section 12.3 to give the roots r_i = 0.4556, 3.1800, 7.7510, 11.8740 and 15.2190. The natural frequencies are given by

$$\omega_i = \frac{1}{2\pi}\sqrt{r_i \frac{c^*}{I^*}}$$

3398, 8975, 14000, 17350 and 19650 cps.

12.6 The Holzer Method (Close Coupled Systems) in Transfer Matrix Form

Holzer's method in Section 11.6 advanced a step ahead in the early 1940s by expressing it in transfer matrix form that is made more easily adaptable for computerization. The transfer of state quantities from station to station here are facilitated through matrices called transfer matrices. Holzer's method is illustrated here in transfer matrix form, which is the simplest, since it has only two state quantities [14]. We can write in matrix form the transfer equations of a discrete lumped mass spring torsional system from station $i-1$ to i as follows:

$$\begin{Bmatrix} \theta \\ M \end{Bmatrix}_i^L = \begin{bmatrix} 1 & \frac{1}{K} \\ 0 & 1 \end{bmatrix} \begin{Bmatrix} \theta \\ M \end{Bmatrix}_{i-1}^R$$

$$\{S\}_i^L = [F]_i \{S\}_{i-1}^R \tag{12.59}$$

where the state quantities are transferred from the right of station $i-1$ to the left of station i across the field represented by field matrix $[F]_i$. Similarly the point matrix $[P]_i$ across the mass at station i is

$$\begin{Bmatrix} \theta \\ M \end{Bmatrix}_i^R = \begin{bmatrix} 1 & 0 \\ -m\omega^2 & 1 \end{bmatrix} \begin{Bmatrix} \theta \\ M \end{Bmatrix}_i^L \tag{12.60}$$
$$\{S\}_i^R = [P]_i \{S\}_i^L$$

The above two equations are combined to give the transfer matrix $[T]_i$:

$$\begin{aligned} \{S\}_i^R &= [P]_i [F]_i \{S\}_{i-1}^R \\ &= [T]_i \{S\}_{i-1}^R \end{aligned} \tag{12.61}$$

where

$$\{T\}_i = \begin{bmatrix} 1 & \frac{1}{K} \\ -m\omega^2 & 1 - \frac{m\omega^2}{K} \end{bmatrix}_i \tag{12.62}$$

Now let us take the example of a locomotive drive train in Table 12.2 and show its solution by the transfer matrix method. We start from the free end station 1 with Point matrix $[P]_1$ and transfer matrix for station 2 and keep transferring the state quantities to reach station 12, another free end where we satisfy the boundary conditions. This builds up to an Overall Transfer Matrix as follows. The frequency assumed in the calculations is $\omega = 86$ rad/s.

$$[T]_2\,[P]_1 = \begin{bmatrix} 1 & 6.2539 \times 10^{-6} \\ -1538.2 & 0.9904 \end{bmatrix} \begin{bmatrix} 1 & 0 \\ -2735.3 & 1 \end{bmatrix} = \begin{bmatrix} 0.9829 & - \\ -4247.14 & - \end{bmatrix}$$

$$[T]_3\,[T]_2\,[P]_1 = \begin{bmatrix} 1 & 36.4033 \times 10^{-6} \\ -457.09 & 0.9834 \end{bmatrix} \begin{bmatrix} 10.9829 & - \\ -4247.14 & - \end{bmatrix} = \begin{bmatrix} 0.8283 & - \\ -4265.7 & - \end{bmatrix}$$

$$[T]_4 \cdots [P]_1 = \begin{bmatrix} 1 & 5.8584 \times 10^{-6} \\ -1044.76 & 0.9939 \end{bmatrix} \begin{bmatrix} 0.8283 & - \\ -4265.7 & - \end{bmatrix} = \begin{bmatrix} 0.8012 & - \\ -5462.74 & - \end{bmatrix}$$

$$[T]_5 \cdots [P]_1 = \begin{bmatrix} 1 & 0.9572 \times 10^{-6} \\ -28296.36 & 0.9729 \end{bmatrix} \begin{bmatrix} 0.8012 & - \\ -5462.74 & - \end{bmatrix} = \begin{bmatrix} 0.7959 & - \\ -27985.3 & - \end{bmatrix}$$

$$[T]_6 \cdots [P]_1 = \begin{bmatrix} 1 & 0.5727 \times 10^{-6} \\ -6769.34 & 0.9996 \end{bmatrix} \begin{bmatrix} 0.7959 & - \\ -27985.3 & - \end{bmatrix} = \begin{bmatrix} 0.7943 & - \\ -33362.46 & - \end{bmatrix}$$

$$[T]_7 \cdots [P]_1 = \begin{bmatrix} 1 & 0.1699 \times 10^{-6} \\ -5927.72 & 0.999 \end{bmatrix} \begin{bmatrix} 0.7943 & - \\ -33362.46 & - \end{bmatrix} = \begin{bmatrix} 0.7887 & - \\ -38037.8 & - \end{bmatrix}$$

$$[T]_8 \cdots [P]_1 = \begin{bmatrix} 1 & 0.1699 \times 10^{-6} \\ -5927.72 & 0.999 \end{bmatrix} \begin{bmatrix} 0.7887 & - \\ -38037.8 & - \end{bmatrix} = \begin{bmatrix} 0.7822 & - \\ -42674.84 & - \end{bmatrix}$$

$$[T]_9 \cdots [P]_1 = \begin{bmatrix} 1 & 0.1699 \times 10^{-6} \\ -6769.34 & 0.9988 \end{bmatrix} \begin{bmatrix} 0.7822 & - \\ -42674.84 & - \end{bmatrix} = \begin{bmatrix} 0.775 & - \\ -47920.88 & - \end{bmatrix}$$

$$[T]_{10} \cdots [P]_1 = \begin{bmatrix} 1 & 0.1416 \times 10^{-6} \\ -24741.2 & 0.9965 \end{bmatrix} \begin{bmatrix} 0.775 & - \\ -47920.88 & - \end{bmatrix} = \begin{bmatrix} 0.7682 & - \\ -66926.84 & - \end{bmatrix}$$

$$[T]_{11} \cdots [P]_1 = \begin{bmatrix} 1 & 10.7302 \times 10^{-6} \\ -4353.3 & 0.9533 \end{bmatrix} \begin{bmatrix} 0.7682 & - \\ -66926.84 & - \end{bmatrix} = \begin{bmatrix} 0.0501 & - \\ -67144.7 & - \end{bmatrix}$$

$$[T]_{12} \cdots [P]_1 = \begin{bmatrix} 1 & 10.7302 \times 10^{-6} \\ -99668.5 & -0.0695 \end{bmatrix} \begin{bmatrix} 0.0501 & - \\ -67144.7 & - \end{bmatrix} = \begin{bmatrix} 0.6704 & - \\ -324.4 & - \end{bmatrix} \tag{12.63}$$

$$\{S\}_{12} = \begin{Bmatrix} \theta_{12} \\ 0 \end{Bmatrix} = [T]\,\{S\}_1$$
$$= \begin{bmatrix} 0.6704 & - \\ -324.4 & - \end{bmatrix} \begin{Bmatrix} \theta_1 \\ 0 \end{Bmatrix} \tag{12.64}$$

The term $T21 = -324.4$ can be taken as zero compared to the terms before in the transfer matrices. Therefore 86 rad/s is a natural frequency close to correct value.

Next we can start with one unit amplitude at station 1 and apply equation (12.61) repeatedly to get the mode shape.

Table 12.4 Prieb's scheme of calculations of polynomial frequency equation coefficients

$A_{11}' = I_0'$									
1.360000									
I_1'									
1.460000	$k_1' = 5.4054$								
A_{21}'	$B_{21}' = A_{11}'/k_1'$	$A_{22}' = I_1'B_{21}'$							
2.820000	0.249358	0.364063							
I_2'	A_{21}'	$I_2'B_{21}'$							
1.610000	0.487854	1.186927	$k_2' = 5.7803$						
A_{31}'	B_{31}'	A_{32}'	$B_{32}' = A_{22}'/k_2'$	$A_{33}' = I_2'B_{32}'$					
4.430000	0.737222	1.550990	0.062983	0.101403					
I_3'	A_{31}'/k_3'	$I_3'B_{41}'$	A_{32}'/k_3'	$I_3'B_{42}'$					
1.770000	0.660071	2.473209	0.231098	0.520523	$k_3' = 6.7114$				
A_{41}'	B_{41}'	A_{42}'	B_{42}'	A_{43}'	$B_{43}' = A_{33}'/k_3'$	$A_{44}' = I_3'B_{43}$			
6.200000	1.397293	4.024199	0.294081	0.621926	0.015109	0.026743			
I_4'	A_{41}'/k_4'	$I_4'B_{51}'$	A_{42}'/k_4'	$I_4'B_{52}'$	A_{43}'/k_4'	$I_4'B_{53}'$			
2.020000	0.830799	4.500746	0.539242	1.683312	0.083338	0.198863	$k_4' = 7.4627$		
A_{51}'	B_{51}'	A_{52}'	B_{52}'	A_{53}'	B_{53}'	A_{54}'	$B_{54}' = A_{44}'/k_4'$	$A_{55}' = I_4'B_{54}'$	
8.220000	2.228092	8.524945	0.833323	2.305238	0.098447	0.225606	0.003584	0.007240	
	A_{51}'/k_5'		A_{52}'/k_5'		A_{53}'/k_5'		A_{54}'/k_5'		
	0.558959		0.579696		0.156756		0.015341	$k_5' = 14.7059$	
	B_{61}'		B_{62}'		B_{63}'		B_{64}'		$B_{65}' = A_{55}'/k_5'$
	2.787051		1.413019		0.255203		0.018925		0.000492
	$-\xi$		ξ^2		$-\xi^3$		ξ^4		$-\xi^5$

12.7 Myklestad–Thomson (1949, 1953) – Prohl Methods in Transfer Matrix Form for Far-Coupled Systems

The transfer equations from station to station of the state vector, deflection, slope, bending moment and shear force across a field and point are written for a discretized beam in the x–z plane with point masses in a similar manner as in close coupled systems. First the relations for shear force V_z and bending moment M_y for an ith field of length l_i are

$$V_{z,i}^L = V_{z,i-1}^R$$

$$M_{y,i}^L = M_{y,i-1}^R + V_{z,i}^L l_i \tag{12.65}$$

In deriving the relations for the deflection and slope to the left of station i in terms of equations to the right of station $i - 1$, we use the cantilever relations for an applied shear force V and bending moment M at the station i

$$w = -\frac{Ml^2}{2EI} + \frac{Vl^3}{3EI}$$

$$\theta = \frac{Ml}{EI} + \frac{Vl^2}{2EI}_i \tag{12.66}$$

Now we can write the transfer relations for deflection and slope

$$-w_i^L = -w_{i-1}^R - \theta_{i-1}^R l_i - M_{y,i-1}^R \frac{l_i^2}{2EI_i} + V_{z,i-1}^R \frac{l_i^3}{6EI_i}$$

$$\theta_i^L = \theta_{i-1}^R - M_{y,i-1}^R \frac{l_i}{EI_i} + V_{z,i-1}^R \frac{l_i^2}{2EI_i} \tag{12.67}$$

Equations (12.65) and (12.67) are now put in a field transfer matrix form

$$\begin{Bmatrix} -w \\ \theta \\ M_y \\ V_Z \end{Bmatrix}_i^L = \begin{bmatrix} 1 & l & \frac{l^2}{2EI} & \frac{l^3}{6EI} \\ 0 & 1 & \frac{l}{EI} & \frac{l^2}{2EI} \\ 0 & 0 & 1 & l \\ 0 & 0 & 0 & 1 \end{bmatrix} \begin{Bmatrix} -w \\ \theta \\ M_y \\ V_Z \end{Bmatrix}_{i-1}^R$$

$$\{S\}_i^L = [F]_i \{S\}_{i-1}^R \tag{12.68}$$

Similarly, we can write the point transfer matrix as

$$\begin{Bmatrix} -w \\ \theta \\ M_y \\ V_Z \end{Bmatrix}_R^L = \begin{bmatrix} 1 & 0 & 0 & 0 \\ 0 & 1 & 0 & 0 \\ 0 & 0 & 1 & 0 \\ m\omega^2 & 0 & 0 & 1 \end{bmatrix} \begin{Bmatrix} -w \\ \theta \\ M_y \\ V_Z \end{Bmatrix}_i^L$$

$$\{S\}_i^R = [P]_i \{S\}_i^L \tag{12.69}$$

The transfer matrix relation can now be obtained as

$$\begin{aligned} \{S\}_i^R &= [P]_i \{S\}_i^L \\ &= [P]_i [F]_i \{S\}_{i-1}^R \\ &= [T]_i \{S\}_{i-1}^R \end{aligned} \tag{12.70}$$

where

$$[T]_i = \begin{bmatrix} 1 & l & \frac{l^2}{2EI} & \frac{l^3}{6EI} \\ 0 & 1 & \frac{l}{EI} & \frac{l^2}{2EI} \\ 0 & 0 & 1 & l \\ m\omega^2 & m\omega^2 l & \frac{m\omega^2 l^2}{2EI} & 1 + \frac{m\omega^2 l^3}{6EI} \end{bmatrix} \tag{12.71}$$

Consider a turbine blade modeled as a cantilever with five stations starting from cantilever end 0. Table 12.5 gives the data.

First the transfer matrices for the five beam and mass elements are set up for $\omega =$ 8950 rad/s.

Table 12.5 Cantilever lumped at 5 stations with fixed end 0 and free end at station 5.

i	m_i kg	$(EI)_i$ Nm^2	l_i m
1	0.4493	32659.2	0.04
2	0.3494	15366.3	0.04
3	0.2496	5600.0	0.04
4	0.1497	1209.0	0.04
5	0.0499	44.8	0.04

$$[T]_1 = \begin{bmatrix} 1 & 0.04 & 0.2449 \times 10^{-7} & 0.3266 \times 10^{-9} \\ 0 & 1 & 0.1225 \times 10^{-5} & 0.2449 \times 10^{-7} \\ 0 & 0 & 1 & 0.04 \\ 35.9885 \times 10^{6} & 14.3954 \times 10^{5} & 0.8816 & 1.0117 \end{bmatrix}$$

$$[T]_2 = \begin{bmatrix} 1 & 0.04 & 0.5206 \times 10^{-7} & 0.6942 \times 10^{-9} \\ 0 & 1 & 0.2603 \times 10^{-5} & 0.5206 \times 10^{-7} \\ 0 & 0 & 1 & 0.04 \\ 27.991 \times 10^{6} & 11.1964 \times 10^{5} & 1.4573 & 1.0194 \end{bmatrix}$$

$$[T]_3 = \begin{bmatrix} 1 & 0.04 & 0.1428 \times 10^{-6} & 0.1905 \times 10^{-8} \\ 0 & 1 & 0.7143 \times 10^{-5} & 0.1428 \times 10^{-6} \\ 0 & 0 & 1 & 0.04 \\ 19.9936 \times 10^{6} & 79.9743 \times 10^{5} & 2.8562 & 1.0381 \end{bmatrix}$$

$$[T]_4 = \begin{bmatrix} 1 & 0.04 & 0.6614 \times 10^{-6} & 0.8818 \times 10^{-8} \\ 0 & 1 & 0.3307 \times 10^{-4} & 0.6614 \times 10^{-6} \\ 0 & 0 & 1 & 0.04 \\ 11.9962 \times 10^{6} & 47.9846 \times 10^{4} & 7.934 & 1.1058 \end{bmatrix}$$

$$[T]_5 = \begin{bmatrix} 1 & 0.04 & 0.1786 \times 10^{-4} & 0.2381 \times 10^{-6} \\ 0 & 1 & 0.8929 \times 10^{-3} & 0.1786 \times 10^{-4} \\ 0 & 0 & 1 & 0.04 \\ 39.9872 \times 10^{5} & 15.9949 \times 10^{4} & 71.4067 & 1.9521 \end{bmatrix}$$

The overall transfer matrix is

$$\{S\}_5 = [U]\{S\}_1 = [T]_5 \cdots [T]_1 \{S\}_1$$

$$= \begin{Bmatrix} -w \\ \theta \\ 0 \\ 0 \end{Bmatrix}_5$$

$$= \begin{bmatrix} 191.612 & 11.977 & 0.4366 \times 10^{-4} & 0.4649 \times 10^{-5} \\ 10.0429 \times 10^7 & 633.222 & 0.224 \times 10^{-2} & 0.2394 \times 10^{-3} \\ 14.255 \times 10^6 & 92.35 \times 10^4 & 3.143 & 0.3319 \\ 92.8663 \times 10^7 & 59.3422 \times 10^6 & 209.698 & 22.1226 \end{bmatrix} \begin{Bmatrix} 0 \\ 0 \\ M_y \\ V_z \end{Bmatrix}_1$$

This leads to

$$f\left(\omega^2\right) = \begin{vmatrix} 3.143 & 0.3319 \\ 209.698 & 22.1226 \end{vmatrix} = -0.0657 \approx 0$$

The transfer matrix method was extensively used in rotor dynamics calculations for determining critical speedscritical speeds, unbalance response etc. Rao and Sarma [13] extended the transfer matrix method for transient whirl analysis. The main advantage of the transfer matrix calculations is that they involve simple matrix multiplications and the overall transfer matrix size is always small; e.g., the unbalance response calculations for bending in two planes may involve only 17 × 17 size and requires very little memory. Thus this method was preferred in the early days of the computational era.

12.8 A Brief Note on Computers and Evolution

Carl Sagan [15], a noted physicist, theorized that evolution was connected with extra somatic activity of humans. The earth is roughly 5000 million years old. Viruses, bacteria, and unicellular algae came into existence over 3000 million years ago. Their genetic information in DNA nucleotide pairs per haploid cell ranged from 3×10^4 to 10^8 bits. Protozoa appeared around 1000 million years ago with about 10^9 bits of genetic information. Amphibians appeared around 150 million years ago with about 10^9 bits of genetic information; however the first time brain information was developed in the evolution process involved about 3×10^5 bits. The reptiles that appeared around 100 million years ago broke the ground of developing 10^{10} bits of genetic as well as brain information simultaneously. Mammals that appeared around 60 million years ago increased their brain information to nearly 2×10^{11} bits. The genetic information of mammals remained almost the same as that of reptiles viz., 10^{10} bits. Humans evolved out of mammals around 2.5 million years ago with essentially the same amount of genetic information; however what set them apart from mammals was their brain which can contain 10^{13} bits of information. Thus, the information age essentially began with the evolution of human beings.

There seems to be a dead end in terms of genetic information since the appearance of reptiles. The capability of storing brain information also seems to have reached a dead end with the evolution of humans from mammals. This is attributed to the female pelvic cavity which controls the maximum possible brain size at the time of birth of a human.

Humans with their brain capacity have evolved extra body activities in their living process, like domesticating animals and producing grain around 15,000 BC in India. There was a need to pass on this information to their offspring – e.g., cave paintings of hunting appeared in France around this time. *Vedas* provide a typical example of passing on information from generation to generation in an exact manner. As time passed around 5000–4000 BC, the rudiments of writing were developed. Paper made from pith of papyrus reed was used nearly 4000 years ago to record information. Paper as we know it today, was invented in China between 100 BC and 150 AD. Water mills appeared around 100 BC and windmills in 700 AD. Suddenly there was more information – and a book was sold for two cows in Spain in 800 AD. By the year 1400, a book on medicine was sold at half an ounce of gold. There was a need for making more books at lesser price, and a movable type book printing press was introduced in 1445. In the year 1450, 300 pages could be printed in one day and by 1700, 1250 pages could be printed on one machine per day.

Then began a scientific revolution followed by the industrial revolution in 1780. The typewriter was invented in 1870. This was followed rapidly by Faraday's electromagnetic induction, Morse's telegraphy, and Alexander Graham Bell's telephone in 1876. All these inventions and others till recently seem to be a process of evolution to supplement the internal body capacity to store information. The development of digital computers follows this line of thought that would enable humans to collect information, store it and retrieve it for future processing, for which brain information alone cannot suffice.

The Electronic Numerical Integrator And Computer (ENIAC) was the first general-purpose electronic computer [4]. It was a digital computer capable of being reprogrammed to solve a full range of computing problems and designed to calculate artillery firing tables for the US Army's Ballistic Research Laboratory. It was heralded as a "Giant Brain" and boasted speeds one thousand times faster than electro-mechanical machines, a leap in computing power that no single machine has since matched. This mathematical power, coupled with general-purpose programmability, excited scientists and engineers. Gradually it rendered the development of the line of industrial applications discussed in this section of less importance.

From 1952 into the late 1960s, IBM's 700/7000 series came on the market. The first-generation 700s were based on vacuum tubes, while the later, second-generation 7000s used transistors. These computers were initially without any software, expecting engineers to write their own. Later, IBM provided compilers for the newly developed higher level programming languages such as Fortran. Engineers began writing computer programs to solve their problems rather than using the tabular or numerical methods that were popular until the advent of digital computers.

With the advent of high speed desk top computers with huge memory and parallel computing handling large size matrices, commercial rugged solvers appeared on the market. Coupled with preprocessor and postprocessor technologies, these solvers have revolutionalized the way we handle matrices of sizes arising out of finite element formulations with millions of elements. Suddenly we had the capability of handling high stress concentration areas without depending on factors of safety and strength of materials. The strength of materials approach developed in place of classical theories of elasticity from the science revolution period suddenly disappeared from analysis of complex problems. Classical elasticity and related theories suddenly became appropriate in the form of variational principles or energy methods through which modern finite elements are derived. This way we returned to the theory of elasticity approach and derivation of finite elements for different applications. All this happened in just two decades and tabular methods, transfer matrix methods, etc., lost their place in modern day designs.

References

1. Biezeno, C.B. and Grammel, R. (1939) *Technische Dynamik*, Springer Verlag.
2. Den Hartog, J.P. (1940) *Mechanical Vibration*, McGraw-Hill Book Co.
3. Duncan, W.J. and Collar, A.R. (1934) Solution of Oscillation Problems by Matrices, *Philos. Mag.*, vol. 17, p. 866.
4. Goldstine, H.H. and Goldstine, A. (1946) *The Electronic Numerical Integrator and Computer (ENIAC)*. Reprinted in *The Origins of Digital Computers: Selected Papers*, Springer-Verlag, New York, p. 359.
5. Hahn 1918 Note sur la vitesse critique et la formule du Dunkerley, *Schweiz. Bauztg.*, vol. 72, p. 191.
6. Householder, A.S. (1959) Dandelin, Lobachevskii, or Gräffe? *Amer. Math. Monthly*, vol. 66, p. 464.
7. Jacobsen, L.S. and Ayre, R.S. (1958) *Engineering Vibrations*, McGraw-Hill Book Co., New York.
8. Nestorides, E.J. (1958) *A Handbook of Torsional Vibrations*, Cambridge University Press.
9. Pipes, L.A. (1946) *Applied Mathematics for Engineers and Physicists*, McGraw-Hill Book Co.
10. Priebs, R. (1962) Ein einfaches Rechenschema zur Aufstellung der Frequenzgleichung eines an den Enden freien Drehschwingers, *Ing. Archiv*, vol. 80, no. 2, p. 14,
11. Rao, D.K. and Rao, J.S. (1974) *Computer Programs for Predicting the Torsional Vibration Characteristics of Diesel Engine – Driven Sets*, Garden Reach Workshops Ltd., 43/46 Garden Reach, Calcutta.
12. Rao, J.S. (1964) Vibration of Cantilever Beams in Torsion, *Journal of Science and Engineering Research*, vol. 8, part 2, p. 351.
13. Rao, J.S., Sarma, K.V.B. and Gupta, K. (1987) Transient Analysis of Rotors by Transfer Matrix Method, in *Rotating Machinery Dynamics*, ASME DE-Vol. 2, p. 545.
14. Rao, J.S. and Gupta, K. (1999) *Introductory Course on Theory and Practice of Mechanical Vibrations*, New Age International.
15. Sagan C, (1988) *The Dragons of Eden*, Random House.
16. Scanlan, R.H. and Rosenbaum, R. (1951) *Introduction to the Study of Aircraft Vibration and Flutter*, MacMillan.
17. Sylvester, J.J. (1852) *Philos. Mag.*, vol. 4, p. 138.

18. Thomson, W.T. (1949) Matrix Solution of Vibration of Non-Uniform Beams, ASME Paper 49 A-11.
19. Thomson, W.T. (1953) A Note on Tabular Methods for Flexural Vibrations, *J. Aero. Sci.*, vol. 20, p. 62.
20. Timoshenko, S.P. (1955) *Vibration Problems in Engineering*, D. Van Nostrand Co. Inc.
21. Wilson, W.K. (1949)*Practical Solution of Torsional Vibration Problems*, vol. 1, John Wiley & Sons.
22. Wilson, W.K. (1949) *Practical Solution of Torsional Vibration Problems*, vol. 2, John Wiley & Sons.
23. Wilson, W.K. (1965) *Practical Solution of Torsional Vibration Problems*, vol. 3, Chapman and Hall.

Chapter 13
Finite Element Methods

Once the possibility of computers was foreseen, the engineering community turned its attention to solving complex elasticity and structural analysis problems rather than depending on an approximate strength of materials approach. It also allowed the engineering community to depend less on factors of a safety approach and to remove or reduce to a considerable extent the unknown factors and thus render more accurate designs. Finite element methods allowing more accurate predictions reduced costly experimentation and introduced simulation, thus achieving cheaper but more accurate designs before testing prototypes.

Alexander Hrennikoff (1896–1984), a Russian-Canadian engineer, was considered by many as the primary founder of the finite element method (1941). Richard Courant (1888–1972), a German-American mathematician/engineer, was also considered as the father of this method [5, 6]. Priority aside, both are credited with the beginnings of modern finite element methods for structures. While the approaches used by these pioneers were dramatically different, they shared one essential characteristic: mesh discretization of a continuous domain into a set of discrete subdomains, usually called elements.

Hrennikoff's work discretizes the domain by using a lattice analogy while Courant's approach divides the domain into finite triangular subregions for solution of second order elliptic partial differential equations that arise from the problem of torsion of a cylinder. Courant's contribution was evolutionary, drawing on a large body of earlier results developed by Rayleigh, Ritz, and Galerkin.

Developments of the finite element method began in earnest in the middle to late 1950s for airframe and structural analysis [22], and gathered momentum through the work of Argyris (1913–2004) at Stuttgart and Clough at Berkeley in the 1960s. Clough refers to the finite element methods as "the Argyris method" and considers his work Argyris (1954) to be the most important series of papers ever published in the field of Structural Mechanics. Turner et al. [21] gave a stiffness method; Clough [4] coined the term "finite elements".

By the late 1950s, the key concepts of stiffness matrix and element assembly existed essentially in the form used today. NASA issued a request for proposals for development of the finite element software NASTRAN in 1965. The method was provided with a rigorous mathematical foundation by Strang and Fix [19] and has since been generalized into a branch of applied mathematics for numerical modeling of physical systems in a wide variety of engineering disciplines, e.g., electromagnetism and fluid dynamics.

Some important developments in finite element methods are rather very recent, just about four to five decades ago. These methods have come into vogue as computers have become more common, particularly in the last two to three decades. But the real difference was observed when commercial software became available to industry, gradually replacing old practices, some of which had appeared as recently as a decade ago. Finite element methods are so common today that many young and fresh engineers think they must have been available for ages, somewhat akin to an impression that mobile phones were known to our grandparents. While life has become easier for engineers to quickly make designs through simulation and bring new products to market in record time, there are a number of pitfalls in the procedures. The recent engineering generation tends to assume that answers obtained from available commercial codes are correct and that nothing can go wrong with the solutions obtained. Average engineers today cannot produce a design from fundamentals or basics; this creates problems.

We should remember that finite element analysis is only as good as the element devised and chosen – and the assumptions in the chosen element are not always clearly understood. There is no magic when a computer is used in an engineering solution; all that the computer does is crunch the numbers faster. The way in which it performs the number crunching is simply the way it was told to do so. So it is crucial for the engineer to know how a finite element code operates and the limitations in the codes.

The basic sciences were all well established during the scientific revolution from Newton, Euler, Bernoulli, Lagrange, Hamilton, etc., in the 17th century; the most general tools being energy methods and variational principles. With the advent of rotating machinery towards the end of the 19th century, these energy methods were applied by Rayleigh, Ritz, Galerkin, and Stodola to determine critical speeds or through a strength of materials approach to determine the stress and strain fields of a structure. This classical approach gained importance as it was the only practical way to get an approximate design, since energy methods could not provide practical solutions of structures with stress raisers and discontinuities. However, once the number crunching capability increased rapidly, the design community gradually shifted back to basic energy methods through finite element formulations. This treatment is of course not limited to structural analysis alone, it is applicable to any branch of physics, fluids, electromagnetism, and so on.

Here we will illustrate some simple structural finite elements to show how the subject developed so rapidly. Since this subject is of very recent origin and there are several text books on the market from which to learn the fundamentals. The purpose of the treatment here is to show how the transition took place in industry from the strength of materials approach in to basic science and finite element methods of solution for structures.

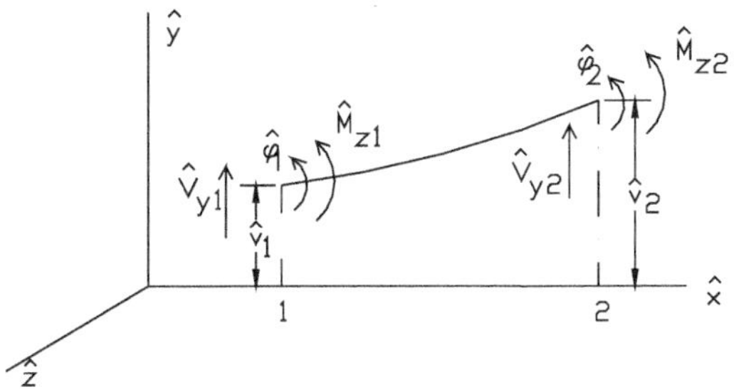

Fig. 13.1 A simple beam element in local coordinates

13.1 Beam Finite Element

We discretize the given beam into several finite elements with one element as shown in Figure 13.1. This is a simple element and has two nodes 1 and 2 and is located in the local coordinate system $\hat{x}\hat{y}\hat{z}$. The nodal deflections $\hat{v}$ are measured positive in the direction of the $\hat{y}$axis. The nodal slopes $\hat{\varphi}$ are positive in the counterclockwise direction according to the right-hand rule and similarly the bending moments $\hat{M}_z$ and shear forces $\hat{V}_y$ are as shown.

Following a Theory of Elasticity approach, we assume the transverse displacement to be a cubic polynomial that satisfies the governing differential equation of a beam. In addition, we also note that the cubic displacement shape function satisfies the continuity condition of both the deflection and slope at the nodes.

$$\hat{v}(\hat{x}) = a_1\hat{x}^3 + a_2\hat{x}^2 + a_3\hat{x} + a_4 \tag{13.1}$$

From the boundary conditions at node 1, we have

$$\begin{aligned} \hat{v}(0) &= \hat{v}_1 = a_4 \\ \frac{d\hat{v}(0)}{d\hat{x}} &= \hat{\varphi}_1 = a_3 \end{aligned} \tag{13.2}$$

From the boundary conditions of the element at node 2, we have

$$\begin{aligned} \hat{v}(L) &= \hat{v}_2 = a_1L^3 + a_2L^2 + a_3L + a_4 \\ \frac{d\hat{v}(L)}{d\hat{x}} &= \hat{\varphi}_2 = 3a_1L^2 + 2a_2L + a_3 \end{aligned} \tag{13.3}$$

The shape function in equation (13.1) can now be written as

$$\begin{aligned} \hat{v}(\hat{x}) = &\left[\frac{2}{L^3}(\hat{v}_1 - \hat{v}_2) + \frac{1}{L^2}(\hat{\varphi}_1 + \hat{\varphi}_2)\right]\hat{x}^3 \\ &- \left[\frac{3}{L^2}(\hat{v}_1 - \hat{v}_2) + \frac{1}{L}(2\hat{\varphi}_1 + \hat{\varphi}_2)\right]\hat{x}^2 + \hat{\varphi}_1\hat{x} + \hat{v}_1 \end{aligned} \tag{13.4}$$

Since there are four boundary conditions, we can express the degrees of freedom in the chosen element as

$$\left\{\hat{d}\right\} = \begin{Bmatrix} \hat{v}_1 \\ \hat{\varphi}_1 \\ \hat{v}_2 \\ \hat{\varphi}_2 \end{Bmatrix} \tag{13.5}$$

Equation (13.4) is now written in matrix form

$$\begin{aligned} \hat{v} &= [N]\left\{\hat{d}\right\} \\ &= \left[N_1 \; N_2 \; N_3 \; N_4 \right] \end{aligned} \tag{13.6}$$

where N_i are called shape functions.

$$\begin{aligned} N_1 &= \frac{1}{L^3}(2\hat{x}^3 - 3\hat{x}^2 L + L^3) \\ N_2 &= \frac{1}{L^3}(\hat{x}^3 L - 2\hat{x}^2 L^2 + \hat{x} L^3) \\ N_3 &= \frac{1}{L^3}(-2\hat{x}^3 + 3\hat{x}^2 L) \\ N_4 &= \frac{1}{L^3}(\hat{x}^3 L - \hat{x}^2 L^2) \end{aligned} \tag{13.7}$$

The strain energy in the system from equation (10.23) is given by

$$\begin{aligned} U &= \frac{1}{2} E I_{zz} \int_0^L \hat{v}^2_{,\hat{x}\hat{x}} d\hat{x} \\ &= \frac{1}{2} E I_{zz} \{\hat{d}\}^T \int_0^L [B]^T [B] \{\hat{d}\} d\hat{x} \end{aligned} \tag{13.8}$$

where

$$[B] = [N_{1,\hat{x}\hat{x}} \quad N_{2,\hat{x}\hat{x}} \quad N_{3,\hat{x}\hat{x}} \quad N_{4,\hat{x}\hat{x}}] \tag{13.9}$$

and

$$\begin{aligned} N_{1,\hat{x}\hat{x}} &= \frac{1}{L^3}(12\hat{x} - 6L) \\ N_{2,\hat{x}\hat{x}} &= \frac{1}{L^3}(6\hat{x}L - 4L^2) \\ N_{3,\hat{x}\hat{x}} &= \frac{1}{L^3}(-12\hat{x} + 6L) \\ N_{4,\hat{x}\hat{x}} &= \frac{1}{L^3}(6\hat{x}L - 2L^2) \end{aligned} \tag{13.10}$$

Following equation (10.75)

$$\frac{\partial U}{d\{\hat{d}\}} = EI_{zz} \int_0^L [B]^T [B]\{\hat{d}\} d\hat{x} = F \tag{13.11}$$

Then the elemental stiffness matrix is obtained as

$$\left[\hat{K}\right] = EI_{zz} \int_0^L \begin{bmatrix} N_{1,\hat{x}\hat{x}} \\ N_{2,\hat{x}\hat{x}} \\ N_{3,\hat{x}\hat{x}} \\ N_{4,\hat{x}\hat{x}} \end{bmatrix} \begin{bmatrix} N_{1,\hat{x}\hat{x}} & N_{2,\hat{x}\hat{x}} & N_{3,\hat{x}\hat{x}} & N_{4,\hat{x}\hat{x}} \end{bmatrix} d\hat{x}$$

$$= \frac{EI_{zz}}{L^3} \begin{bmatrix} 12 & 6L & -12 & 6L \\ & 4L^2 & -6L & 2L^2 \\ & & 12 & -6L \\ & & & 4L^2 \end{bmatrix}_{\text{sym}} \tag{13.12}$$

Let us consider just one element in a cantilever with node 1 fixed and a tip force at P at node 2. Then

$$[K]\{d\} = \{F\}$$

$$\frac{EI}{L^3} \begin{bmatrix} 12 & 6L & -12 & 6L \\ & 4L^2 & -6L & 2L^2 \\ & & 12 & -6L \\ & & & 4L^2 \end{bmatrix} \begin{Bmatrix} v_1 \\ \varphi_1 \\ v_2 \\ \varphi_2 \end{Bmatrix} = \begin{Bmatrix} 0 \\ 0 \\ P \\ 0 \end{Bmatrix}$$

We apply the boundary conditions

$$\frac{EI}{L^3} \begin{bmatrix} 12 & 6L & -12 & 6L \\ & 4L^2 & -6L & 2L^2 \\ & & 12 & -6L \\ & & & 4L^2 \end{bmatrix} \begin{Bmatrix} 0 \\ 0 \\ v_2 \\ \varphi_2 \end{Bmatrix} = \begin{Bmatrix} 0 \\ 0 \\ P \\ 0 \end{Bmatrix}$$

$$\frac{EI}{L^3} \begin{bmatrix} 12 & -6L \\ & 4L^2 \end{bmatrix} \begin{Bmatrix} v_2 \\ \varphi_2 \end{Bmatrix} = \begin{Bmatrix} P \\ 0 \end{Bmatrix}$$

This gives two equations

$$\frac{12EI}{L^3} v_2 - \frac{6EI}{L^2} \varphi_2 = P$$

$$-\frac{6EI}{L^2} v_2 + \frac{4EI}{L} \varphi_2 = 0$$

and the slope and deflections at the tip as

$$\varphi_2 = \frac{3}{2L} v_2$$

$$v_2 = P \frac{L^3}{3EI}$$

We see that even one element gives correct tip deflection and the shape function for the cantilever is

$$\begin{aligned} v(x) &= \left[\frac{2}{L^3} (v_2) + \frac{1}{L^2} (\varphi_2) \right] x^3 - \left[\frac{3}{L^2} (v_2) + \frac{1}{L} (\varphi_2) \right] x^2 \\ &= \left[\frac{2}{L^3} \frac{PL^3}{3EI} + \frac{1}{L^2} \frac{PL^2}{2EI} \right] x^3 - \left[\frac{3}{L^2} \frac{PL^3}{3EI} + \frac{1}{L} \frac{PL^2}{2EI} \right] x^2 \\ &= \frac{Px^2}{6EI} (7x - 9L) \end{aligned}$$

The correct relation for a cantilever is

$$y = \frac{Px^2}{6EI} (x - 3L)$$

The tip deflections are equivalent by the finite element method and are the correct values. The deflection shape is also a polynomial of the same degree in both cases, however between the fixed and free ends; there will be a difference because of approximation in the assumed shape function.

As another example to illustrate the element assembly, let us consider a fixed-fixed beam with two elements. The two elements are assembled together to give a system stiffness matrix as below:

$$\begin{array}{cccccc} v_1 & \varphi_1 & v_2 & \varphi_2 & v_3 & \varphi_3 \end{array}$$

$$\left[\hat{K}\right] = \frac{EI_{zz}}{L^3} \begin{bmatrix} 12 & 6L & -12 & 6L & & \\ & 4L^2 & -6L & 2L^2 & & \\ & & 24 & 0 & -12 & 6L \\ & & & 8L^2 & -6L & 2L^2 \\ & & & & 12 & -6L \\ & & & & & 4L^2 \end{bmatrix}$$

Let the beam be loaded at mid point node 2 and apply boundary conditions

$$\frac{EI}{L^3} \begin{bmatrix} 24 & 0 \\ 0 & 8L^2 \end{bmatrix} \begin{Bmatrix} v_2 \\ \phi_2 \end{Bmatrix} = \begin{Bmatrix} P \\ 0 \end{Bmatrix}$$

$$\frac{24EI}{L^3} v_2 - 0\phi_2 = P$$

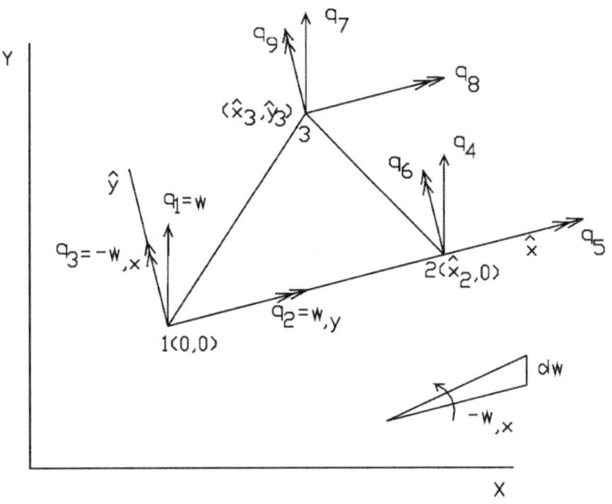

Fig. 13.2 Tocher's triangular plate element.

$$0v_2 + \frac{8EI}{L}\phi_2 = 0 \rightarrow \phi_2 = 0$$

$$v_2 = P\frac{L^3}{24EI}$$

Again we get the correct answer at the midpoint of the beam; however there will be some difference between the node points from the exact solution. However, the advantage of the finite element method is in choosing as many numbers of elements as we can so that polynomials of the same degree spread over several small elements of the beam combine together to give an almost exact solution.

13.2 Tocher Triangular Plate Element (1962)

Consider the element in Figure 13.2 with three nodes in a local coordinate system, each node having three degrees of freedom as given in equation (13.13)

$$q_1 = w\,(0,0)$$

$$q_2 = w_{,y}\,(0,0)$$

$$q_3 = -w_{,x}\,(0,0)$$

$$q_4 = w\,(x_2,0)$$

$$q_5 = w_{,y}(x_2, 0)$$
$$q_6 = -w_{,x}(x_2, 0)$$
$$q_7 = w(x_3, y_3)$$
$$q_8 = w_{,y}(x_3, y_3)$$
$$q_9 = -w_{,x}(x_3, y_3) \tag{13.13}$$

The deflection and two slopes form the shape function of the element. The nodal degrees of freedom vector for the three nodes is

$$\{\hat{d}\} = \{\hat{q}_1 \quad \hat{q}_2 \quad \hat{q}_3 \quad \hat{q}_4 \quad \hat{q}_5 \quad \hat{q}_6 \quad \hat{q}_7 \quad \hat{q}_8 \quad \hat{q}_9\}^T \tag{13.14}$$

The displacement function in the element is defined by

$$\begin{aligned} w(\hat{x}, \hat{y}) &= a_1 + a_2\hat{x} + a_3\hat{y} + a_4\hat{x}^2 \\ &\quad + a_5\hat{x}\hat{y} + a_6\hat{y}^2 + a_7\hat{x}^3 + a_8(\hat{x}^2\hat{y} + \hat{x}\hat{y}^2) + a_9\hat{y}^3 \\ &= \begin{bmatrix} 1 & \hat{x} & \hat{y} & \hat{x}^2 & \hat{x}\hat{y} & \hat{y}^2 & \hat{x}^3 & (\hat{x}^2\hat{y} + \hat{x}\hat{y}^2) & \hat{y}^3 \end{bmatrix} \begin{Bmatrix} a_1 \\ a_2 \\ a_3 \\ a_4 \\ a_5 \\ a_6 \\ a_7 \\ a_8 \\ a_9 \end{Bmatrix} \\ &= [N]\{a\} \end{aligned} \tag{13.15}$$

where $[N]$ is the shape function with nine arbitrary parameters a_1 to a_9.

The above shape function function does not represent a complete third-degree polynomial as it has only nine terms instead of ten. Two terms $(\hat{x}^2\hat{y} + \hat{x}\hat{y}^2)$ are combined into one so that the nine arbitrary parameters a_1 to a_9 can be evaluated. Deflections and slopes are not continuous across the elements through a common node. Therefore, the shape function above produces a non-conformal element.

At any point in the triangular element, the three quantities of the state, the deflection and two slopes are given by

$$\begin{Bmatrix} w \\ w_{,y} \\ -w_{,x} \end{Bmatrix} = \begin{bmatrix} 1 & \hat{x} & \hat{y} & \hat{x}^2 & \hat{x}\hat{y} & \hat{y}^2 & \hat{x}^3 & (\hat{x}^2\hat{y} + \hat{x}\hat{y}^2) & \hat{y}^3 \\ 0 & 0 & 1 & 0 & \hat{x} & 2\hat{y} & 0 & (\hat{x}^2 + 2\hat{x}\hat{y}) & 3\hat{y}^2 \\ 0 & -1 & 0 & -2\hat{x} & -\hat{y} & 0 & -3\hat{x}^2 & (2\hat{x}\hat{y} + \hat{y}^2) & 0 \end{bmatrix} \{a\} \tag{13.16}$$

Substituting the nodal coordinates for all the three nodes, the above equation gives the following nine equations in (13.17):

$$\begin{bmatrix} 1 & 0 & 0 & 0 & 0 & 0 & 0 & 0 & 0 \\ 0 & 0 & 1 & 0 & 0 & 0 & 0 & 0 & 0 \\ 0 & -1 & 0 & 0 & 0 & 0 & 0 & 0 & 0 \\ 1 & x_2 & 0 & x_2^2 & 0 & 0 & x_2^3 & 0 & 0 \\ 0 & 0 & 1 & 0 & x_2 & 0 & 0 & x_2^2 & 0 \\ 0 & -1 & 0 & -2x_2 & 0 & 0 & -3x_2^2 & 0 & 0 \\ 1 & x_3 & y_3 & x_3^2 & x_3y_3 & y_3^2 & x_3^3 & (x_3^2y_3 + x_3y_3^2) & y_3^3 \\ 0 & 0 & 1 & 0 & x_3 & 2y_3 & 0 & (x_3^2 + 2x_3y_3) & 3y_3^2 \\ 0 & -1 & 0 & -2x_3 & -y_3 & 0 & -3x_3^2 & -(2x_3y_3 + y_3^2) & 0 \end{bmatrix} \begin{Bmatrix} a_1 \\ a_2 \\ a_3 \\ a_4 \\ a_5 \\ a_6 \\ a_7 \\ a_8 \\ a_9 \end{Bmatrix} = \begin{Bmatrix} \hat{q}_1 \\ \hat{q}_2 \\ \hat{q}_3 \\ \hat{q}_4 \\ \hat{q}_5 \\ \hat{q}_6 \\ \hat{q}_7 \\ \hat{q}_8 \\ \hat{q}_9 \end{Bmatrix} \tag{13.17}$$

Writing the above in matrix form

$$[\hat{A}]\{a\} = \{\hat{q}\} \tag{13.18}$$

Solving for $\{a\}$

$$\{a\} = [\hat{A}]^{-1}\{\hat{q}\} \tag{13.19}$$

Now equation (13.15) can be written as

$$w = [N][\hat{A}]^{-1}\{\hat{q}\} \tag{13.20}$$

Note that the element becomes singular when $x_2 - 2x_3 - y_3 = 0$ and it should be so chosen to avoid this condition. The strain field is obtained as follows:

$$\begin{aligned} \varepsilon_{xx} &= -zw,_{xx} \\ \varepsilon_{yy} &= -zw,_{yy} \\ \varepsilon_{zz} &= -zw,_{xy} \end{aligned} \tag{13.21}$$

$$\begin{Bmatrix} w,_{xx} \\ w,_{yy} \\ -w,_{xy} \end{Bmatrix} = \begin{bmatrix} 0 & 0 & 0 & 2 & 0 & 0 & 6\hat{x} & 2\hat{y} & 0 \\ 0 & 0 & 0 & 0 & 0 & 2 & 0 & 2\hat{x} & 6\hat{y} \\ 0 & 0 & 0 & 0 & 1 & 0 & 0 & 2\left(\hat{x} + \hat{y}\right) & 0 \end{bmatrix} \{a\}$$

$$\begin{aligned} \left\{\hat{\varepsilon}\right\} &= \begin{bmatrix} 0 & 0 & 0 & 2 & 0 & 0 & 6\hat{x} & 2\hat{y} & 0 \\ 0 & 0 & 0 & 0 & 0 & 2 & 0 & 2\hat{x} & 6\hat{y} \\ 0 & 0 & 0 & 0 & 2 & 0 & 0 & 4\left(\hat{x} + \hat{y}\right) & 0 \end{bmatrix} \{a\} \\ &= \left[\bar{B}\right]\{a\} = \left[\bar{B}\right]\left[\hat{A}\right]^{-1}\left\{\hat{q}\right\} = [B]\left\{\hat{q}\right\} \end{aligned} \tag{13.22}$$

where $[B]$ is the matrix relating strain and the nodal degrees of freedom. The stress field is determined from Hooke's law for plane stress condition

$$\tau_{xx} = \frac{E}{1-\nu^2}\left(\varepsilon_{xx} + \nu\varepsilon_{yy}\right)$$

$$\tau_{yy} = \frac{E}{1-\nu^2}\left(\varepsilon_{yy} + \nu\varepsilon_{xx}\right)$$

$$\tau_{xy} = G\gamma_{xy} = \frac{E\,(1-\nu)}{2\left(1-\nu^2\right)}\gamma_{xy}$$

$$\begin{aligned}\{\hat{\tau}\} &= \frac{E}{(1-\nu^2)}\begin{bmatrix} 1 & \nu & 0 \\ \nu & 1 & 0 \\ 0 & 0 & \frac{(1-\nu)}{2} \end{bmatrix}\{\hat{\varepsilon}\} \\ &= [D]\{\hat{\varepsilon}\} \end{aligned} \tag{13.23}$$

The matrix relating stress and strain $[D]$ is

$$[D] = \frac{E}{\left(1-\nu^2\right)}\begin{bmatrix} 1 & \nu & 0 \\ \nu & 1 & 0 \\ 0 & 0 & \frac{(1-\nu)}{2} \end{bmatrix} \tag{13.24}$$

The strain energy U in the plate is now written as

$$\begin{aligned}\hat{U} &= \frac{1}{2}\iiint_V \{\hat{\varepsilon}\}^T\{\hat{\tau}\}dV \\ &= \frac{1}{2}\iiint_V \{\hat{\varepsilon}\}^T[D]\{\hat{\varepsilon}\}dV \\ &= \frac{1}{2}\iiint_V \{\hat{q}\}^T[B]^T[D]\{\hat{q}\}dV \end{aligned} \tag{13.25}$$

Let $Q(x, y)$ be the lateral force applied on the surface of the plate, then

$$\begin{aligned} W &= \iint_A wQ\,(x, y)\,dA \\ &= \iint_A \{\hat{q}\}^T[N]^T([\hat{A}]^{-1})^T Q\,(x, y)\,dA \end{aligned} \tag{13.26}$$

The potential functional is now written as

$$\begin{aligned} \pi &= \frac{1}{2}\{\hat{q}\}^T\left[\iiint_V [B]^T[D][B]\,dV\right]\{\hat{q}\} \\ &\quad - \iint_A \{\hat{q}\}^T[N]^T([\hat{A}]^{-1})^T Q\,(x, y)\,dA \end{aligned} \tag{13.27}$$

Let us use Ritz minimization to derive this element to evaluate the stiffness matrix as follows:

$$\frac{\partial \pi}{\partial \{\hat{q}\}^T} = 0$$

$$\left[\iiint_V [B]^T [D] [B] \, dV\right] \{\hat{q}\} = \iint_A [N]^T ([\hat{A}]^{-1})^T Q(x, y) \, dA$$

$$[\hat{K}]\{\hat{q}\} = \{\hat{Q}\}$$

$$[\hat{K}] = \left[\iiint_V [B]^T [D] [B] \, dV\right]$$

$$\{\hat{Q}\} = \iint_A [N]^T ([\hat{A}]^{-1})^T Q(x, y) \, dA$$

$$\begin{aligned} [\hat{K}] &= ([\hat{A}]^{-1})^T \left[\iint_A dA \int_{-h/2}^{+h/2} [B]^T [D] \left[\bar{B}\right] dz\right] [\hat{A}]^{-1} \\ &= ([\hat{A}]^{-1})^T [\bar{K}][\hat{A}]^{-1} \end{aligned} \tag{13.28}$$

In the above $\{\hat{Q}\}$ is a nodal force vector. The elemental stiffness matrix can now be evaluated as

$$\begin{aligned} [\bar{K}] &= \iint_A dA \int_{-h/2}^{+h/2} [B]^T [D] [\bar{B}] dz \\ &= \frac{Eh^3}{12\left(1 - \nu^2\right)} \iint_A \begin{bmatrix} 0 & 0 & 0 \\ 0 & 0 & 0 \\ 0 & 0 & 0 \\ 2 & 0 & 0 \\ 0 & 0 & 2 \\ 0 & 2 & 0 \\ 6\hat{x} & 0 & 0 \\ 2\hat{y} & 2\hat{x} & 4(\hat{x} + \hat{y}) \\ 0 & 6\hat{y} & 0 \end{bmatrix} \\ &\times \begin{bmatrix} 1 & \nu & 0 \\ \nu & 1 & 0 \\ 0 & 0 & \frac{1-\nu}{2} \end{bmatrix} \begin{bmatrix} 0 & 0 & 0 & 2 & 0 & 0 & 6\hat{x} & 2\hat{y} & 0 \\ 0 & 0 & 0 & 0 & 0 & 2 & 0 & 2\hat{x} & 6\hat{y} \\ 0 & 0 & 0 & 0 & 2 & 0 & 0 & 4(\hat{x} + \hat{y}) & 0 \end{bmatrix} d\hat{x} d\hat{y} \end{aligned} \tag{13.29}$$

to give

$$
\left[\bar{K}\right] = \frac{Eh^3}{12\left(1-\nu^2\right)}
\iint_A \begin{bmatrix}
0 & 0 & 0 & 0 & 0 & 0 & 0 & 0 & 0 \\
 & 0 & 0 & 0 & 0 & 0 & 0 & 0 & 0 \\
 & & 0 & 0 & 0 & 0 & 0 & 0 & 0 \\
 & & & 4 & 0 & 4\nu & 12\hat{x} & 4(\nu\hat{x}+\hat{y}) & 12\nu\hat{y} \\
 & & & & 2(1-\nu) & 0 & 0 & 4(1-\nu)(\hat{x}+\hat{y}) & 0 \\
 & & & & & 4 & 12\nu\hat{x} & 4(\hat{x}+\nu\hat{y}) & 12\hat{y} \\
 & & & & & & 36\hat{x}^2 & 12(\nu\hat{x}^2+\hat{x}\hat{y}) & 36\nu\hat{x}\hat{y} \\
 & & & & & & & \left\{\begin{matrix}(12-8\nu)(\hat{x}+\hat{y})^2 \\ -8(1-\nu)\hat{x}\hat{y}\end{matrix}\right\} & 12(\nu\hat{y}^2+\hat{x}\hat{y}) \\
 & & & & & & & & 36\hat{y}^2
\end{bmatrix} d\hat{x}d\hat{y}
\tag{13.30}
$$

The following integrals are used in evaluating the above:

$$
\iint_A x^m y^n dA = \sum_{r=0}^{m+1}\sum_{s=0}^{r} \frac{(-1)^{r+s} m!}{(m+1-r)!(r-s)!s!(n+r+1)} x_2^{m+1-s} x_3^s y_3^{n+1} - \frac{x_3^{m+1} y_3^{n+1}}{(m+1)(m+n+2)}
$$

for $x_3 \neq 0, x_3 \neq x_2$

$$
\iint_A x^m y^n dA = \sum_{r=0}^{m+1} \frac{(-1)^r \, m!}{(m+1-r)!r! \, (n+r+1)} x_2^{m+1} y_3^{n+1}
$$

for $x_3 = 0$

$$
\iint_A x^m y^n dA = \sum_{r=0}^{m+1} \frac{1}{(n+1)\,(m+n+2)} x_2^{m+1} y_3^{n+1}
$$

for $x_3 = x_2$ (13.31)

The elemental degrees of freedom can be expressed in global coordinates through the following transformation:

$$
\begin{Bmatrix} w \\ w,_y \\ w,_x \end{Bmatrix} = \begin{bmatrix} 1 & 0 & 0 \\ 0 & \cos(x,X) & \cos(x,Y) \\ 0 & \cos(y,X) & \cos(y,Y) \end{bmatrix} \begin{Bmatrix} w \\ w,_Y \\ w,_X \end{Bmatrix}
= [L] \begin{Bmatrix} w \\ w,_Y \\ w,_X \end{Bmatrix}
\tag{13.32}
$$

The directional cosines are given by

$$\cos(x, X) = \frac{X_2 - X_1}{\sqrt{(X_2 - X_1)^2 + (Y_2 - Y_1)^2}}$$

$$\cos(x, Y) = \frac{Y_2 - Y_1}{\sqrt{(X_2 - X_1)^2 + (Y_2 - Y_1)^2}}$$

$$\cos(y, X) = \frac{Y_2 - Y_1}{\sqrt{(X_2 - X_1)^2 + (Y_2 - Y_1)^2}}$$

$$\cos(y, Y) = \frac{X_2 - X_1}{\sqrt{(X_2 - X_1)^2 + (Y_2 - Y_1)^2}} \tag{13.33}$$

The transformation matrix from elemental to global coordinates is given by

$$\{\hat{q}\} = [R]\{q\}$$

$$[R] = \begin{bmatrix} [L] & & \\ & [L] & \\ & & [L] \end{bmatrix} \tag{13.34}$$

The potential functional in global coordinates is

$$\pi = \frac{1}{2}[R]^T\{q\}^T\left[\iiint_V [B]^T[D][B]\,dV\right][R]\{q\} - \iint_A [R]^T\{q\}^T[N]^T\left(\left[\hat{A}\right]^{-1}\right)^T Q(x, y)\,dA \tag{13.35}$$

The stiffness matrix, force vector and elemental equation in global coordinates are

$$[K] = [R]^T[\hat{K}][R]$$

$$\{Q\} = [R]^T\{\hat{Q}\}$$

$$[K]\{q\} = \{Q\} \tag{13.36}$$

Within two years of Clough's coining the word "finite elements", there were several authors developing finite elements of different classes of structures, the earliest one being from Tocher [20] for plates using triangular elements.

We consider here an example of a square fixed-fixed plate given by Rao [14].

Typical finite element idealization is shown in Figure 13.3 for $n = 1$, $n = 2$, $n = 4$, etc., for a quarter of a plate. Because of symmetry it is sufficient to consider a quarter of a plate, with fixed boundary conditions on the two edges and the slopes $w_{,x}$ and $w_{,y}$ zero on the edges 2–4 and 3–4 respectively. The deflection coefficient $10^3 D/qa^4$ is plotted in Figure 13.4 as a function of the mesh size used in Figure 13.3. It should be noted that monotonic convergence is not always obtained with non-conforming elements.

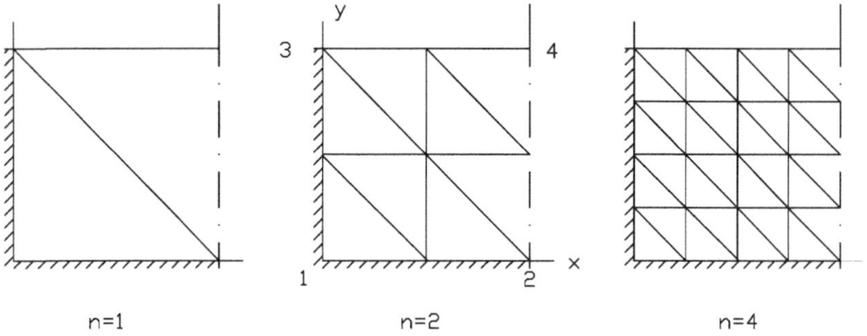

Fig. 13.3 A square fixed-fixed plate

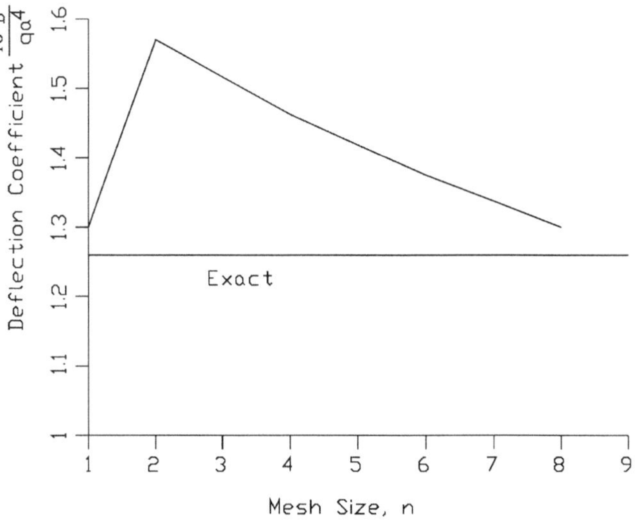

Fig. 13.4 Deflection as a function of mesh size

We will now extend from beams and plates to one more case shell and look at how the finite element methods are developed for structures.

13.3 Shell Element

Shells are plates with their middle surface having a curvature. External faces of the shell are curved, while the sections across the thickness are generated by straight lines. There are many kinds of shells used in structures, simple cylindrical shells, paraboloids, hyperbolic shape shells or any structures whose middle surface is not in a plane. These shells can also be modeled as finite elements. We will consider

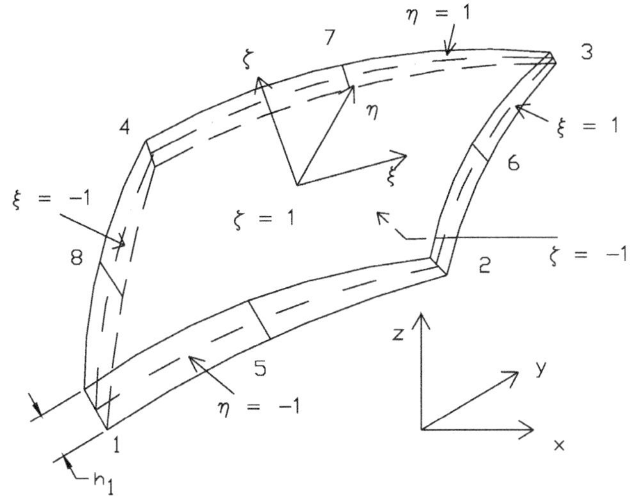

Fig. 13.5 Ahmad's shell element

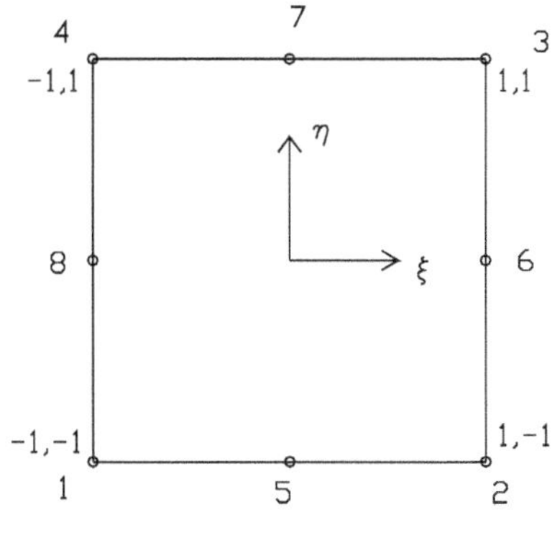

Fig. 13.6 Mid-surface of parent element in natural coordinates

a shell element as derived by Ahmad et al. [1] within a decade of plate elements derivation. The geometry of the element, see Figure 13.5, is described by a pair of points, $\hat{i}_{\text{top}}, \hat{i}_{\text{bottom}}$ each with given Cartesian coordinates. ξ, η are curvilinear coordinates in the middle plane of the shell and ζ is a linear coordinate in the thickness direction. The mid-surface of the parent element in natural coordinates is shown in Figure 13.6.

Shape Functions: ξ, η and ζ vary between -1 and $+1$ and Cartesian coordinates of any point on the shell are expressed in terms of geometric shape functions, as

$$\begin{Bmatrix} x \\ y \\ z \end{Bmatrix} = \sum_{i=1}^{8} N_i\left(\xi,\eta\right) \frac{1+\zeta}{2} \begin{Bmatrix} x \\ y \\ z \end{Bmatrix}_{\text{top}} + \sum_{i=1}^{8} N_i\left(\xi,\eta\right) \frac{1-\zeta}{2} \begin{Bmatrix} x \\ y \\ z \end{Bmatrix}_{\text{bottom}} \tag{13.37}$$

In the above the shape functions N are assumed as

$$N_1(\xi,\eta) = \frac{1}{4}(1-\xi)(1-\eta)(-\xi-\eta-1)$$
$$N_2(\xi,\eta) = \frac{1}{4}(1+\xi)(1-\eta)(\xi-\eta-1)$$
$$N_3(\xi,\eta) = \frac{1}{4}(1+\xi)(1+\eta)(\xi+\eta-1)$$
$$N_4(\xi,\eta) = \frac{1}{4}(1-\xi)(1+\eta)(-\xi+\eta-1)$$

$$N_5(\xi,\eta) = \frac{1}{2}(1-\xi^2)(1-\eta)$$
$$N_6(\xi,\eta) = \frac{1}{2}(1+\xi)(1-\eta^2)$$
$$N_7(\xi,\eta) = \frac{1}{2}(1-\xi^2)(1+\eta)$$
$$N_8(\xi,\eta) = \frac{1}{2}(1-\xi)(1-\eta^2) \tag{13.38}$$

Note that $N_i\ (\xi,\eta)$ takes a value of unity at node i and zero at all other nodes. Equation (13.38) can also be written in terms of the mid-surface coordinates

$$\begin{Bmatrix} x \\ y \\ z \end{Bmatrix} = \sum_{i=1}^{8} N_i\ (\xi,\eta) \begin{Bmatrix} x \\ y \\ z \end{Bmatrix}_{\text{mid}} + \sum_{i=1}^{8} N_i\ (\xi,\eta)\, \frac{\zeta}{2} \tilde{V}_{3i}$$

$$\tilde{V}_{3i} = \begin{Bmatrix} x_i \\ y_i \\ z_i \end{Bmatrix}_{\text{top}} - \begin{Bmatrix} x_i \\ y_i \\ z_i \end{Bmatrix}_{\text{bottom}} \tag{13.37a}$$

Displacement Field: The displacement field is defined by the three displacements of the mid-surface nodes and two rotations of the nodal vector $\tilde{V}_{3i}$ about its own orthogonal directions as in Figure 13.7

$$\begin{Bmatrix} u \\ v \\ w \end{Bmatrix} = \sum_{i=1}^{8} N_i\ (\xi,\eta) \begin{Bmatrix} u_i \\ v_i \\ w_i \end{Bmatrix} + \sum_{i=1}^{8} N_i\ (\xi,\eta)\, \frac{\zeta h_i}{2} \begin{bmatrix} V_{1x}^i & -V_{2x}^i \\ V_{1y}^i & -V_{2y}^i \\ V_{1z}^i & -V_{2z}^i \end{bmatrix} \begin{Bmatrix} -\beta_i \\ \alpha_i \end{Bmatrix} \tag{13.39}$$

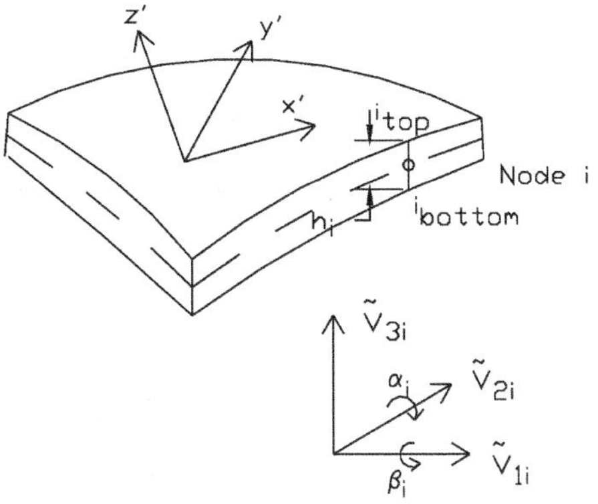

Fig. 13.7 Displacement field

$$\tilde{V}_{1i} = V^i_{1x}\hat{e}_x + V^i_{1y}\hat{e}_y + V^i_{1z}\hat{e}_z = \frac{\hat{e}_x \times \tilde{V}_{3i}}{|\hat{e}_x \times \tilde{V}_{3i}|}$$

$$\tilde{V}_{2i} = V^i_{2x}\hat{e}_x + V^i_{2y}\hat{e}_y + V^i_{2z}\hat{e}_z = \tilde{V}_{3i} \times \tilde{V}_{1i}$$

u_i, v_i, w_i are the displacements at the mid surface nodes and α_i, β_i are the two rotations about the directions $\tilde{V}_{2i}$, $\tilde{V}_{1i}$respectively. There are five degrees of freedom per each node given by

$$\left\{\hat{d}_i\right\} = \begin{Bmatrix} u_i \\ v_i \\ w_i \\ \alpha_i \\ \beta_i \end{Bmatrix} \tag{13.40}$$

and 40 degrees of freedom for the element. Equation (13.39) can be written as

$$\begin{Bmatrix} u \\ v \\ w \end{Bmatrix} = [N] \begin{Bmatrix} \hat{d}_1 \\ \hat{d}_2 \\ \cdots \\ \cdots \\ \hat{d}_8 \end{Bmatrix} = [N]\{\hat{d}\} \tag{13.41}$$

where

$$[N] = \left[[N]_1 \; [N]_2 \cdots [N]_8 \right]$$

$$[N]_1 = \begin{bmatrix} N_1 & 0 & 0 & -\frac{1}{2}V^1_{2x}N_1h_1\zeta & -\frac{1}{2}V^1_{1x}N_1h_1\zeta \\ 0 & N_1 & 0 & -\frac{1}{2}V^1_{2y}N_1h_1\zeta & -\frac{1}{2}V^1_{1y}N_1h_1\zeta \\ 0 & 0 & N_1 & -\frac{1}{2}V^1_{2z}N_1h_1\zeta & -\frac{1}{2}V^1_{1z}N_1h_1\zeta \end{bmatrix}$$

$$[N]_2 = \begin{bmatrix} N_2 & 0 & 0 & -\frac{1}{2}V_{2x}^2 N_2 h_2 \zeta & -\frac{1}{2}V_{1x}^2 N_2 h_2 \zeta \\ 0 & N_2 & 0 & -\frac{1}{2}V_{2y}^2 N_2 h_2 \zeta & -\frac{1}{2}V_{1y}^2 N_2 h_2 \zeta \\ 0 & 0 & N_2 & -\frac{1}{2}V_{2z}^2 N_2 h_2 \zeta & -\frac{1}{2}V_{1z}^2 N_2 h_2 \zeta \end{bmatrix} \tag{13.42}$$

...

The total number of degrees of freedom of the element is 40, whereas, the definition contains only 24 displacements, therefore, the element is of the super-parametric kind.

Strain-Displacement Relations: Only the components of strains and stresses in directions of orthogonal axes related to the surface ζ = constant are considered. Let z' be a normal to this surface at any point and let the other two orthogonal axes be x' and y' which are tangent to z', see Figure 13.7. The strain components are defined by

$$\{\varepsilon'\} = \begin{Bmatrix} \varepsilon'_{xx} \\ \varepsilon'_{yy} \\ \gamma_{x'y'} \\ \gamma_{y'z'} \\ \gamma_{x'z'} \end{Bmatrix} = \begin{Bmatrix} \frac{\partial u'}{\partial x'} \\ \frac{\partial v'}{\partial y'} \\ \frac{\partial u'}{\partial y'} + \frac{\partial v'}{\partial x'} \\ \frac{\partial v'}{\partial z'} + \frac{\partial w'}{\partial y'} \\ \frac{\partial u'}{\partial z'} + \frac{\partial w'}{\partial x'} \end{Bmatrix} = [H] \begin{Bmatrix} \frac{\partial u'}{\partial x'} \\ \frac{\partial u'}{\partial y'} \\ \frac{\partial u'}{\partial z'} \\ \frac{\partial v'}{\partial x'} \\ \frac{\partial v'}{\partial y'} \\ \frac{\partial v'}{\partial z'} \\ \frac{\partial w'}{\partial x'} \\ \frac{\partial w'}{\partial y'} \\ \frac{\partial w'}{\partial z'} \end{Bmatrix} \tag{13.43}$$

where

$$[H] = \begin{bmatrix} 1 & 0 & 0 & 0 & 0 & 0 & 0 & 0 & 0 \\ 0 & 0 & 0 & 0 & 1 & 0 & 0 & 0 & 0 \\ 0 & 1 & 0 & 1 & 0 & 0 & 0 & 0 & 0 \\ 0 & 0 & 0 & 0 & 0 & 1 & 0 & 1 & 0 \\ 0 & 0 & 1 & 0 & 0 & 0 & 1 & 0 & 0 \end{bmatrix} \tag{13.44}$$

Stress-Strain Relation: The stress strain relation is taken as

$$\{\sigma'\} = [D]\{\varepsilon'\}\{\sigma'\} = [D]\{\varepsilon'\} \tag{13.45}$$

where

$$[D] = \begin{bmatrix} 1 & \nu & 0 & 0 & 0 \\ \nu & 1 & 0 & 0 & 0 \\ 0 & 0 & \frac{1-\nu}{2} & 0 & 0 \\ 0 & 0 & 0 & \frac{1-\nu}{2k} & 0 \\ 0 & 0 & 0 & 0 & \frac{1-\nu}{2k} \end{bmatrix} \tag{13.46}$$

The factor k in the above equation (13.46) is taken as 1.2 to improve the shear displacement approximation.

We need now a transformation from the global coordinates x, y, z to local coordinates x', y', z'. We first define a vector normal to the surface ζ = constant, by taking the vector product of any two vectors tangent to the surface,

$$\tilde{V}_{z'} = \begin{Bmatrix} \frac{\partial x}{\partial \xi} \\ \frac{\partial y}{\partial \xi} \\ \frac{\partial z}{\partial \xi} \end{Bmatrix} \times \begin{Bmatrix} \frac{\partial x}{\partial \eta} \\ \frac{\partial y}{\partial \eta} \\ \frac{\partial z}{\partial \eta} \end{Bmatrix} \tag{13.47}$$

The other two vectors $\tilde{V}_{x'}$, $\tilde{V}_{y'}$ along x' and y' directions can also be obtained by the process given above and reducing these to unit magnitudes, we can construct a matrix of unit vectors in x', y', z' directions

$$\begin{aligned} [\theta] &= \begin{bmatrix} \tilde{V}_{x'} & \tilde{V}_{y'} & \tilde{V}_{z'} \end{bmatrix} \\ &= \begin{bmatrix} l_1 & l_2 & l_3 \\ m_1 & m_2 & m_3 \\ n_1 & n_2 & n_3 \end{bmatrix} \end{aligned} \tag{13.48}$$

We can now write

$$\begin{bmatrix} \frac{\partial u'}{\partial x'} & \frac{\partial v'}{\partial x'} & \frac{\partial w'}{\partial x'} \\ \frac{\partial u'}{\partial y'} & \frac{\partial v'}{\partial y'} & \frac{\partial w'}{\partial y'} \\ \frac{\partial u'}{\partial z'} & \frac{\partial v'}{\partial z'} & \frac{\partial w'}{\partial z'} \end{bmatrix} = [\theta]^T \begin{bmatrix} \frac{\partial u}{\partial x} & \frac{\partial v}{\partial x} & \frac{\partial w}{\partial x} \\ \frac{\partial u}{\partial y} & \frac{\partial v}{\partial y} & \frac{\partial w}{\partial y} \\ \frac{\partial u}{\partial z} & \frac{\partial v}{\partial z} & \frac{\partial w}{\partial z} \end{bmatrix} [\theta] \tag{13.49}$$

i.e.,

$$\begin{Bmatrix} \frac{\partial u'}{\partial x'} \\ \frac{\partial u'}{\partial y'} \\ \frac{\partial u'}{\partial z'} \\ \frac{\partial v'}{\partial x'} \\ \frac{\partial v'}{\partial y'} \\ \frac{\partial v'}{\partial z'} \\ \frac{\partial w'}{\partial x'} \\ \frac{\partial w'}{\partial y'} \\ \frac{\partial w'}{\partial z'} \end{Bmatrix} = [T] \begin{Bmatrix} \frac{\partial u}{\partial x} \\ \frac{\partial u}{\partial y} \\ \frac{\partial u}{\partial z} \\ \frac{\partial v}{\partial x} \\ \frac{\partial v}{\partial y} \\ \frac{\partial v}{\partial z} \\ \frac{\partial w}{\partial x} \\ \frac{\partial w}{\partial y} \\ \frac{\partial w}{\partial z} \end{Bmatrix} \tag{13.50}$$

where

$$[T] = \begin{bmatrix} l_1\,[\theta]^T & m_1\,[\theta]^T & n_1\,[\theta]^T \\ l_2\,[\theta]^T & m_2\,[\theta]^T & n_2\,[\theta]^T \\ l_3\,[\theta]^T & m_3\,[\theta]^T & n_3\,[\theta]^T \end{bmatrix} \tag{13.51}$$

and $l_1, l_2, \ldots,$ are directional derivatives. The derivatives of the global displacements u, v, and w with respect to the curvilinear coordinates are related to the derivatives with respect to Cartesian coordinates through the Jacobian by

$$\begin{bmatrix} \frac{\partial u}{\partial x} & \frac{\partial v}{\partial x} & \frac{\partial w}{\partial x} \\ \frac{\partial u}{\partial y} & \frac{\partial v}{\partial y} & \frac{\partial w}{\partial y} \\ \frac{\partial u}{\partial z} & \frac{\partial v}{\partial z} & \frac{\partial w}{\partial z} \end{bmatrix} = [J]^{-1} \begin{bmatrix} \frac{\partial u}{\partial \xi} & \frac{\partial v}{\partial \xi} & \frac{\partial w}{\partial \xi} \\ \frac{\partial u}{\partial \eta} & \frac{\partial v}{\partial \eta} & \frac{\partial w}{\partial \eta} \\ \frac{\partial u}{\partial \zeta} & \frac{\partial v}{\partial \zeta} & \frac{\partial w}{\partial \zeta} \end{bmatrix} \tag{13.52}$$

Therefore,

$$\begin{Bmatrix} \frac{\partial u}{\partial x} \\ \frac{\partial u}{\partial y} \\ \frac{\partial u}{\partial z} \\ \frac{\partial v}{\partial x} \\ \frac{\partial v}{\partial y} \\ \frac{\partial v}{\partial z} \\ \frac{\partial w}{\partial x} \\ \frac{\partial w}{\partial y} \\ \frac{\partial w}{\partial z} \end{Bmatrix} = [\Gamma] \begin{Bmatrix} \frac{\partial u}{\partial \xi} \\ \frac{\partial u}{\partial \eta} \\ \frac{\partial u}{\partial \zeta} \\ \frac{\partial v}{\partial \xi} \\ \frac{\partial v}{\partial \eta} \\ \frac{\partial v}{\partial \zeta} \\ \frac{\partial w}{\partial \xi} \\ \frac{\partial w}{\partial \eta} \\ \frac{\partial w}{\partial \zeta} \end{Bmatrix} \tag{13.53}$$

where

$$[\Gamma] = \begin{bmatrix} [J]^{-1} & 0 & 0 \\ 0 & [J]^{-1} & 0 \\ 0 & 0 & [J]^{-1} \end{bmatrix} \tag{13.54}$$

The Jacobian is

$$[J] = \begin{bmatrix} \frac{\partial x}{\partial \xi} & \frac{\partial y}{\partial \xi} & \frac{\partial z}{\partial \xi} \\ \frac{\partial x}{\partial \eta} & \frac{\partial y}{\partial \eta} & \frac{\partial z}{\partial \eta} \\ \frac{\partial x}{\partial \zeta} & \frac{\partial y}{\partial \zeta} & \frac{\partial z}{\partial \zeta} \end{bmatrix} \tag{13.55}$$

where

$$\frac{\partial x}{\partial \xi} = \sum_{i=1}^{8} \frac{\partial N_i}{\partial \xi} x_{\text{mid}}^i + \sum_{i=1}^{8} \frac{1}{2} \frac{\partial N_i}{\partial \xi} h_i \zeta V_{3ix}$$

$$\frac{\partial y}{\partial \xi} = \sum_{i=1}^{8} \frac{\partial N_i}{\partial \xi} y_{\text{mid}}^i + \sum_{i=1}^{8} \frac{1}{2} \frac{\partial N_i}{\partial \xi} h_i \zeta V_{3iy}$$

$$\frac{\partial z}{\partial \xi} = \sum_{i=1}^{8} \frac{\partial N_i}{\partial \xi} z_{\text{mid}}^i + \sum_{i=1}^{8} \frac{1}{2} \frac{\partial N_i}{\partial \xi} h_i \zeta V_{3iz}$$

$$\frac{\partial x}{\partial \eta} = \sum_{i=1}^{8} \frac{\partial N_i}{\partial \eta} x^i_{\mathrm{mid}} + \sum_{i=1}^{8} \frac{1}{2}\frac{\partial N_i}{\partial \eta} h_i \zeta V_{3ix}$$

$$\frac{\partial y}{\partial \eta} = \sum_{i=1}^{8} \frac{\partial N_i}{\partial \eta} y^i_{\mathrm{mid}} + \sum_{i=1}^{8} \frac{1}{2}\frac{\partial N_i}{\partial \eta} h_i \zeta V_{3iy}$$

$$\frac{\partial z}{\partial \eta} = \sum_{i=1}^{8} \frac{\partial N_i}{\partial \eta} z^i_{\mathrm{mid}} + \sum_{i=1}^{8} \frac{1}{2}\frac{\partial N_i}{\partial \eta} h_i \zeta V_{3iz}$$

$$\frac{\partial x}{\partial \zeta} = \sum_{i=1}^{8} \frac{1}{2} N_i h_i \zeta V_{3ix}$$

$$\frac{\partial y}{\partial \zeta} = \sum_{i=1}^{8} \frac{1}{2} N_i h_i \zeta V_{3iy}$$

$$\frac{\partial z}{\partial \zeta} = \sum_{i=1}^{8} \frac{1}{2} N_i h_i \zeta V_{3iz} \tag{13.56}$$

From (13.39), we can obtain

$$\left\{ \begin{array}{c} \frac{\partial u}{\partial \xi} \\ \frac{\partial u}{\partial \eta} \\ \frac{\partial u}{\partial \zeta} \\ \frac{\partial v}{\partial \xi} \\ \frac{\partial v}{\partial \eta} \\ \frac{\partial v}{\partial \zeta} \\ \frac{\partial w}{\partial \xi} \\ \frac{\partial w}{\partial \eta} \\ \frac{\partial w}{\partial \zeta} \end{array} \right\} = [L] \left\{ \begin{array}{c} \hat{d}_1 \\ \hat{d}_2 \\ \cdots \\ \cdots \\ \hat{d}_8 \end{array} \right\} \tag{13.57}$$

where

$$[L] = \begin{bmatrix} [L_{11}] & [L_{12}] & \cdots & [L_{18}] \\ [L_{21}] & [L_{22}] & \cdots & [L_{28}] \\ [L_{31}] & [L_{32}] & \cdots & [L_{38}] \end{bmatrix} \tag{13.58}$$

$$[L_{11}] = \begin{bmatrix} \frac{\partial N_1}{\partial \xi} & 0 & 0 & -\frac{1}{2}V_{2x}^1 \frac{\partial N_1}{\partial \xi} h_1 \zeta & -\frac{1}{2}V_{1x}^1 \frac{\partial N_1}{\partial \xi} h_1 \zeta \\ \frac{\partial N_1}{\partial \eta} & 0 & 0 & -\frac{1}{2}V_{2x}^1 \frac{\partial N_1}{\partial \eta} h_1 \zeta & -\frac{1}{2}V_{1x}^1 \frac{\partial N_1}{\partial \eta} h_1 \zeta \\ 0 & 0 & 0 & -\frac{1}{2}V_{2x}^1 N_1 h_1 & -\frac{1}{2}V_{1x}^1 N_1 h_1 \end{bmatrix}$$

$$[L_{21}] = \begin{bmatrix} 0 & \frac{\partial N_1}{\partial \xi} & 0 & -\frac{1}{2}V_{2y}^1 \frac{\partial N_1}{\partial \xi} h_1 \zeta & -\frac{1}{2}V_{1y}^1 \frac{\partial N_1}{\partial \xi} h_1 \zeta \\ 0 & \frac{\partial N_1}{\partial \eta} & 0 & -\frac{1}{2}V_{2y}^1 \frac{\partial N_1}{\partial \eta} h_1 \zeta & -\frac{1}{2}V_{1y}^1 \frac{\partial N_1}{\partial \eta} h_1 \zeta \\ 0 & 0 & 0 & -\frac{1}{2}V_{2y}^1 N_1 h_1 & -\frac{1}{2}V_{1y}^1 N_1 h_1 \end{bmatrix}$$

$$[L_{31}] = \begin{bmatrix} 0 & 0 & \frac{\partial N_1}{\partial \xi} & -\frac{1}{2}V_{2z}^1 \frac{\partial N_1}{\partial \xi} h_1 \zeta & -\frac{1}{2}V_{1z}^1 \frac{\partial N_1}{\partial \xi} h_1 \zeta \\ 0 & 0 & \frac{\partial N_1}{\partial \eta} & -\frac{1}{2}V_{2z}^1 \frac{\partial N_1}{\partial \eta} h_1 \zeta & -\frac{1}{2}V_{1z}^1 \frac{\partial N_1}{\partial \eta} h_1 \zeta \\ 0 & 0 & 0 & -\frac{1}{2}V_{2z}^1 N_1 h_1 & -\frac{1}{2}V_{1z}^1 N_1 h_1 \end{bmatrix}$$

$$[L_{12}] = \begin{bmatrix} \frac{\partial N_2}{\partial \xi} & 0 & 0 & -\frac{1}{2}V_{2x}^2 \frac{\partial N_2}{\partial \xi} h_2 \zeta & -\frac{1}{2}V_{1x}^2 \frac{\partial N_2}{\partial \xi} h_2 \zeta \\ \frac{\partial N_2}{\partial \eta} & 0 & 0 & -\frac{1}{2}V_{2x}^2 \frac{\partial N_2}{\partial \eta} h_2 \zeta & -\frac{1}{2}V_{1x}^2 \frac{\partial N_2}{\partial \eta} h_2 \zeta \\ 0 & 0 & 0 & -\frac{1}{2}V_{2x}^2 N_2 h_2 & -\frac{1}{2}V_{1x}^2 N_2 h_2 \end{bmatrix}$$

...

From (13.44), (13.50), (13.53) and (13.58), we now obtain

$$\{\varepsilon'\} = [B] \begin{Bmatrix} \hat{d}_1 \\ \hat{d}_2 \\ \cdots \\ \cdots \\ \hat{d}_8 \end{Bmatrix} \tag{13.59}$$

where

$$[B] = [H][T][\Gamma][L] \tag{13.60}$$

The elemental stiffness matrix is then given by

$$\begin{aligned} \left[\hat{K}\right] &= \int_V [B]^T [D] [B] \, dxdydz \\ &= \int_V [B]^T [D] [B] \, |J| \, d\xi d\eta d\zeta \end{aligned} \tag{13.61}$$

Assembling the elemental matrices and solving the problem follows the same directions as we discussed before.

We will close here the discussion on finite element methods by stating that several authors following different energy methods have derived several elements in the last three to four decades [23].

Once several of these elements were published a number of commercial codes appeared using such libraries of elements. Before the commercial codes, when computers were fairly slow and with less memory, the practice was to manually mesh the chosen elements, number the nodes and make sure that the mesh had no discontinuities. Today, commercial preprocessors perform the mesh work automatically and prepare the finite element model in a short period of time. Today the mesh work can be morphed without remeshing, a fact that becomes useful in making metamodels for Design of Experiments study and optimization.

What we call today solvers in commercial codes were written as subroutines or subprograms in Fortran or Algol and designated as eigen-values, matrix multiplications etc. as discussed with respect to beam or plate elements.

The engineer is also required to bring the displacement field to different nodes and use formulae to determine the strain and stress fields or any other information required in the design like principal stresses and strains, von Mises stress etc. It was required to make the plots manually or write special programs that can give a printout that is made to resemble the plots. These steps can be seen in papers as recently as 1990, see Rao et al. [16] to study the friction damping effects from the blade disk root region using contact elements.

13.4 Interface Damping through Finite Element Analysis

Interfacial damping arising out of friction between two surfaces moving relative to each other is a complex analytical problem, particularly for the type of blade structures with blade root junctions, shroud bands, lacing wires and platform dampers. Here we discuss how such a problem was tackled by finite contact elements [16].

Static Contact Problem: When there are two separate bodies such as a blade and a disk with the corresponding root, the contact under static loading should be first established before modeling the dynamic contact problem. In the case of a blade root system, the contact can take place with a separation and friction, which is a nonlinear and irreversible case. The quasi-static conditions corresponding to the effect of centrifugal load on the blade and the disk represent the equilibrium state about which oscillations would occur. Thus it is necessary to solve the static contact problem before arriving at initial conditions for the forced vibration problem.

The structure is assumed to be an assembly of a large number of finite elements as shown in Figure 13.8. We can use quadratic Isoparametric elements for the blade and root structure. These elements are conveniently handled by normalized coordinates. Figure 13.9 shows the parent element and the transformed element.

Figure 13.10 shows two elastic bodies, A and B, before and after they come into contact. The body B is considered to have fixed boundary conditions and the body A is acted upon by a force vector $\{f\}$. Elastically the problem is the same if the contact is brought about by rigid body movement of the support of body B and the force vector $\{f\}$ results in the body A at the boundary points. The rigid body movement of the supports of the body B is given by the vector

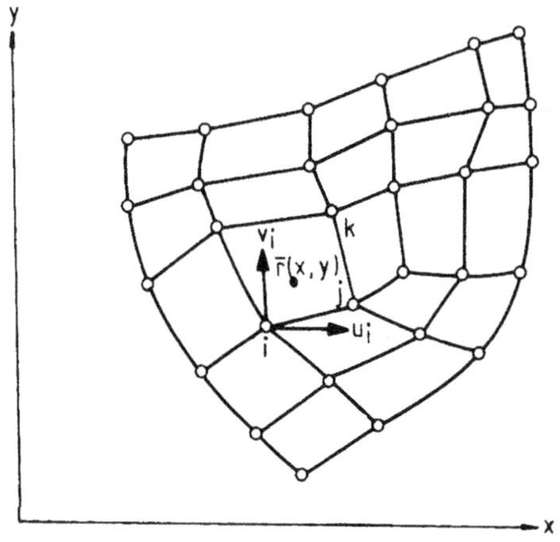

Fig. 13.8 Finite element idealization of a plane region

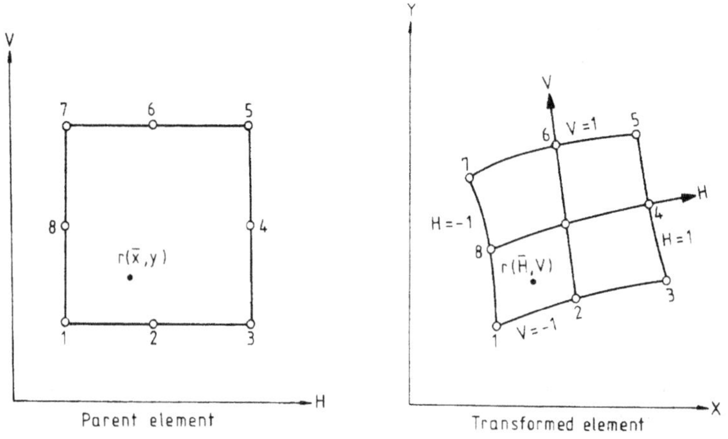

Fig. 13.9 Coordinate transformation

$$\{\Delta\} = \begin{Bmatrix} U \\ V \\ \theta \end{Bmatrix} \tag{13.62}$$

Consider the series of boundary points i_A and i_B ($i = 1$ to n) in Figure 13.10 at which possible contact can take place. It is assumed that the nodal contact forces are $\{p_j^n\}$ and $\{p_j^t\}$ on body A and B respectively. It is to be noted that these vectors have two components, one in the tangential and the other in normal directions as below:

$$\{p_j\} = \begin{Bmatrix} \{p_j^t\} \\ \{p_j^n\} \end{Bmatrix} \quad \text{and} \quad \{p_j'\} = \begin{Bmatrix} \{p_j'^t\} \\ \{p_j'^n\} \end{Bmatrix} \tag{13.63}$$

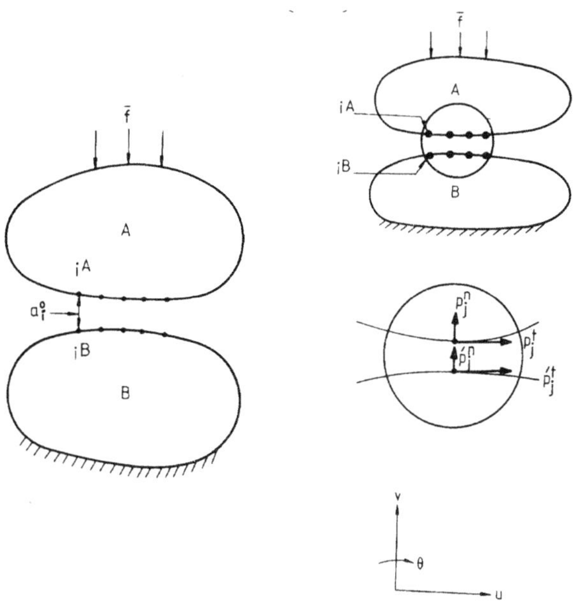

Fig. 13.10 Two bodies A and B before and after contact

If the body B has fixed supports and the only forces acting are the contact forces, the displacements of points i in the body B can be written as

$$\{a_i^B\} = \sum_{j=1}^{n} [C_{ij}^B]\{p_j'\} \tag{13.64}$$

where $\left[C_{ij}^B\right]$ above represents a 2×2 submatrix of flexibility coefficients corresponding to the tangential and normal deflections, respectively at the node i due to the force at j.

The submatrix of flexibility coefficients is obtained by eliminating all the nodes except those where possible contact can take place and inverting the so formed condensed stiffness matrix. If the supports of the body B are given the rigid displacements given in equation (13.62), the total displacement of the point i is given by

$$\{a_j^B\} = \sum_{j=1}^{n} [C_{ij}^B]\{p_j'\} + [\sigma_i]\{\Delta\} \tag{13.65}$$

where

$$[\sigma_i] = \begin{bmatrix} \cos\phi & \sin\phi & -(y\cos\phi - x\sin\phi) \\ -\sin\phi & \cos\phi & -(x\cos\phi + y\sin\phi) \end{bmatrix} \tag{13.66}$$

is the kinematic transformation matrix, see Figure 13.11.

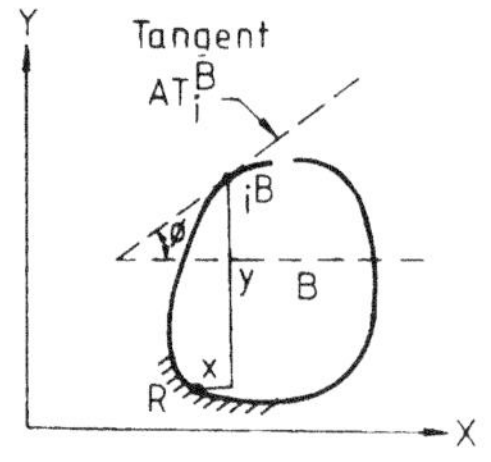

Fig. 13.11 Computation of transformation matrix

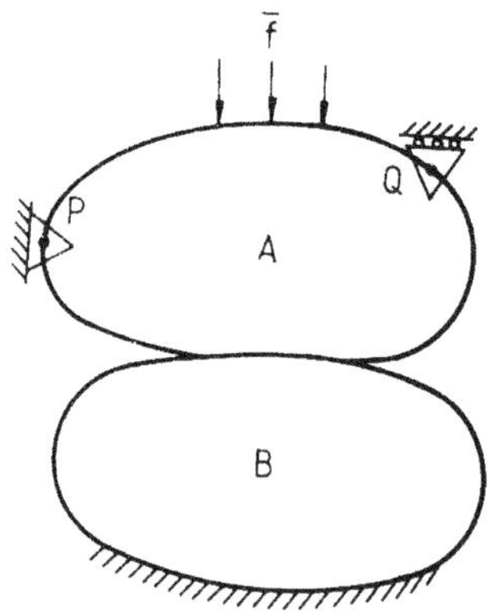

Fig. 13.12 Body A with imaginary support points P and Q

As the body A has a force boundary condition, it is free to have a rigid body movement. Hence a slightly different method is to be used for getting the flexibility coefficient for this. The rigid body movement of A is restricted by assuming two imaginary supports as shown in Figure 13.12. Displacements are assumed zero in both the directions at one of the points and at the other point in one direction only. Hence three degrees of freedom are restrained. All the nodes except those at the imaginary reaction points and the likely contact points and those with the external forces are eliminated.

The condensed system of equations in general can be written as

$$\begin{bmatrix} K_{11} & K_{12} & K_{13} \\ K_{21} & K_{22} & K_{23} \\ K_{31} & K_{32} & K_{33} \end{bmatrix} \begin{Bmatrix} a^R \\ a^C \\ a^f \end{Bmatrix} = \begin{Bmatrix} R \\ p^C \\ f \end{Bmatrix} \tag{13.67}$$

where K are submatrices of the condensed stiffness matrix of body A, $\{a^R\}$ is the vector of displacement of reaction points, $\{a^C\}$ is the vector of displacements at contact points, $\{a^f\}$ is the vector of displacements at points where external forces are acting on A. $\{R\}$ is the vector of reaction forces; $\{p^C\}$ and $\{f\}$ are the vectors of forces at the contact points and at the points where external forces are applied. Since the displacements at the reaction points are zero,

$$\begin{bmatrix} K_{12} & K_{13} \end{bmatrix} \begin{Bmatrix} \{a^C\} \\ \{a^f\} \end{Bmatrix} = \{R\} \tag{13.68}$$

and

$$\begin{bmatrix} K_{22} & K_{23} \\ K_{32} & K_{33} \end{bmatrix} \begin{Bmatrix} \{a^C\} \\ \{a^f\} \end{Bmatrix} = \begin{Bmatrix} p^C \\ \{f\} \end{Bmatrix} \tag{13.69}$$

From the above two equations we get,

$$\begin{bmatrix} K_{12} & K_{13} \end{bmatrix} \begin{bmatrix} K_{22} & K_{23} \\ K_{32} & K_{33} \end{bmatrix}^{-1} \begin{Bmatrix} \{p^C\} \\ \{f\} \end{Bmatrix} = \{R\} = 0 \tag{13.70}$$

since in reality there are no external forces present in this case.

The displacement of points i in the direction normal to the surface are given by the expression

$$\{a_i^A\} = \sum_{j=1}^{n} [C_{ij}^A]\{p_j\} + \sum_{k=1}^{e} [C_{i,n+k}]\{f_k\} \tag{13.71}$$

where e is the number of nodes with external forces and $\{f^k\}$ is the vector of external forces at the kth loaded node such that

$$[f_k] = \begin{Bmatrix} \{f_k^t\} \\ \{f_k^n\} \end{Bmatrix} \tag{13.72}$$

where $\{f_k^t\}$ and $\{f_k^n\}$ are the components in the tangential and normal directions respectively. It may be noted that

$$[C_{ij}] = \begin{bmatrix} K_{22} & K_{23} \\ K_{32} & K_{33} \end{bmatrix}^{-1} \tag{13.73}$$

This is easily accomplished by inverting the condensed matrix obtained after imposing boundary conditions for the fictitious nodes. When the two bodies are in contact the following compatibility equation must be satisfied:

$$\{a_i^A\} = \{a_i^B\} + \{a_i^0\} \tag{13.74}$$

where $\{a_i^0\}$ is the vector of clearances. Hence from equations (13.65) and (13.71)

$$\sum_{j=1}^{n} [C_{ij}^A]\{p_j\} + \sum_{k=1}^{e} [C_{i,n+k}]\{f_k\} = \sum_{j=1}^{n} [C_{ij}^B]\{p'_j\} + \{\sigma_i\}\{\Delta\} + \{a_i^0\} \tag{13.75}$$

At the contact nodes $\{p_j\} = -\{p'_j\}$ and hence

$$\sum_{j=1}^{n} ([C_{ij}^A] + [C_{ij}^B])\{p_j\} - \{\sigma_i\}\{\Delta\} = -\sum_{k=1}^{e} [C_{i,n+k}]\{f_k\} + \{a_i^0\} \tag{13.76}$$

where n is the number of nodes in contact and e is the number of nodes with external forces.

For the case when the nodes slip relative to each other in the direction parallel to the contact surface, the compatibility condition (13.74) is not valid for the tangential direction. Hence the above equation will not be valid along the tangential direction and the corresponding equations are replaced by the slip condition

$$\{p_j^t\} = \mu\{p_j^n\} \tag{13.77}$$

where μ is the coefficient of dynamic friction and the subscript j represents the node in contact. If m nodes are in contact, the above two equations constitute $2m$ equations in $2m+3$ unknowns and the three additional equations are the equations of equilibrium suitably represented by (13.70). This system is linear and the solution is straight-forward if the contact zone is known apriori and friction is sufficient to prevent slipping. However, in practice the area of contact is not known and some nodes are observed to be slipping. Hence the following steps are to be followed:

1. Assume a set of nodal points to be in contact and solve equation (13.76) for a prescribed load.
2. Delete all those points where nodal forces become negative from the possible contact zone.
3. If the ratio of nodal force in tangential direction to the nodal force in the normal direction is greater than the coefficient of friction, the nodes are slipping, then the portion of equation (13.76) corresponding to the tangential direction is replaced by equation (13.77).
4. The above steps are repeated until all normal forces in the contact zone come out to be positive only and the ratio of tangential force to normal force for all the nodes in contact is either less than or equal to the coefficient of friction.

Now consider a quasi-static problem of blade root-disk deformation under centrifugal loading with a T root interface joint. Figure 13.13 shows the finite element model of T root of the 19th stage of a 210 MW steam turbine. Body A has been modeled with 34 elements having 282 degrees of freedom while the body B has 26 elements with 230 degrees of freedom. Eight noded quadratic isoparametric elements [10, 23], are used in the analysis.

For simplicity, the entire centrifugal load is assumed to act at the tip of the blade. The body has been fixed along the edge AB, while the body A has been imposed with an axial force of prescribed value along the surface. Fictitious nodes for restraining rigid body motion have been assumed for body A at two nodes on opposite faces of the blade. Figure 13.14 shows the enlarged details of pairing of contact nodes for the T root junction. The deformation of the two-body assembly under the action of axial force is shown in Figure 13.15. The deformation is symmetric about the central line. For the given mesh two nodes lift and then are found to slip when an axial load of 1000 kg is applied. The deformations shown are enlarged 400 times. When the axial force is increased to 2000 kg, the deformed shape indicates that two nodes lift and eight are slipping. The dovetail is nearly locked and the lips of the

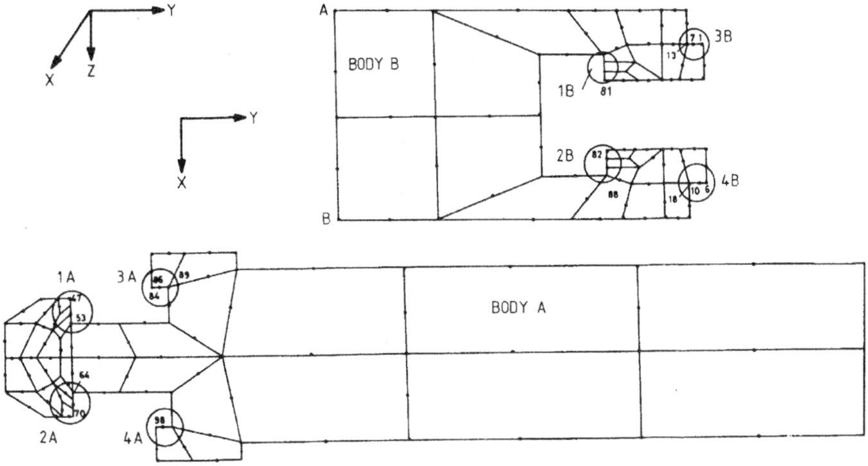

Fig. 13.13 Finite element model of T root blade and disk

disk open out causing slip on the two side surfaces. The displacements shown in Figure 13.16 are enlarged 300 times.

The nodal displacements thus determined in the quasi-static problem constitute the initial condition for the dynamic problem.

Dynamic Contact Problem: The system equation for a single blade assembly in the absence of material damping is

$$[M]\{\ddot{a}\} + [K]\{a\} = \{f\} \tag{13.78}$$

The behavior of the contact zone between A and B is nonlinear on account of opening and slipping which are to be suitably modeled. Thus partitioning the displacement vector

$$\{a_t\}^A = \begin{Bmatrix} \{a_t^C\} \\ \{a_t^A\} \end{Bmatrix}$$

and (13.79)

$$\{a_t\}^B = \begin{Bmatrix} \{a_t^{C'}\} \\ \{a_t^B\} \end{Bmatrix}$$

we can write the equations of motion for both the bodies as

$$[M^B]\begin{Bmatrix} \{\bar{a}_t^{C'}\} \\ \{\bar{a}_t^B\} \end{Bmatrix} + [K^B]\begin{Bmatrix} \{a_t^{C'}\} \\ \{a_t^B\} \end{Bmatrix} = \begin{Bmatrix} \{f_t^{C'}\} \\ \{f_t^B\} \end{Bmatrix} \tag{13.80}$$

$$[M^A]\begin{Bmatrix} \{\bar{a}_t^C\} \\ \{\bar{a}_t^A\} \end{Bmatrix} + [K^A]\begin{Bmatrix} \{a_t^C\} \\ \{a_t^A\} \end{Bmatrix} = \begin{Bmatrix} \{f_t^C\} \\ \{f_t^A\} \end{Bmatrix} \tag{13.81}$$

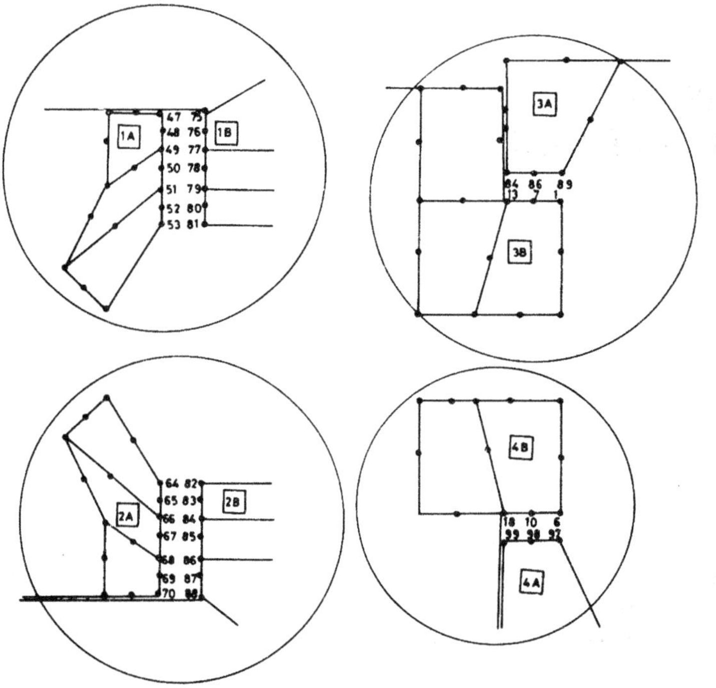

Fig. 13.14 Details of finite element mesh in the neighborhood of contact regions

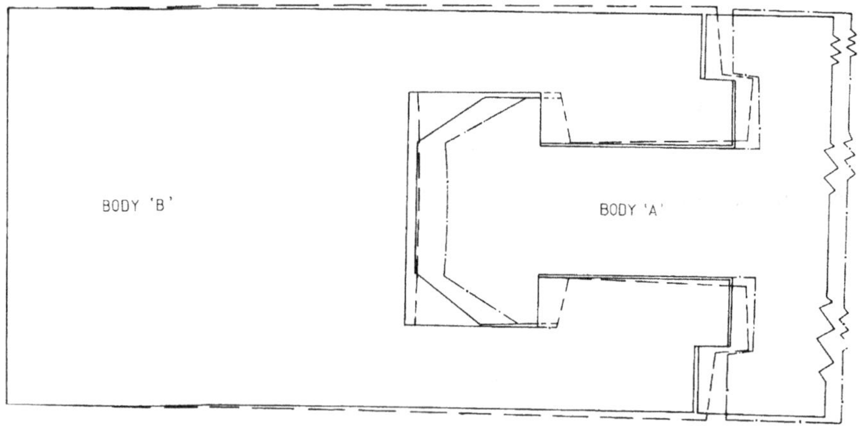

Fig. 13.15 Deformation pattern of *T* root blade/disk junction under a centrifugal load of 1000 kg

On account of the nonlinearities, modal methods are not found suitable, instead direct integration algorithms will be useful.

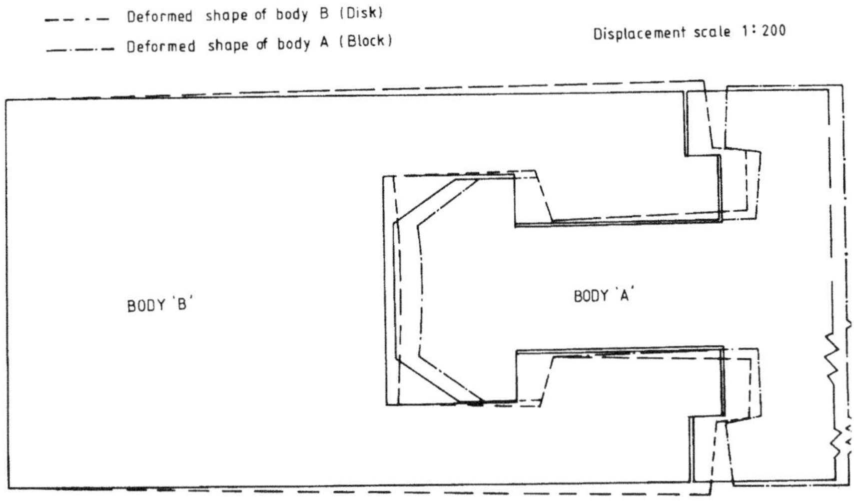

Fig. 13.16 Deformation pattern of T root blade/disk junction under a centrifugal load of 2000 kg

The Wilson-θ method, discussed extensively by Bath and Wilson (1978), is an implicit integration method, in which the acceleration varies linearly over the time interval from t to $(t + \theta\Delta t)$ where $\theta \geq 1$ is determined to obtain optimum stability and accuracy characteristics. θ is usually taken to be 1.4. Equation (13.78) is written as

$$[M]\{\ddot{a}_{t+\theta\Delta t}\} + [K]\{a_{t+\theta\Delta t}\} = \{f_{t+\theta\Delta t}\} \tag{13.82}$$

where

$$\{f_{t+\theta\Delta t}\} = \{f_t\} + \theta(\{f_{t+\Delta t}\} - \{f_t\})$$

For adopting the Wilson-θ scheme, the following procedure is used:

Initial calculations:

1. Form stiffness matrix $[K]$ and mass matrix $[M]$.
2. Initialize $\{a_0\}$, $\{\dot{a}_0\}$, $\{\ddot{a}_0\}$.
3. Select time step Δt.

$$A0 = \frac{6}{(\theta\Delta t)^2}; \quad A2 = \frac{6}{\theta\Delta t}; \quad A4 = \frac{\Delta\theta}{\theta}; \quad A5 = \frac{A2}{\theta};$$

$$A6 = 1 - \frac{3}{\theta}; \quad A7 = \frac{\Delta t}{2}; \quad A8 = \frac{(\Delta t)^2}{6} \tag{13.83}$$

4. Form the effective stiffness matrix

$$[\hat{K}] = [K] + A0[M] \tag{13.84}$$

5. Carry out reduction of the equation to upper triangular form $[\tilde{K}]$
6. For each time step, the effective load at time $t + \theta\Delta t$ is calculated as

$$\{\hat{f}_{t+\theta\Delta t}\} = \{f_t\} + \theta(\{f_{t+\Delta t}\} - \{f_t\}) + [M](A0\{a_t\} + A2\{\dot{a}_t\} + 2\{\ddot{a}_t\}) \quad (13.85)$$

7. Form the following equation to determine $\{a_{t+\theta\Delta t}\}$:

$$[\tilde{K}]\{a_{t+\theta\Delta t}\} = \{\hat{f}_{t+\theta\Delta t}\} \quad (13.86)$$

8. Determine accelerations, velocities and displacements at time $t + \Delta t$.

$$\begin{aligned}
\{\ddot{a}_{t+\Delta t}\} &= A4(\{a_{t+\theta\Delta t}\} - \{a_t\}) + A5\{\dot{a}_t\} + A6\{\ddot{a}_t\} \\
\{\dot{a}_{t+\Delta t}\} &= \{\dot{a}_t\} + A7(\{\ddot{a}_{t+\theta\Delta t}\} - \{\ddot{a}_t\}) \\
\{a_{t+\Delta t}\} &= \{a_t\} + \Delta t\{\dot{a}_t\} + A8(\{\ddot{a}_{t+\theta\Delta t}\} + 2\{\ddot{a}_t\}) \\
\{\tilde{a}_{t+\Delta t}\} &= \frac{6}{(\theta\Delta t)^2}(\{a_{t+\theta\Delta t}\} - \{a_t\}) - \frac{6}{\theta\Delta t}\{\dot{a}_t\} - 2\{\ddot{a}_t\}
\end{aligned} \quad (13.87)$$

To solve the transient dynamic contact problem which is nonlinear, the dynamic equations of motion for the two bodies are considered separately. Each region further consists of a general domain and a contact zone. The dynamic equations are condensed out for the contact zone only for the two bodies A and B. Equations of equilibrium and compatibility along with the constitutive relations for contact behavior are set up to obtain the solution for the behavior through a heuristic iterative procedure.

Consider equation (13.80) at $t + \theta\Delta t$, then from the last equation of (13.87), we get

$$[P^B]\begin{Bmatrix} \{a^{C'}_{t+\theta\Delta t}\} \\ \{a^{B}_{t+\theta\Delta t}\} \end{Bmatrix} = [Q^B]\begin{Bmatrix} \{a^{C'}_t\} + \theta\Delta t\{\dot{a}^{C'}_t\} + \frac{(\theta\Delta t)^2}{3}\{\ddot{a}^{C'}_t\} \\ \{a^{B}_t\} + \theta\Delta t\{\dot{a}^{B}_t\} + \frac{(\theta\Delta t)^2}{3}\{\ddot{a}^{B}_t\} \end{Bmatrix} + \begin{Bmatrix} \{f^{C'}_{t+\theta\Delta t}\} \\ \{f^{B}_{t+\theta\Delta t}\} \end{Bmatrix} \quad (13.88)$$

where

$$[P^B] = \frac{6}{(\theta\Delta t)^2}[M^B] + [K^B] = \begin{bmatrix} P'_{CC} & P_{CB} \\ P_{BC} & P_{BB} \end{bmatrix}$$

$$[Q^B] = \frac{6}{(\theta\Delta t)^2}[M^B] = \begin{bmatrix} Q'_{CC} & Q_{CB} \\ Q_{BC} & Q_{BB} \end{bmatrix}$$

For body A, similarly, we get

$$[P^A]\begin{Bmatrix} \{a^{C'}_{t+\theta\Delta t}\} \\ \{a^{AQ}_{t+\theta\Delta t}\} \end{Bmatrix} = [Q^A]\begin{Bmatrix} \{a^{C}_t\} + \theta\Delta t\{\dot{a}^{C}_t\} + \frac{(\theta\Delta t)^2}{3}\{\ddot{a}^{C}_t\} \\ \{a^{A}_t\} + \theta\Delta t\{\dot{a}^{A}_t\} + \frac{(\theta\Delta t)^2}{3}\{\ddot{a}^{A}_t\} \end{Bmatrix} + \begin{Bmatrix} \{f^{C}_{t+\theta\Delta t}\} \\ \{f^{A}_{t+\theta\Delta t}\} \end{Bmatrix} \quad (13.89)$$

where

$$[P^A] = \begin{bmatrix} P_{CC} & P_{CA} \\ P_{AC} & P_{AA} \end{bmatrix}$$

$$[Q^A] = \begin{bmatrix} Q_{CC} & Q_{CA} \\ Q_{AC} & Q_{AA} \end{bmatrix}$$

If the pairing nodes i_A and i_B are in contact, then

$$\begin{aligned} \{a^C_{t+\theta\Delta t}\} &= \{a^{C'}_{t+\theta\Delta t}\} \\ \{f^C_{t+\theta\Delta t}\} &= -\{f^{C'}_{t+\theta\Delta t}\} = \{p^C_{t+\theta\Delta t}\} \end{aligned} \tag{13.90}$$

Imposing the above conditions on equations (13.88) and (13.89), we get

$$\begin{aligned} [C^*_B + C^*_A]\{p^C_{t+\theta\Delta t}\} &= [C^*_A S^*_1 - C^*_B Q^*_1]\left\{\{a^C_t\} + \theta\Delta t\{\dot{a}^C_t\} + \frac{(\theta\Delta t)^2}{3}\{\ddot{a}^C_t\}\right\} \\ &+ C^*_A S^*_2\left\{\{a^A_t\} + \theta\Delta t\{\dot{a}^A_t\} + \frac{(\theta\Delta t)^2}{3}\{\ddot{a}^A_t\}\right\} \\ &- C^*_B Q^*_2\left\{\{a^B_t\} + \theta\Delta t\{\dot{a}^B_t\} + \frac{(\theta\Delta t)^2}{3}\{\ddot{a}^B_t\}\right\} - C^*_A P_{CA} P^{-1}_{AA}\{f^A_{t+\theta\Delta t}\} \end{aligned} \tag{13.91}$$

where

$$\begin{aligned} C^*_A &= [P_{CC} - P_{CA}P^{-1}_{AA}P_{AC}]^{-1}, \quad C^*_B = [P'_{CC} - P_{CB}P^{-1}_{BB}P_{BC}]^{-1} \\ Q^*_1 &= Q_{CC} - P_{CA}P^{-1}_{AA}Q_{AC}, \quad Q^*_2 = Q_{CA} - P_{CA}P^{-1}_{AA}Q_{AA} \\ S^*_1 &= Q'_{CC} - P_{CB}P^{-1}_{BB}Q_{BC}, \quad S^*_2 = Q_{CB} - P_{CB}P^{-1}_{BB}Q_{BC} \end{aligned}$$

The solution of linear equation (13.91) directly yields the unknown contact forces at the nodes. This can be used at any particular time instant to obtain the nodal contact forces, if the nodes in contact are known apriori. However, this is usually unknown and some assumptions are necessary. These assumptions are to be further confirmed by the signs of interface forces which have to be compressive. If some of these are found to be tensile, node opening occurs and these nodes have to be excluded from the contact zone during which further redistribution of contact pressures occur. The contact nodes have to be further checked for slip and the modeling of slip has to be done accurately.

Simulation of the Opening: At a particular time instant, given the right-hand side of equation (13.91), the nodal contact forces can be first computed by solving for $p^C_{t+\theta\Delta t}$. If some of the nodes exhibit tensile forces, these cannot be supported by the contact zone. Hence the nodal forces have to be recomputed by deleting these nodes from the list of original contact nodes and the interface forces have to be computed again.

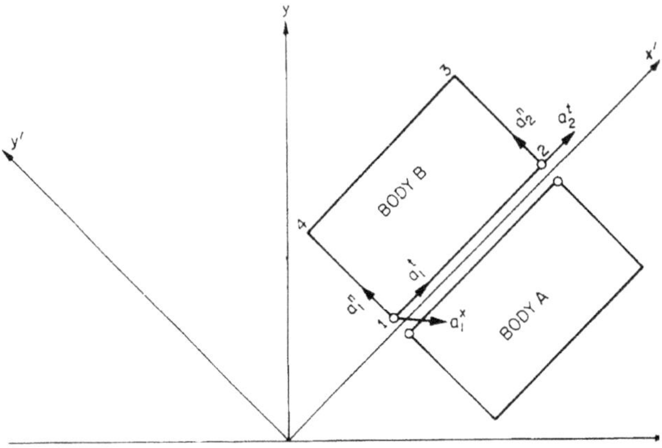

Fig. 13.17 Computation of transformation matrix

Simulation of Slip: At any particular time instant, after the nodes in contact are firmly established, the contact nodes have to be checked for conditions of slip. Thus for all the nodes, the ratio of the tangential to the normal load force is to be computed. If this ratio exceeds the prescribed coefficient of dynamic friction then the no-slip condition assumed is not valid. Hence the compatibility condition along the tangential direction is to be replaced by equation (13.77), for the particular node under consideration. Then equation (13.91) gives rise to $2m$ equations corresponding to m nodes in contact in $2m$ variables, viz., the unknown contact forces. When certain nodes are found to slip, slip conditions have to be imposed, thereby giving rise to as many equations as there are unknowns and an iterative procedure is to be used to determine the contact forces accurately.

Contact along Inclined Surfaces: If the contact is along an inclined surface, the conditions of compatibility and equilibrium have to be imposed along the direction $x'y'$. This is easily accomplished by transforming the stiffness matrix corresponding to these nodes along the directions $x'y'$, see Figure 13.17. The transformed stiffness matrices are then assembled and condensed in arriving at the final form of equation (13.91), to determine the contact forces along the tangential and normal directions. It is obvious that an efficient computing procedure should be developed to reduce the number of iterations.

Consider the T root blade example discussed earlier for the static contact problem, shown in Figure 13.13. The contact zone extends over four different surfaces shown in Figure 13.14. No contact is expected along other surfaces on account of the ample clearances present in the assembly. The contact region on the surface 1A and 2A is modeled using six elements, while the contact on surfaces 3A and 4A is modeled with two elements only.

The displacements obtained in a quasi-static analysis constitute initial conditions for the dynamic problem. The initial velocities and accelerations are assumed to be zero. The blade in Figure 13.13 is considered with unit thickness and Young's mod-

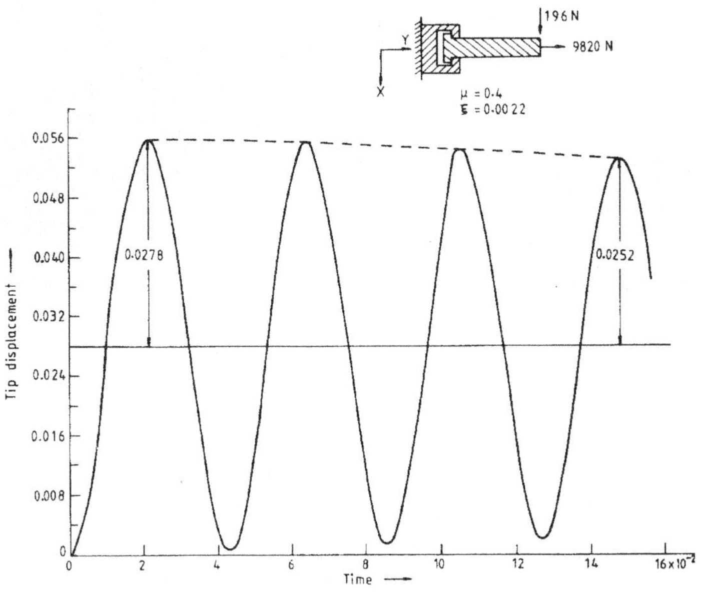

Fig. 13.18 Decay in tip response

ulus is 2.1×10^6 kg/cm^2, Poisson's ratio is 0.3 and coefficient of dynamic friction is 0.4. The blade has an overall length of 29 cm and width 5 cm. The excitation due to a suddenly applied load is as shown in Figure 13.18. For an axial load of 1000 kg and a transverse load of 20 kg, the tip displacement is obtained as a function of time. A time step of 0.0002 sec was used so that at least 20 such steps are included in one cycle of response. The natural period was found to be 0.00432 sec. Figure 13.18 shows that the decay in response is clearly due to the Coulomb friction model. The average damping ratio is estimated as 0.0022.

When the axial load is increased to 1200 kg, the equivalent viscous damping was found to be 0.00158, showing thereby, that the blade is becoming locked in the root of the disk. At 1500 kg axial load, very little difference in the successive amplitudes was found, showing that there is practically no slip between the blade and the root. When the axial load was further increased to 2000 kg the value of damping ratio actually increased to 0.0033. This behavior is attributed to the deformation pattern of the joint assembly as shown in Figure 13.16. While the surfaces 1A and 2A, which are perpendicular to the centrifugal load and therefore get locked under increased axial pull, the two surfaces 3A and 4A being parallel to the application of axial pull, now slip and contribute to additional damping in the system.

Today there are several commercial codes dealing with contact problems with a preprocessor for preparing the finite element model and the results from the solver are all processed in a postprocessor to obtain a user-friendly form of the results.

We will discuss such an application using a commercial code in the next section that has taken away the drudgery of preparing a dedicated code and processing the results from a design engineer and produces the desired results in a quick turn around time.

13.5 Illustration of Turbomachine Blade Analysis using Commercial Codes

This work is reported from Rao [15]. The blade data is as follows:

- Normal operating speed of machine = 8500 RPM
- Blade material $X_{20}Cr_{13}/X_{20}CrMo_{13}$;
- Density = 7700 kg/m^3;
- Poisson's ratio = 0.3;
- Ultimate tensile strength S_u = 880 MPa (at operating temp, 50.4° C);
- Yield stress $S_y = 575$ MPa (at operating temp, 50.4° C);
- Young's Modulus $E = 218000$ MPa (at operating temp, 50.4° C);
- Lazan's coefficient $J = 30$ kNm/m^3/cycle;
- Lazan's exponent = 2.4.

The blade geometric model is shown in Figures 13.19 and 13.20 is the FE model. Kardestuncer and Norrie [11] gave several finite elements and several commercial codes provide libraries. Here Solid 45 elements are chosen for the blade vane, dovetail and Solid 92 for the connection between for the blade vane and dovetail. The total number of elements used is 198267 and the total nodes are 252868.

Steady steam pressures at 10% intervals from hub to tip on pressure and suction surfaces at 10 points across each airfoil are given in Table 13.1.

Static stress analysis was carried out for 8500 RPM with the static steam pressure loads given in Table 13.1. The boundary conditions are given in Figure 13.21. Steam Pressure Loads are mapped as illustrated in Figure 13.22. Pressure loads are applied on sections from hub to tip. Half of the elements above and below each of these sections receive the same pressure.

Static Steam loads are negligible and the maximum stress observed is 4 MPa only. The centrifugal load causes maximum stress and the peak value is 2470 MPa far beyond the yield as shown in Figure 13.23. The blade otherwise is globally elastic with an average stress 343 MPa. The plastic region is small in depth and elastic analysis is not applicable here, which is why we have a highly unrealistic value. The plastic region beyond yield 585 MPa is shown in Figure 13.24. The depth of the max stress is one element and the element depth is 0.37 mm.

One can perform an elasto-plastic analysis and find that the maximum stress will follow the material plastic law. However the stress in the singular elements with a sudden rise in stress at the node usually represents a condition where the finite element method fails to predict the correct stress values. The true stress is governed by local strain concentration and Neuber's law gives the true strain and stress condition. We will discuss this later. An elasto plastic analysiselasto plastic

Table 13.1 Steady steam pressures (in Pa)

Pressure Surfaces

HUB ... TIP

x/c	0% Span	10% Span	20% Span	30% Span	40% Span	50% Span	60% Span	70% Span	80% Span	90% Span	100% Span
0	9000	9060	9120	9180	9240	9300	9620	9940	10260	10580	10900
0.1	7900	7920	7940	7960	7980	8000	8280	8560	8840	9120	9400
0.2	8060	8168	8276	8384	8432	8600	8940	9280	9620	9960	10300
0.3	8220	8356	8432	8628	8764	8900	9200	9500	9800	10100	10400
0.4	8350	8500	8650	8800	8950	9100	9360	9620	9880	10140	10400
0.5	8400	8580	8760	8940	9120	9300	9520	9740	9960	10810	10400
0.6	8440	8606	8772	8938	9104	9270	9436	9722	9948	10174	10400
0.7	8480	8634	8788	8942	9096	9250	9480	9710	9940	10170	10400
0.8	8520	8661	8802	8943	9084	9225	9440	9655	9870	10085	10300
0.9	8560	8688	8816	8944	9072	9200	9400	9600	9800	10000	10200
1	8600	8716	8832	8948	9064	9180	9354	9528	9702	9876	10050

Suction Surfaces

HUB ... TIP

x/c	0% Span	10% Span	20% Span	30% Span	40% Span	50% Span	60% Span	70% Span	80% Span	90% Span	100% Span
0	9000	9060	9120	9180	9240	9300	9620	9940	10260	10580	10900
0.1	8300	8700	9100	9500	9900	10300	10760	11220	11680	12140	12600
0.2	8150	8380	8610	9070	9070	9300	9600	9900	10200	10500	10800
0.3	8100	8260	8420	8580	8740	8900	9080	9260	9440	9620	9800
0.4	8150	8280	8410	8540	8670	8800	9000	9200	9400	9600	9800
0.5	8200	8320	8440	8560	8680	8800	9000	9200	9400	9600	9800
0.6	8290	8392	8494	8596	8698	8800	9000	9200	9400	9600	9800
0.7	8370	8456	8542	8628	8714	8800	9000	9200	9400	9600	9800
0.8	8460	8552	8644	8736	8828	8920	9096	9272	9448	9624	9800
0.9	8530	8634	8738	8842	8946	9050	9225	9400	9575	9750	9925
1	8600	8716	8832	8948	9064	9180	9354	9528	9702	9876	10050

analysis will not change the global elastic region, the limited plastic region results will be limited to be just above yield.

A free vibration analysis is carried out and from the Campbell diagram it is found that, 344.50707 Hz 1F is the critical natural frequency. The unsteady pressures are defined in the form given from measurements

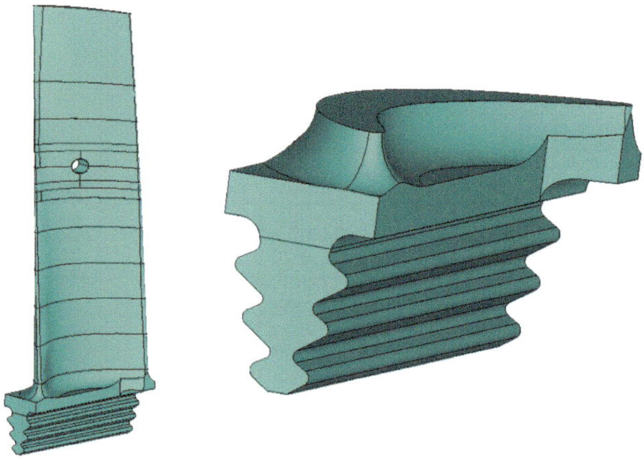

Fig. 13.19 Blade geometric model

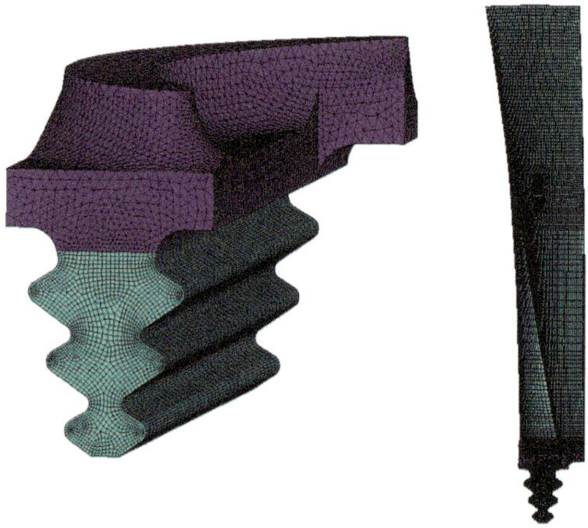

Fig. 13.20 FE model of the blade

$$\frac{dp}{P_{o1}} \cos(\omega t + \psi)$$

$$P_{o1} = 2000\,\mathrm{N/m^2}$$

$$\psi = \text{Phase angle}$$

Figure 13.25 gives the harmonic variation of the pressure at the hub. Similarly pressure values are prepared along the blade span at 10% intervals and across the aerofoil section at 10 points from leading edge to trailing edge. Alternating pressure is applied in the form

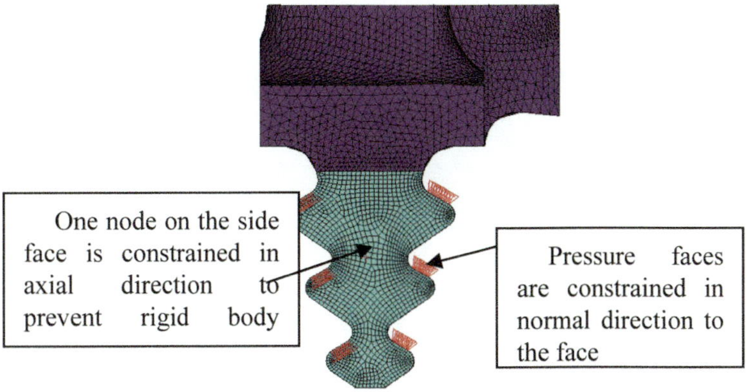

Fig. 13.21 Boundary conditions

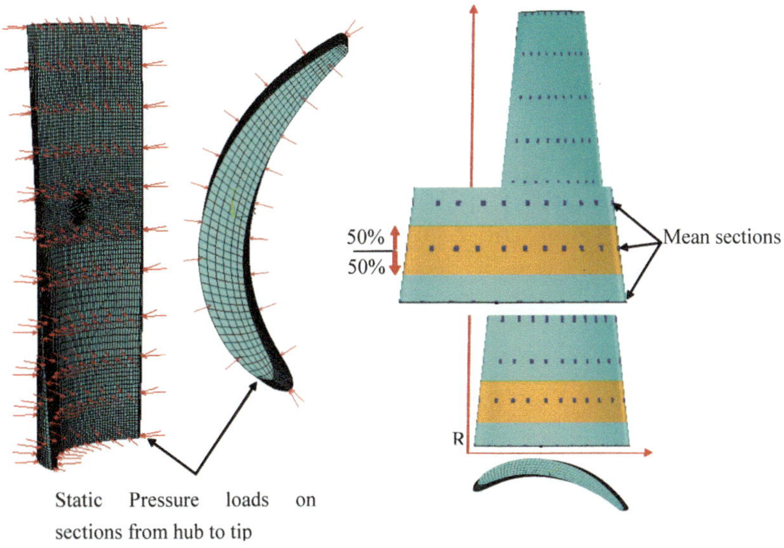

Fig. 13.22 Mapping of steam pressure loads

$$\frac{dp}{P_{o1}} \cos(\omega t + \psi) 2000 \times 10^{-6}$$

where $P_{o1} = 2000\ \text{N/m}^2$.

Table 13.2 gives the real and imaginary parts of the pressures that take into account the phase at the hub as given in Figure 13.25. These pressures are calculated on similar lines and given at different spans of the blade from root to tip and also from leading edge to trailing edge of an aerofoil profile. These values are applied on the full span of blade by considering 50% elements above and below mean sections of the blade span from root to tip as shown in Figure 13.22.

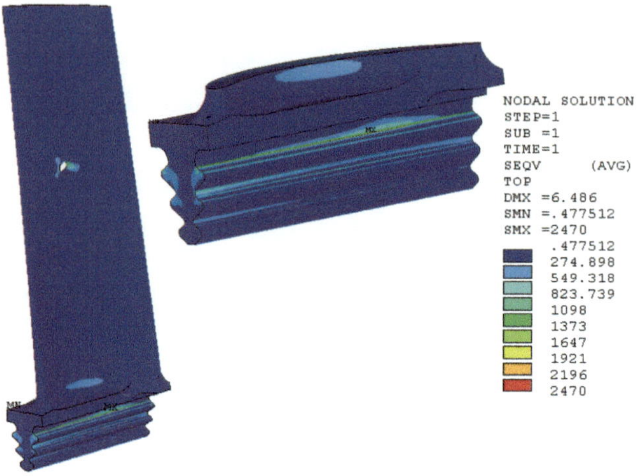

Fig. 13.23 Von Mises stress plot due to centrifugal load

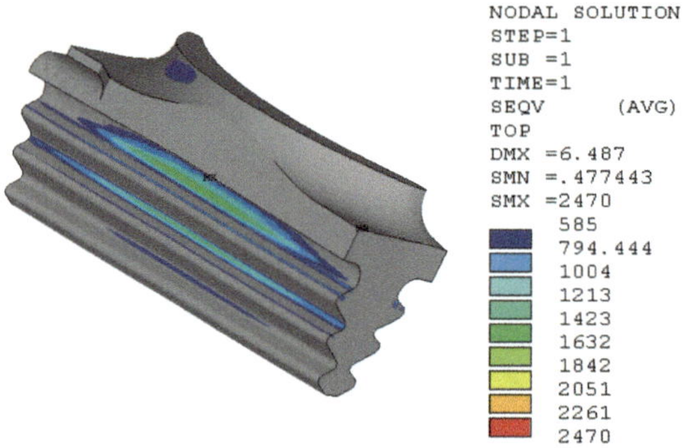

Fig. 13.24 Plastic region beyond yield 585 MPa of Figure 13.23

In order to obtain the equivalent static response with appropriate phases taken into account, a harmonic analysis is performed at very low frequency close to zero i.e. is 0.01 Hz. The average section stress is 0.13373 MPa that controls the dynamic stress value for the purpose of strain-based life estimation adopted in the following sections. The peak stress at the stress raiser is 0.9547 MPa.

The equivalent damping ratio at resonance is taken as 0.00226, 0.226% (see Section for a discussion on damping models). Therefore the magnification factor is $1/2\xi = 221.239$. The peak steady stress is 0.9547 MPa and therefore the peak dynamic stress is $0.9547 \times 221.239 = 211.2168$ MPa.

Even prior to Laval's impulse turbine in 1883, Rayleigh (1877) gave an approximate method based on the energy conservation principle to determine the fundamental mode of a cantilever blade or critical speed of a rotor. Soon the need to

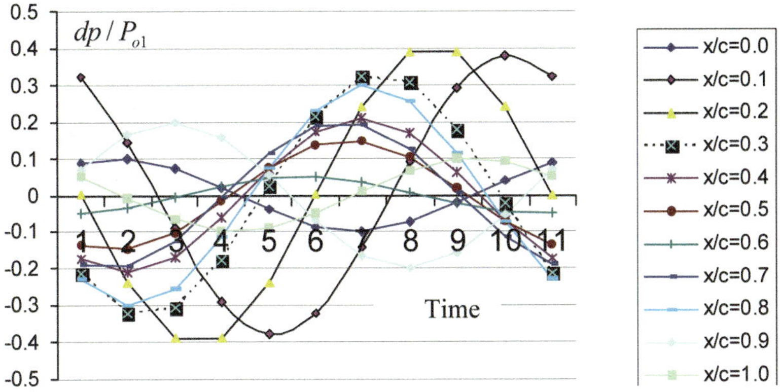

Fig. 13.25 Unsteady pressure

Table 13.2 Real and imaginary pressure components at the Hub corresponding to Chord locations in Figure 13.22

dp/P_{o1}(max)	$dp = \frac{dp}{P_{o1}} 2000 \times 10^{-6}$	Ψ	Real part $A \cos \Psi$	Imaginary part $A \sin \Psi$
0.1	0.0002	330	0.000173	–0.000100004
0.38	0.00076	32	0.000645	0.000402737
0.41	0.00082	90	4.96E-09	0.00082
0.33	0.00066	130	–0.00042	0.000505593
0.21	0.00042	145	–0.00034	0.000240905
0.15	0.0003	155	–0.00027	0.000126788
0.05	0.0001	190	–9.8E-05	–1.73636E-05
0.2	0.0004	160	–0.00038	0.000136812
0.3	0.0006	140	–0.00046	0.000385677
0.2	0.0004	290	0.000137	–0.00037588
0.1	0.0002	60	0.0001	0.000173205

determine natural frequencies or critical speedscritical speeds became an essential feature in design. Graphical and tabular methods etc. were developed by Stodola and Viannello [18], and then came tabular methods such the ones by Dunkerley [7], Holzer [8], Myklestad [12], Prohl [13] amongst others gave way ultimately to finite element methods as described above. Because of the ability to model a full structure with all the discontinuities and arrive at results faster, the finite element methods are a norm today for industrial practice – all this happened in a span of 100 years.

As an engineer using commercial codes for strength analysis, one should remember the following:

1. Basic theory of elasticity is well established; for any structure it involves 15 coupled partial differential equations with 15 unknowns of displacements, strains and stresses.
2. Unfortunately, the solutions of these complete sets of equations are difficult to accomplish.

3. It has been found useful to adopt basic energy methods by making assumptions on the displacement field, thus allowing evaluation of strain and stress fields and determining the solutions; however they are limited to the extent of a few terms in the assumed solutions to be able to achieve the desired solutions. Also, they cannot handle the engineering aspects of structures such as cut outs, grooves, dovetails, fillets, etc.
4. While mathematicians attempted solutions of specific classes of elasticity problems, engineers found it convenient to formulate problems from an equilibrium approach, called *Strength of Materials*, that allowed practical solutions for idealized structures such as beams and plates.
5. The Strength of Materials approach is however inadequate to account for intricate practical geometry of stationary and rotating structures and thus engineering practices evolved by adopting *Stress Concentration*, *Factor of Safety*, etc., to be able to achieve required designs.
6. Thanks to electronic valves, transistors and silicon chips, digital computation has come of age gradually over the last five decades which has ushered in the *finite element method*; in this method the structure is divided into several small finite structures or elements which are satisfied by the energy methods. The finite element has brought back the approximate way of solving problems in the Theory of Elasticity starting from shape functions representing the displacement field.
7. Today the task of engineers is simplified; most structural problems are solved by using one of several commercial codes;

   a. CAD Model;
   b. Meshed FE Model;
   c. Loads and Boundary conditions;
   d. Solver;
   e. Post process results.

8. While it appears that anyone can obtain a structure problem solution, the engineering aspect remains the same; ensure validity of the result, interpret the displacement, strains, stresses, and ensure structural integrity.
9. Another major task of engineering activity today is to estimate the load itself; this might arise out of flow, thermal considerations, electromagnetic fields – we will consider them separately.
10. Another significant design aspect is in understanding the dynamic characteristics of structures, resonance and stresses under resonance and the design for life based on mean and alternating stresses.

While structures advanced to finite element methods, rotors and their analysis lagged behind. The difference between a stator and rotor began to be understood only in 1919. They were treated as a separate class of problems and modeled as beams until the beginning of the 21st century when we learnt modeling of solid rotors. The centrifugal affects, spin softening affects on rotors have become understood only recently.

Although rotor dynamics is basically a vibration subject, it is treated separately because of some special effects such as oil film supports, misalignment, etc., and we will explore these developments in the next section.

References

1. Ahmad, S., Irons, B.M. and Zeinkiewicz, O.C. (1971) Analysis of Thick and Shell Structures by Curved Finite Elements, *International Journal Numerical Methods in Engineering*, vol. 3, p. 575 .
2. Argyris, J.H. (1954) Energy Theorems and Structural Analysis: A Generalized Discourse with Applications on Energy Principles of Structural Analysis Including the Effects of Temperature and Non-Linear Stress-Strain Relations, *Aircraft Engineering and Aerospace Technology*, vol. 26, no. 10, p. 347.
3. Bath, K.J. and Wilson, E.L. (1978) *Numerical Methods in Finite Element Analysis*, Prentice Hall.
4. Clough, R.W. (1960) The Finite Element Method in Plane Stress Analysis, in *Proceedings Second ASCE Conference on Electronic Computation*, Pittsburgh, PA, p. 345.
5. Courant, R. (1941) Variational Methods for the Solution of Boundary Value Problems, in *Symposium on the Rayleigh–Ritz Method and Its Applications*, American Mathematical Society, May 2–3, Washington D.C.
6. Courant, R. (1943) Variational Methods for the Solution of Problems of Equilibrium and Vibrations, *Bulletin of the American Mathematical Society*, vol. 49, p. 1.
7. Dunkerley, S. (1894) On the Whirling of Vibration of Shafts, *Philosophical Transactions of the Royal Society, Series A*, vol. 185, p. 279.
8. Holzer. H. (1922) Tabular Method for Torsional Vibration Analysis of Multiple-Rotor Shaft Systems, *Machine Design*, May, p. 141.
9. Hrennikoff, A. (1941) Solution of Problems of Elasticity by the Frame-Work Method, *ASME Journal of Applied Mechanics*, vol. 8, A619–A715.
10. Huebner, K.H. (1975) *The Finite Element Method for Engineers*, John Wiley.
11. Kardestuncer, H. and Norrie, D.H. (1987) *Finite Element Handbook*, McGraw-Hill.
12. Myklestad, N.O. (1944) *Vibration Analysis*, McGraw-Hill Book Co.
13. Prohl, M.A. (1945) A General Method of Calculating Critical Speeds of Flexible Rotors, *Journal of Applied Mechanics, Trans. ASME, Series E*, vol. 67, p. 142.
14. Rao, J.S. (1998) *Dynamics of Plates*, Marcel Dekker.
15. Rao, J.S., Peraiah, K.Ch. and Udai Kumar Singh (2009) Estimation of Dynamic Stresses in Last Stage Steam Turbine Blades under Reverse Flow Conditions, Advances in Vibration Engineering, *Journal of Vibration Institute of India*, vol. 8, no. 1, p. 71.
16. Rao, J.S., Usmani, M.A.W. and Ramakrishnan, C.V. (1990) Interface Damping in Blade Attachment Region, in *Proceedings 3rd International Conference Rotor DynamicsRotor Dynamics*, Lyon, p. 185.
17. Rayleigh, J.W.S. (1945) *Theory of Sound*, Macmillan, London, 1877, Dover Publication.
18. Stodola, A. (1910) *Dampf- und Gasturbinen*, Springer, Berlin. Translation (1927) *Steam and Gas Turbines*, McGraw-Hill.
19. Strang, G. and Fix, G.J. (1973) *An Analysis of the Finite Element Method*, Prentice Hall, Englewood Cliffs, NJ.
20. Tocher, J.L. (1962) Analysis of Plate Bending Using Triangular Elements, PhD Thesis, University of California, Berkeley.
21. Turner, M.J., Clough, R.C., Martin, H.C. and Topp, L.J. (1956) Stiffness and Deflection Analysis of Complex Structures, *J. Aero Sci.*, vol. 23, p. 805.
22. Weaver, W. and Gere, J.M. (1966) *Matrix Analysis Of Framed Structures*, Springer-Verlag, New York.
23. Zeinkiewicz, O.C. (1977) *The Finite Element Method*, McGraw-Hill.

Chapter 14
Rotor Dynamics Methods

The industrial revolution began with reciprocating steam engines as devised by James Watt in 1780, and the 19th century witnessed a rapid expansion in various industrial sectors. Unfortunately, the reciprocating steam engine had several problems because of external combustion and excessive alternating load due to reciprocating masses that limited speeds and capacities. The industry was looking for non-reciprocating systems, purely rotating systems that could usher in an era of so-called "Vibration Free" engines. The dynamics of rotating structures are different from those of stationary structures. Basically, all the vibration phenomena will be valid, however, there are several differences and we have to set up new procedures for handling rotors and their vibratory phenomena.

The beginnings of idealization of a structure as a beam were in the renaissance period, well before the beginning of the Scientific revolution. Leonardo da Vinci (1452–1519) made fundamental contributions to solid mechanics and understood the central idea of bending of a beam. Galileo (1564–1642) followed in pursuit of problems involving beams; his mathematical treatment of acceleration and his concept of inertia both reflect earlier medieval analyses of motion. The scientific revolution occurred about two centuries after Leonardo da Vinci's contributions to the study of beams and bending, when Isaac Newton (1642–1727) invented Calculus and Calculus of Variations that enabled Euler and Bernoulli in 1750 to formulate the first comprehensive theory of beams. This was three centuries after Leonardo da Vinci suggested the basic idea of beam bending. Thus, beam theory is embedded in the Scientific Revolution period.

There were practically no known attempts to understand vibrations of a rotating structure or a rotor for over a century after beam theory was well understood and expanded to other structures, e.g., Plates by Sophie Germain in 1815. William John Macquorn Rankine (1820–1872) made significant contributions to Thermodynamics, particularly Steam Engines and his publication in 1859 was the first attempt at a practical approach to steam-engine theory. The Rankine cycle is a thermodynamic sequence of events and is still used as a standard for rating steam power plant performance. Ten years later in 1869, the first attempts were made to understand Rotor Dynamics when Rankine performed the first analysis of a spinning shaft. He con-

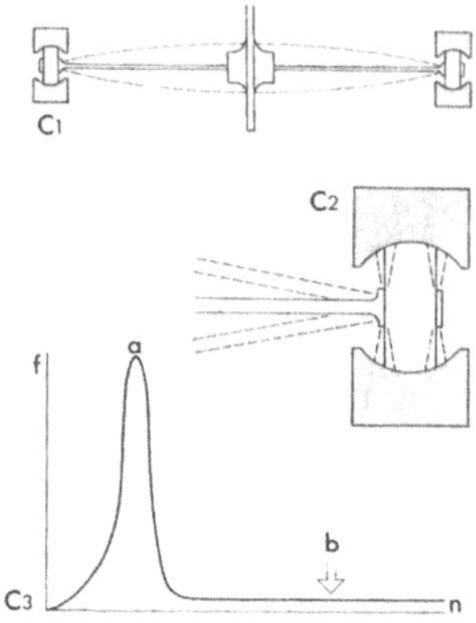

Fig. 14.1 De Laval rotor

sidered centrifugal whirling and showed the existence of critical speeds of a rotor. Rankine correctly concluded that the whirling speed was the same as the frequency at which the shaft vibrated transversely if struck, and he set out to calculate this critical speed. He defined this as a limit of speed for centrifugal whirling. There were many doubts whether a rotor could cross this limit. It was presumed that it would be unstable after exceeding the critical speed. He chose an unfortunate model and predicted that; beyond a certain spin speed "... the shaft is considerably bent and whirls around in this bent form." He defined this certain speed as the "whirling speed" of the shaft. In fact, it can be shown that beyond this whirling speed the radial deflection of Rankine's model increases without limit. Rankine did add the term "whirling" to the rotor dynamics vocabulary. Some believe now that he may have been responsible for setting back the science of rotor dynamics by nearly 50 years [39].

14.1 De Laval Model

Though the basics of rotor dynamics were not yet fully understood, Laval built the first impulse turbine in 1883 which ran successfully at 40000 RPM! From simple equilibrium conditions, he derived a correct relation for the whirl radius y (though whirl and spin have not been clearly differentiated), in his own notation, it is

$$y = \frac{\omega^2 \delta}{\frac{Fg}{W} - \omega^2} \tag{14.1}$$

where δ is the eccentricity, F is stiffness and W is the rotor weight.

When the denominator is zero, the whirl radius is infinity, defining the critical speed. Laval proved that one can exceed the critical speed and have stable operation at high speeds as y approaches $-\delta$. Essentially, this is the first attempt to identify the equivalence of a single degree of freedom vibrating system for a rotor and therefore some prefer to call one-disk rotor models "Laval Rotors". Föppl [20] discussed Laval's rotor. Rayleigh [76], using the energy principle, provided an approximate (upper bound) method to determine the first critical speed of a rotor considering it as a stationary beam. Dunkerley [14] derived an empirical relation for estimating the lower bound value of critical speed. The demonstration that a shaft can have several critical speeds is more than a century behind Lagrange's work for stationary systems. Stodola [82] presented a graphical method to determine the critical speeds of practical rotors. This method continued to be widely used for over five decades until transfer matrix methods and digital computers became available.

Though not directly related to vibrations, a significant experiment was made by British Rail Road Engineer Tower [92], whose work on friction of lubricated bearings by Osborne Reynolds (1842–1912) in 1886, led to the discovery of hydrodynamic bearings that form a load bearing member in the system and thus has stiffness and damping. These bearings play an important role in rotor dynamic behavior, a subject of intense investigations in the 20th century. Actually, Petroff [61] a Russian scientist, had already discussed bearings with centralized journals, i.e., concentrically operating, when Tower discovered pressure in oil film bearings in his railroad experiments. Sommerfeld [81] gave the first solution of Reynolds' equation for long bearings.

Charles Parsons soon after Laval's impulse turbine in 1883 built his reaction turbine in 1884, the precursor of modern rotating machinery for power plants. The invention of the dynamo by Thomas Alva Edison (1847–1931) in 1879 has already introduced the first power plant driven by a steam engine in 1882 at Pearl Street Station in New York producing 100 KW of electricity. The invention of turbine has changed the rotating machinery scenario in 20th century witnessing a rapid development from 1 to 1500 MW of power generation.

The first half of 20th century developed through logarithmic tables, mechanical desk calculators, slide rules and drawing tables with instruments. The world witnessed tremendous changes to satisfy the energy needs of the modern man; these changes were made possible through semiconductor devices as against electronic valves and at best a short exposure of transistors. The vibration engineer transformed the basic science to a rapid tool for fast designs reducing testing times through simulation, accurately estimating life and providing optimal designs all helping reduce design cycle and bringing the advanced machine to market – we have learnt in this century the *Shortest Distance from Concept to Reality*.

Some of the earlier rotor failures belong to propeller shafts in torsion of steam driven war ships during I world war. The story goes thus: When a propeller shaft failed, it was felt that designers did not provide sufficient diameter of the shaft to take care of the transmitted torque, therefore its diameter was increased by 10%. The modified shaft however failed in less than half time of the previous shaft failure. Then the designers began taking rotor failures seriously to adopt dynamic design. It was found that by increasing the diameter of the shaft the natural frequency became closer to the excitation harmonic resulting in an earlier failure. Holzer [33,34] presented a tabular method to determine the torsional natural frequencies of systems, which can be discretized in the form of several rigid inertias, connected by massless torsional springs. This is a simple method in which the inertia torque and torsional amplitude of each disk are calculated sequentially beginning from one end with amplitude equal to unity and arbitrarily chosen frequency. After completing the calculations till the end of the train, the boundary condition was checked – in this case the total inertia of the system in free vibration to be equal to zero. Obviously the first attempt will not yield the assumed frequency to be the correct natural frequency; therefore Holzer proposed an iteration method by varying frequency until a satisfactory answer is obtained. The method works well for a reasonable number of rotors in a simple table – it is so simple that it lasted into the computer age.

14.2 Jeffcott Rotor Analysis

Even with the general knowledge of critical speeds, the shaft behavior at any general speed was still unclear until Jeffcott [37] formulated the rotor problem as one of forced vibration. He showed for the first time that the shaft did not primarily rotate about its rest position, but about its own centerline. The whirl of the rotor corresponds to free or forced vibration of a stationary structure. This is a significant development in the understanding of rotor dynamic behavior. Rotors, modeled as a single disk on a flexible massless shaft, similar to a mass on a spring single degree freedom model are named after Jeffcott (see Figure 14.2).

Here we show basic Jeffcott rotor analysis including rolling element and oil film supports. In this model, the shaft is assumed mass less and the disk to be rigid. The total mass M of the rotor is put as disk and the stiffness K is represented as shaft. The eccentricity is denoted by $EG = a$. The rotor spins about its own axis with an angular velocity ω and whirls with angular velocity ν. Jeffcott considered synchronous whirl, i.e., $\nu = \omega$.

O is the bearing centerline, E is the disk geometric center, G is the mass center, $OE = R$ is the whirl radius about the bearing centerline. The disk rotates/spins about E with an angular velocity of ω in ccw direction and the whirl is assumed synchronous with spin. The whirl is lagging in phase by an angle ϕ from the unbalance force vector in direction of EG. Write down the inertia forces, stiffness forces and damping forces in the respective directions, etc.

$$M\frac{d^2}{dt^2}(z + a\cos\omega t) + C\frac{dz}{dt} + Kz = 0$$

$$M\frac{d^2}{dt^2}(y + a\sin\omega t) + C\frac{dy}{dt} + Ky = 0 \tag{14.2}$$

With $r = z + iy$, the above equations can be reduced to

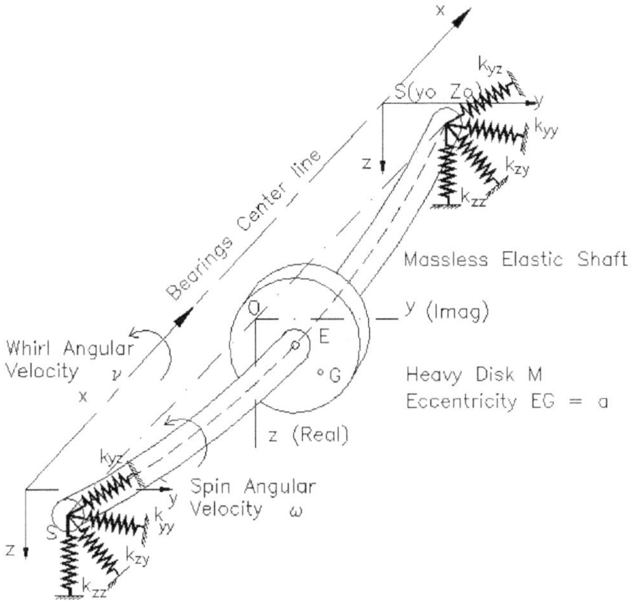

Fig. 14.2 Jeffcott rotor model

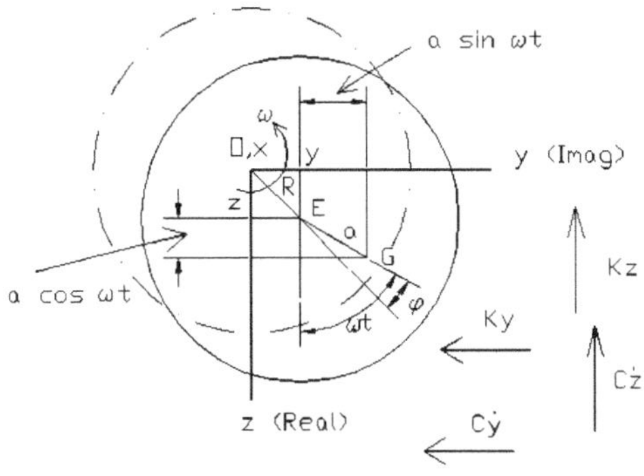

Fig. 14.3 Disk equilibrium relations

$$M\frac{d^2r}{dt^2} + C\frac{dr}{dt} + Kr = Ma\omega^2 e^{i\omega t} \tag{14.3}$$

Thus the Jeffcott rotor equation is same as a single degree of freedom mass system in all respects except that the excitation magnitude is a function of ω^2. Here is where many fail to understand why a Jeffcott rotor is Jeffcott rotor. We need to look at the whirl radius $r = z + iy$ is the solution instead of a displacement in a spring mass stationary system. The solution of (14.3) is of two parts. The complementary function or free vibration in this case is whirl radius at frequency $p = \sqrt{K/M}$ the same as free vibration at natural frequency p of a stationary rotor or for a single degree of freedom system. The forced vibration amplitude, or steady state vibration is

$$\bar{R} = \frac{R}{a} = \frac{\Omega^2}{\sqrt{(1-\Omega^2)^2 + (2\xi\Omega)^2}} \tag{14.4}$$

at frequency ω as in a stationary system but this amplitude of whirl R increases with the square of the spin speed of the rotor ($\Omega = \omega/p$). The phase ϕ here is the angle between the line connecting mass and geometric centers of shaft and the response whirl OE.

$$\phi = \frac{2\xi\Omega}{1-\Omega^2} \tag{14.5}$$

The subtle understanding of the rotor begins with the appreciation that all rotors are unbalanced, i.e., there is always an eccentricity a either due to material inhomogeneity and/or manufacturing errors. This is the source of excitation which is always there unlike a stationary mass spring system where an external force is necessary to be applied. This aspect intrigued many as to why rotors whirl where there is no excitation. When disturbed a rotor spinning at ω responds in free and forced vibration, the free vibration whirl ν at natural frequency p dies due to damping and the forced vibration with a whirl of radius r at whirl frequency $\nu = \omega$ in case of synchronous whirl. Whereas we have both free and forced vibration responses in stationary and rotor systems, Jeffcott was able to explain the subtle differences that set apart the stationary systems vibration to rotor vibrations. That is why we call a single degree of freedom rotor as a Jeffcott rotor. The Jeffcott rotor results can be used in modal analysis multimass rotors in the same way stationary system modal analysis.

We may also note here that the critical speeds of rotors (same as stationary system natural frequencies) are determined from stationary systems, e.g., Prohl's method originally developed for rotor applications is same as Myklestad's method for aircraft wing structure applications. For critical speed determination of rotors, all we need is the natural frequency of stationary rotors. The response in case of rotors is essentially due to unbalance and therefore we call rotor forced vibration as unbalance response (some prefer to call imbalance response). The methods of determining unbalance response actually set rotor dynamics apart from stationary structure forced vibration.

14.3 Fluid Film Bearings

There are several physics phenomena that set rotors apart from stationary structures. One of the main differences is fluid film supports. More than a century ago, when we were beginning to understand rotor dynamics, it was generally believed that a lubricant in the bearing cavity will decrease friction and therefore minimize losses. It was accidentally discovered that the fluid film in the cavity is doing much more than just reducing friction losses.

(a)

(b)

(c)

(d)

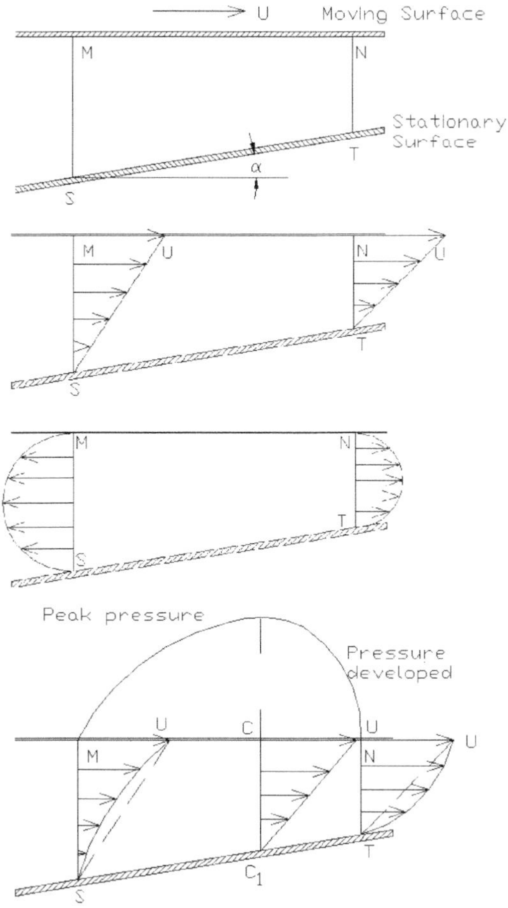

Fig. 14.4 Pressure development in a wedge

Beauchamp Tower [92] was testing for British Rail, to determine the best location for a lubricant inlet hole. He drilled several holes in the bearing and plugged them with just one kept open during the tests. He was unable to control the leakage despite all precautions and multiple pressure gauges; he discovered the familiar wedge action as explained subsequently by Osborne Reynolds. This discovery has completely changed the scenario in rotor dynamics, particularly after Jeffcott's work.

Osborne Reynolds [77] explained Tower's experiments; the flow through the wedge in Figure 14.4a is divided into two parts:

1. A velocity induced flow, with the oil film hugging the stationary surface and moving with the same velocity of the moving surface and linearly varying in between, as shown in Figure 14.4b.

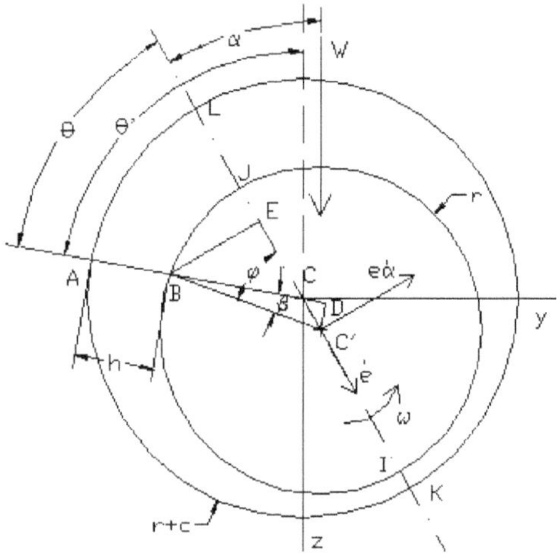

Fig. 14.5 A journal bearing

2. The moving surface cannot support any load in this flow; this flow by itself cannot exist alone as it does not satisfy the mass conservation, unless the film is compressible.
   On the other hand a pure pressure induced flow can squeeze the film out and support a vertical load. This flow is depicted in Figure 14.4c.

A combined velocity and pressure induced flow, shown in Figure 14.4d, explains how an incompressible oil film can satisfy mass conservation inside the cavity. At a section in between, there is no pressure induced flow and the pressure peaks here. The load carried by the moving plate is equal to the net area of the pressure curve.

This wedge action can be caused by several other aspects, e.g., elastic deformation in the bearing shell, temperature gradient in film, and so on.

A journal bearing geometry is shown in Figure 14.5.

The bearing center C and the journal center C' form the attitude of the bearing, which makes an angle α with the vertical load W. The maximum clearance is LJ and minimum clearance is IK and the clearance h varying between these two values forms a wedge, which plays the major role in bearing behavior.

In terms of radial clearance c and eccentricity ratio, $n = e/c$, the film thickness h at any angle is

$$h = c(1 + n\cos\theta) \tag{14.6}$$

The governing equation for pressure p at any angle θ, depends on the journal radius r, film thickness, h, viscosity, μ and the speed ω.

$$\frac{1}{r^2}\frac{\partial}{\partial\theta'}\left[h^3\frac{\partial p}{\partial\theta'}\right] + h^3\frac{\partial^2 p}{\partial x^2} = 6\mu\omega\frac{\partial h}{\partial\theta'} + 12\mu\frac{\partial h}{\partial t} \tag{14.7}$$

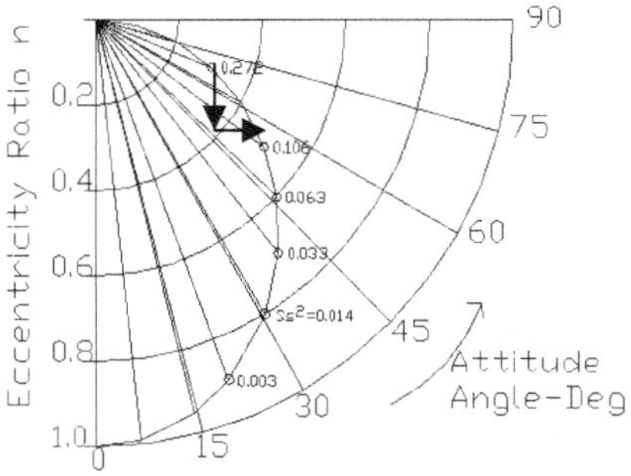

Fig. 14.6 Typical journal locus plot

For a given bearing geometry and speed, the pressure, eccentricity and attitude angle are related. The eccentricity ratio and attitude angle being non-dimensional, one can derive, a bearing non-dimensional load parameter (i.e., pressure), rather than deal with pressure units. Sommerfeld [81], see also [62], derived such a parameter and this parameter is called a Sommerfeld number S.

$$S = \frac{\mu DLN}{W}\left(\frac{r}{c}\right)^2 \tag{14.8}$$

The Sommerfeld number S for a bearing depends on viscosity μ, diameter D, length L, speed in rev/sec, N, radial load, W, radius r and clearance c. For a given bearing operating at a load and speed, we can calculate S. The position taken by the journal, i.e., the eccentricity ratio $n = e/c$ and attitude angle α are determined from Reynolds' steady state equation; they are given in graphical form to help the designers.

Present day numerical methods can simulate and obtain three-dimensional steady state bearing solutions very accurately; in the beginning, Sommerfeld considered a long bearing, $L \gg D$, approximation; later Ocvirk solved for short bearing, $L \ll D$. Ocvirk solutions [59] are closer to general bearings, and these results are shown in Figure 14.6.

The Sommerfeld solution brought out hydrodynamic bearing properties hitherto unknown. From the Ocvirk bearing solution for the short bearings given in Figure 14.6, one can quickly observe the special cross-coupled stiffness (so also damping if a vertical velocity is imposed) properties; when the journal is displaced downwards in an oil film bearing it resists like an isotropic spring but in addition also yields a lateral displacement to seek a steady state solution of equilibrium.

The bearing has four linear stiffness coefficients and similarly four damping coefficients and therefore, an oil film can be represented as an 8-coefficient bearing. These coefficients can be determined mathematically and using numerical methods.

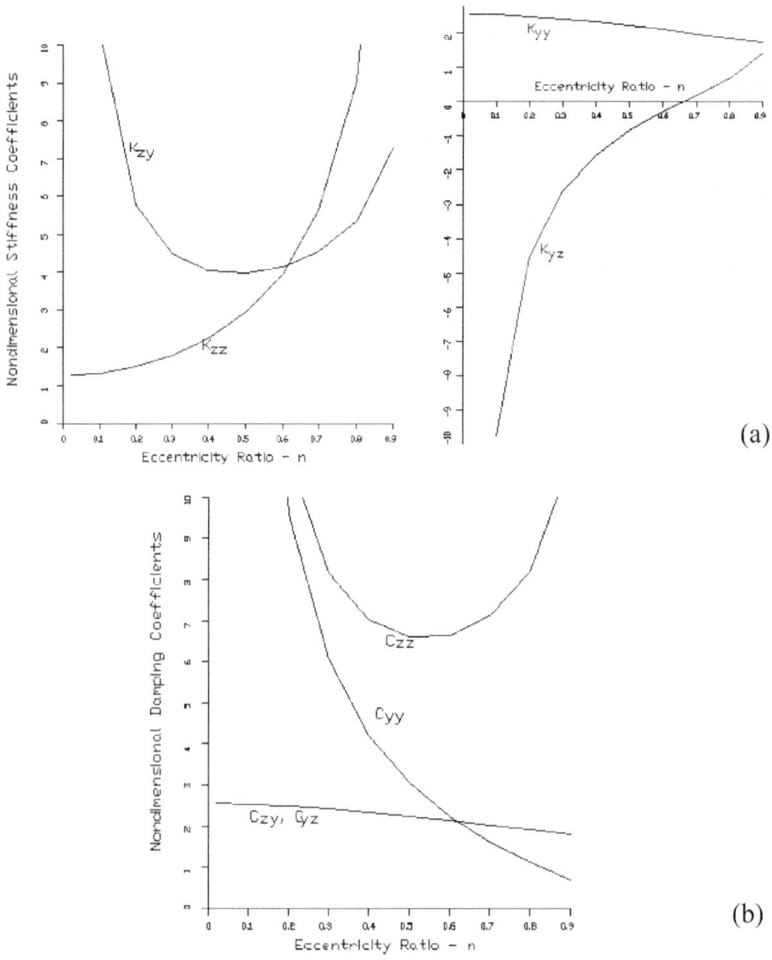

Fig. 14.7 Stiffness and damping coefficients of a short bearing

$$W_z = W_{z0} + W_{z,z}\Delta z + W_{z,y}\Delta y + W_{z,\dot{z}}\Delta\dot{z} + W_{z,\dot{y}}\Delta\dot{y}$$
$$W_y = W_{y0} + W_{y,z}\Delta z + W_{y,y}\Delta y + W_{y,\dot{z}}\Delta\dot{z} + W_{y,\dot{y}}\Delta\dot{y}$$
$$W_z = W_{z0} + K_{zz}\Delta z + K_{zy}\Delta y + C_{zz}\Delta\dot{z} + C_{zy}\Delta\dot{y}$$
$$W_y = W_{y0} + K_{yz}\Delta z + K_{yy}\Delta y + C_{yz}\Delta\dot{z} + C_{yy}\Delta\dot{y}$$
$$K_{zz} = \left(\frac{\partial W_z}{\partial z}\right)_{z=z_0, y=y_0} = (W_{z,z})_{z=z_0, y=y_0} \tag{14.9}$$

For a short bearing [59] the direct and cross-coupled stiffness and damping coefficients can be obtained and are given in Figures 14.7a and b.

The force relations on the bearing are expressed as

$$\begin{Bmatrix} F_z \\ F_y \end{Bmatrix} = \begin{bmatrix} C_{zz} & C_{zy} \\ C_{yz} & C_{yy} \end{bmatrix} \begin{Bmatrix} \dot{z} \\ \dot{y} \end{Bmatrix} + \begin{bmatrix} K_{zz} & K_{zy} \\ K_{yz} & K_{yy} \end{bmatrix} \begin{Bmatrix} z \\ y \end{Bmatrix} \tag{14.10}$$

The main difference between structural springs and oil film supports is in the cross-coupling properties. Notice K_{yz} is negative for certain eccentricities.

Jeffcott equations (14.2) for a rotor on oil film supports without damping in Figure 14.2 gets modified to

$$\begin{aligned} &M\frac{d^2}{dt^2}(z + a\cos\omega t) + K(z - z_0) = 0 \\ &M\frac{d^2}{dt^2}(y + a\sin\omega t) + K(y - y_0) = 0 \\ &K(z - z_0) = 2K_{zz}z_0 + 2K_{zy}y_0 \\ &K(y - y_0) = 2K_{yz}z_0 + 2K_{yy}y_0 \end{aligned} \tag{14.11}$$

Consider first the case of rolling element bearings without accounting for damping – remember the cross-coupled coefficients are zero for a rolling element bearing, the shaft stiffness and bearing stiffness can be considered in series in both vertical and horizontal directions, then, the rotor has two split natural frequencies or critical speeds given by

$$\begin{aligned} K_z &= \frac{2K_{zz}K}{2K_{zz} + K} \\ K_y &= \frac{2K_{yy}K}{2K_{yy} + K} \\ p_1 &= \sqrt{\frac{2K_{zz}K}{(2K_{zz} + K)m}} \\ p_2 &= \sqrt{\frac{2K_{yy}K}{(2K_{yy} + K)m}} \end{aligned} \tag{14.12}$$

The natural frequencies of equation (14.11) can be obtained as

$$p_{1,2}^2 = \frac{1}{2}(\omega_1^2 + \omega_2^2) \pm \sqrt{(\omega_1^2 - \omega_2^2)\mu_1\mu_2\omega_1^2\omega_2^2} \tag{14.13}$$

where

$$\omega_1^2 = \frac{K_1}{M}; \quad \omega_2^2 = \frac{K_2}{M}$$

$$\mu_1 = \frac{K_{12}}{K_1}; \quad \mu_2 = \frac{K_{21}}{K_2}$$

and

$$K_1 = \frac{K[2K_{zz}(2K_{yy} + K) - 4K_{zy}K_{yz}]}{(2K_{zz} + K)(2K_{yy} + K) - 4K_{zy}K_{yz}}$$

$$K_2 = \frac{K[2K_{yy}(2K_{zz} + K) - 4K_{zy}K_{yz}]}{(2K_{zz} + K)(2K_{yy} + K) - 4K_{zy}K_{yz}}$$

$$K_{12} = \frac{2K_{zy}K^2}{(2K_{zz} + K)(2K_{yy} + K) - 4K_{zy}K_{yz}}$$

$$K_{21} = \frac{2K_{yz}K^2}{(2K_{zz} + K)(2K_{yy} + K) - 4K_{zy}K_{yz}}$$

We note that the cross-coupled stiffness K_{yz} can be negative; in such cases, K_{21} can be negative (μ_2) and it is possible that the terms inside the square root of (14.13) together can be negative, creating a problem of non-existence of the critical speeds. (This also indicates instability.) There have been several cases reported of such cases, where a critical speed disappears even when there is no damping considered. This is one of the peculiar behaviors characteristic of rotors which one will not come across in structures [53].

Since the rotor response is expressed as whirl, we will find it convenient to get the solution in the form of forward or backward whirl. The whirl radius from an eccentricity a of the rotor can be obtained from a forced vibration solution which has both cosine and sine terms. These trigonometric terms are expressed in exponential form.

$$\cos \omega t = \frac{1}{2}(e^{i\omega t} + e^{-i\omega t})$$

$$\sin \omega t = -\frac{i}{2}(e^{i\omega t} - e^{-i\omega t})$$

Then the non-dimensional response is determined in terms of forward whirl r^+ and backward whirl r^-.

$$\bar{r} = r^+ e^{i\omega t} + r^- e^{-i\omega t}$$

where

$$r^+ = \frac{1}{2}\omega^2 \left[\frac{(\omega_1^2 + \omega_2^2 - 2\omega^2) - i(\mu_2\omega_2^2 - \mu_1\omega_1^2)}{(\omega_1^2 - \omega^2)(\omega_2^2 - \omega^2) - \mu_1\mu_2\omega_1^2\omega_2^2} \right]$$

$$r^- = -\frac{1}{2}\omega^2 \left[\frac{(\omega_1^2 - \omega_2^2) + i(\mu_2\omega_2^2 + \mu_1\omega_1^2)}{(\omega_1^2 - \omega^2)(\omega_2^2 - \omega^2) - \mu_1\mu_2\omega_1^2\omega_2^2} \right]$$

Starting from time $t = 0$, we can add both the forward and backward whirl components to give the total response. Unlike in structures, we get the sum of these components giving rise to an elliptic whirl, because of the split frequencies and

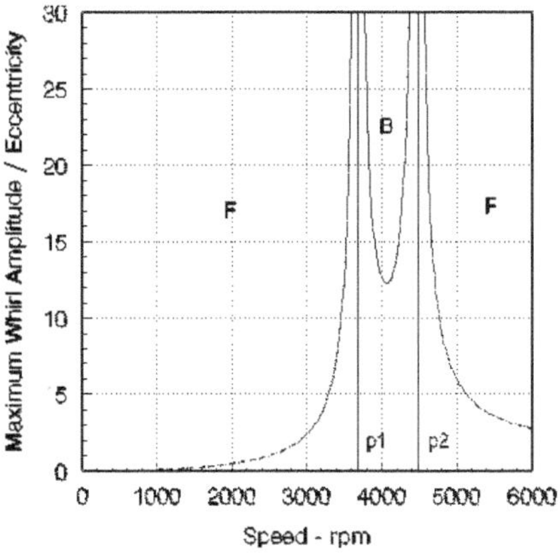

Fig. 14.8 Gunter rotor response

difference in stiffnesses in two perpendicular directions. If the forward whirl component is more than that of backward whirl, the net whirl is forward in the same direction of spin; otherwise, we get a backward whirl, with whirl in a direction opposite to spin, typical of rotors on bearing supports [45, 65].

Gunter [31] considered the following rotor: $M = 54.432$ kg; $K = 1.387 \times 10^7$ N/m; rigid bearing critical speed is 4820 RPM; $K_{zz} = 4.16 \times 10^7$ N/m; $K_{yy} = 1.01 \times 10^7$ N/m; $K_{zy} = 3.12 \times 10^7$ N/m; $K_{yz} = 4.16 \times 10^5$ N/m.

There are two distinct critical speeds obtained p_1 and p_2 given in Figure 14.8 [31]. Between the two criticals, we have backward whirl. The whirl orbits are shown in Figure 14.9. This backward whirl has been seen clearly in laboratory tests.

Morton's compressor rotor is modeled as a shaft of dia 19.05 cm with total mass $M = 453.6$ kg. The bearings are plain cylindrical type 10.16 cm dia, 5.08 cm long with 0.01016 cm diametral clearance. The lubricant viscosity at operating temperature is 0.00568 Nsec/m^2. Rigid bearing critical speed is 8600 RPM. This gives $K = 3.68 \times 10^8$ N/m. The following stiffnesses are obtained for this bearing: $K_1 = 1.839 \times 10^8$ N/m; $K_2 = 2.004 \times 10^8$ N/m; $K_{12} = 1.340 \times 10^8$ N/m; $K_{21} = -1.026 \times 10^8$ N/m. Notice that K_{21} is negative.

The unbalance response obtained is shown in Figure 14.10.

This rotor with a negative cross-coupled stiffness does not show two distinct critical speeds even in the absence of any damping. Initially it was thought that the two critical speeds are so close that they coalesced. Incidentally, the fluid film bearings are physical systems with a negative stiffness, thus showing that negative stiffness need not be simply a concept, but reality. The rotor in the present case does not exhibit any backward whirl.

Subbaiah et al. [86] designed an experimental rig to observe the backward whirl between two critical speeds. The 9.07 kg rotor is mounted centrally on a shaft with stiffness 884000 N/m and mounted on 2.54 cm dia and 2.54 cm long plain cylindrical bearings with 0.0188 cm clearance with a 24 cp lubricant. The backward

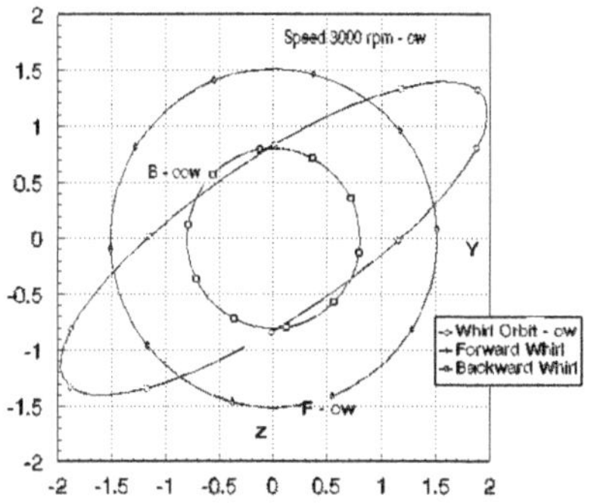

(a) Forward Whirl Orbit below I Critical

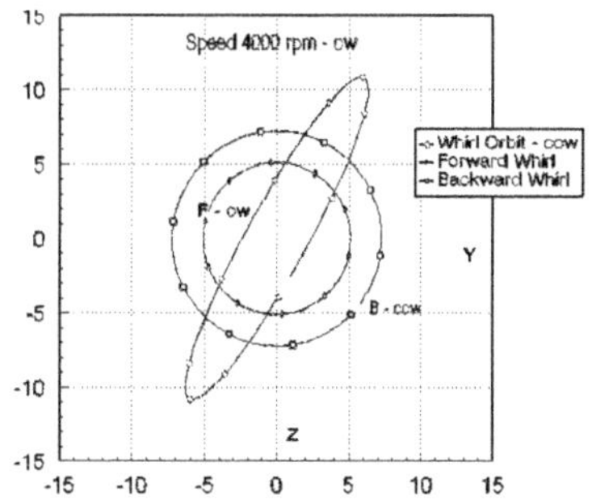

(b) Backward Whirl Orbit between Criticals

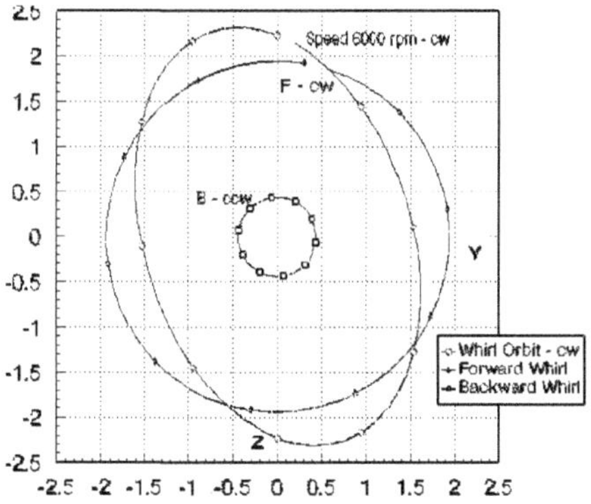

(c) Forward Whirl beyond II Critical

Fig. 14.9 Whirl orbits of Gunter's rotor

whirl between two critical speeds was observed in the laboratory as shown in Figure 14.11.

Because the bearings play a significant role, optimization is proposed in the design by Bhat et al. [3] by using the Vanderplaats [94] method.

14.4 Oil Film Instabilities

The instability of a rotor is a self excited vibration arising out of fluid film forces and is distinct from large amplitudes of whirl caused by residual unbalance. This phenomenon is also known as oil whirl and oil whip and was first observed by Newkirk and Taylor [56]. Robertson [78], Pinkus and Sternlicht [62], Morrison and Peterson [52] discussed this phenomenon and laid down some simple design rules. Lund [45, 46] presented a complete methodology for oil film bearing design and stability; see also [66, 67].

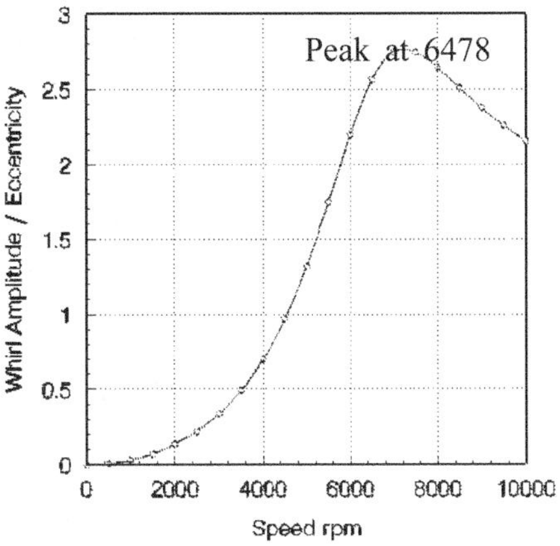

Fig. 14.10 Unbalance response of Morton's compressor rotor with negative cross-coupled stiffness of the plain cylindrical bearings

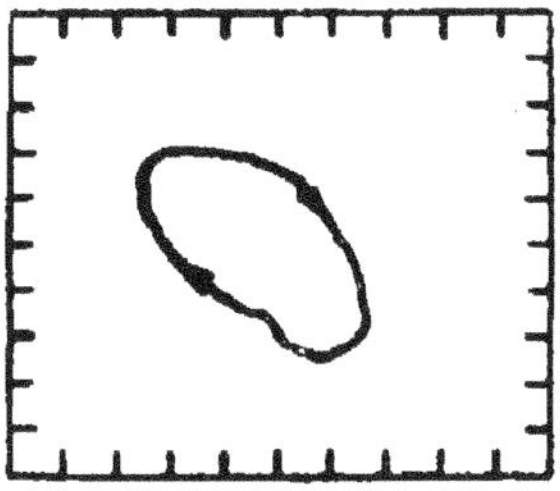

Fig. 14.11 Clockwise whirl of a rotor spinning counterclockwise at 2500 RPM

Newkirk and Taylor [56] found that oil film bearings were the cause of large whirls and instability, called oil whip. We have seen earlier that the load carrying film has two components of flow, velocity induced and pressure induced. The pressure induced flow is essential for the load carrying ability and if this component is small, the bearing becomes lightly loaded and if it disappears, the bearing becomes inactive.

For incompressible flow, velocity induced flow alone causes excess fluid coming into the wedge; this makes the journal lift from the attitude position to provide more space for the excess oil. A continuous lift due to excess oil coming in makes the journal center whirl around the bearing center and such a whirl is called oil whirl. A flow balance gives the oil whirl frequency ν to be exactly half the rotational speed ω.

Figure 14.12 shows the bearing operating only with velocity induced flow. The flow balance equation is

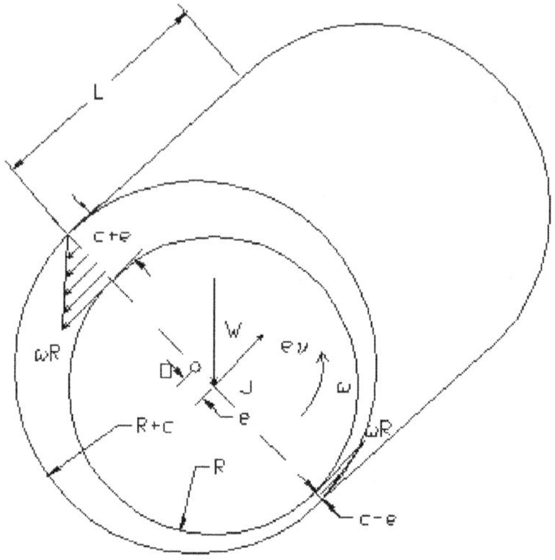

Fig. 14.12 Lightly loaded bearing

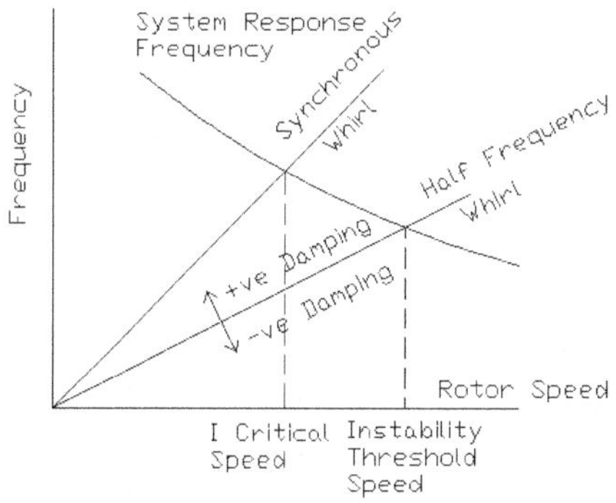

Fig. 14.13 Campbell diagram with half frequency whirl

$$\text{Flow in } F_i \quad = \text{Flow out } F_o \quad + \text{Whirl Flow } F_w$$

$$\frac{1}{2}LR\omega(C+e) = \frac{1}{2}LR\omega(C-e) + 2LR(e\nu)$$

$$\therefore \text{Oil Whirl Frequency} \quad \nu = \frac{1}{2}\omega \tag{14.14}$$

We can explain the instability from the Campbell diagram given in Figure 14.13. The synchronous line is 1 per rev line, giving conventional unbalance excitation and definition of critical speed.

The half frequency whirl or $1/2\times$ line is also drawn. At any rotational speed, when we disturb the rotor, it whirls at a corresponding natural frequency. After exceeding the critical speed, under transient conditions, the rotor gets excited at the

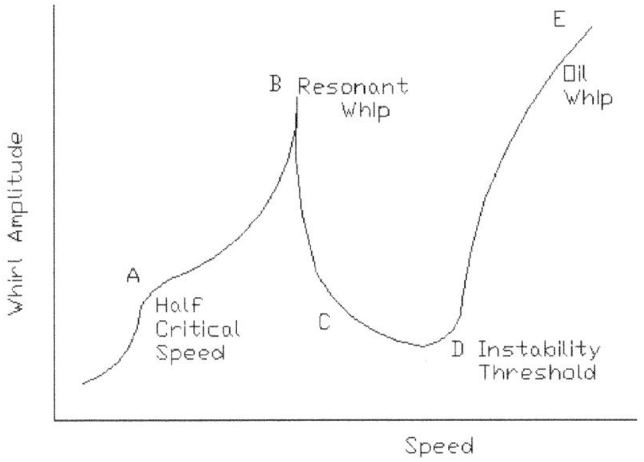

Fig. 14.14 Whirl response

natural frequency which is more than half rotational speed, until the point where the half frequency line and natural frequency intersect. At this speed, the rotor whirl is half the rotational speed and is called Instability Threshold Speed. Before this speed, the transient motion dies, becomes damped and beyond this the whirl becomes uncontrollable and we say the system is negatively damped, feeding energy into the system. The whirls become uncontrollable as we increase the speed further.

Typical whirl response in a coast-up test is shown in Figure 14.14. The oil film in general is nonlinear and therefore has higher harmonic response at about half the critical speed, point A, this is a minor peak observed if you slowly go through this speed.

The conventional unbalance response gives resonance at the critical speed and the rotor gets whipped and this whip is called Resonant Whip, point B. After the resonant whip, the rotor becomes silent from C and should continue to be so under normal conditions with conventional support springs. At D, the instability begins due to oil whirl, the bearing loses its load-carrying capability and the whirl begins to increase. E denotes a condition of large whirls, called Oil Whip, which is only due to oil film. This frequency of whirl is $< 0.5\omega$. Under unstable conditions, the whirl frequency is invariably around 0.47ω. The presence of such a frequency in the vibration spectrum is due to oil whip or due to bearing looseness, etc., causing oil whip conditions.

Rao et al. [71] performed a laboratory experiment to observe the oil whip phenomenon; the rig was designed with analog instrumentation designed for this as shown in Figure 14.15 and the whirl response amplitude identified the points noted in Figure 14.14.

$$M\ddot{z} + K_1 z + K_{12} y = 0$$

$$M\ddot{y} + K_2 y + K_{21} z = 0 \tag{14.15}$$

Fig. 14.15 Laboratory experiment on oil whip

Let us now write the governing equations to determine the complex eigen-values and the whirl response of a Jeffcott rotor with negative cross-coupled stiffness [68].

The complex eigen-values $z = e^{\lambda t}$ of this equation are obtained as follows:

$$\begin{vmatrix} K_1 + \lambda^2 M & K_{12} \\ K_{21} & K_2 + \lambda^2 M \end{vmatrix} = 0$$

$$\lambda^4 + \lambda^2 \frac{K_1 + K_2}{M} + \frac{K_1 K_2 - K_{12} K_{21}}{M^2} = 0 \tag{14.16}$$

If the real part of the root is positive, the rotor whirl amplitudes upon transient disturbance increase and instability occurs. If the real part is negative, the transient disturbance dies and the system remains stable. For Morton's rotor the unbalance response had not shown the critical speeds, and its eigen-values are calculated as

$$\lambda^4 + 0.847 \times 10^6 \lambda^2 + 0.246 \times 10^{12} = 0$$

$$\lambda_{1,2}^2 = \frac{1}{2}\left[-0.847 \times 10^6 \pm \sqrt{0.718 \times 10^{12} - 0.984 \times 10^{12}}\right]$$

$$= -0.424 \times 10^6 \pm 0.258 \times 10^6 i$$

$$\lambda_{1,2} = 190.166 \pm 678.353i \tag{14.17}$$

The imaginary part of the eigen-value gives the natural frequency; here it is 678.353 rad/s, i.e., 6478 RPM, where the peak occurs, however, the real part here is positive and therefore the rotor is susceptible to instability, unless there is some other source of damping to suppress this instability. The transient response is

$$z = e^{190.166t}(Ae^{i678.353t} + Be^{-i678.353t}) \tag{14.18}$$

The peak occurs at 6478 RPM, since the real part of the solution is unstable. Though the unbalance response determined ignoring the transient response shows a low amplitude whirl in these conditions, it is the transient response that should be determined.

Simple design methods, Lund [45] and Rao [66, 67] are devised to find threshold instability speed of a rigid rotor and flexibility rotor due to oil whirl.

$$M\ddot{z} + C_{zz}\dot{z} + C_{zy}\dot{y} + K_{zz}z + K_{zy}y = 0$$
$$M\ddot{z} + C_{yy}\dot{y} + C_{yz}\dot{z} + K_{yy}y + K_{yz}z = 0 \tag{14.19}$$

The threshold condition is determined by when the real part of the eigen-values is zero, therefore we can write

$$z = e^{i\nu t}$$
$$y = e^{i\nu t} \tag{14.20}$$

Hence

$$\begin{vmatrix} K_{zz} - \nu^2 M + i\nu C_{zz} & K_{zy} + i\nu C_{zy} \\ K_{yz} + i\nu C_{yz} & K_{yy} - \nu^2 M + i\nu C_{yy} \end{vmatrix} = 0 \tag{14.21}$$

Cross-multiply and separate the real and imaginary parts to give

$$\frac{(K_{zz} - \nu^2 M)(K_{yy} - \nu^2 M) - K_{yz}K_{zy}}{C_{zz}C_{yy} - C_{zy}C_{yz}} = \nu^2$$
$$\frac{(C_{zz}K_{yy} + C_{yy}K_{zz}) - C_{yz}K_{zy} - C_{zy}K_{yz}}{C_{zz} + C_{yy}} = M\nu^2 \tag{14.22}$$

We can non-dimensionalize and write

$$\frac{(\bar{K}_{zz} - \frac{M\nu^2 C}{W})(\bar{K}_{yy} - \frac{M\nu^2 C}{W}) - \bar{K}_{yz}\bar{K}_{zy}}{\bar{C}_{zz}\bar{C}_{yy} - \bar{C}_{zy}\bar{C}_{yz}} = \frac{\nu^2}{\omega^2}$$
$$\frac{(\bar{C}_{zz}\bar{K}_{yy} + \bar{C}_{yy}\bar{K}_{zz}) - \bar{C}_{yz}\bar{K}_{zy} - \bar{C}_{zy}\bar{K}_{yz}}{\bar{C}_{zz} + \bar{C}_{yy}} = \frac{M\nu^2 C}{W} = \text{constant } K \text{ for a given } S$$
$$\bar{K}_{zz} = \frac{K_{zz}C}{W} \cdots \bar{C}_{zz} = \frac{C_{zz}C\omega}{W} \cdots \tag{14.23}$$

Using (14.8) the first equation in (14.23) is now written as

$$\frac{(\bar{K}_{zz} - K)(\bar{K}_{yy} - K) - \bar{K}_{yz}\bar{K}_{zy}}{\bar{C}_{zz}\bar{C}_{yy} - \bar{C}_{zy}\bar{C}_{yz}} = \frac{KW}{MC\omega^2}$$
$$= \frac{K\mu^2 D^2 L^2}{MC4\pi^2 W S^2}\left(\frac{R}{C}\right)^4 \tag{14.24}$$

Defining the rigid rotor stability threshold parameter χ, we have the criterion to determine the threshold parameter $S_{\text{threshold}}$

$$\chi = \frac{\sqrt{CMW}}{\mu DL(\frac{R}{C})^2} = \frac{1}{2\pi S_{\text{threshold}}}\sqrt{\frac{K(\bar{C}_{zz}\bar{C}_{yy} - \bar{C}_{zy}\bar{C}_{yz})}{(\bar{K}_{zz} - K)(\bar{K}_{yy} - K) - \bar{K}_{yz}\bar{K}_{zy}}} \tag{14.25}$$

Lund [45] gave a procedure for flexible rotors by devising an equivalent two-disk Jeffcott rotor and Rao [66, 67] used this procedure to illustrate the methodology. We have now methods for analysis of distributed rotors for finding the transient response and instability analysis which will be briefly discussed later.

14.5 Quality Factor

It is clear that hydrodynamic bearings play a significant role in the dynamic behavior of rotors. The bearings are eight coefficient bearings with linear properties of stiffness and damping coefficients. We have seen that the cross coupled stiffness terms can dampen the rotors and limit the amplitudes of whirl in unbalance response. The direct and cross-coupled dampings are also present in all hydrodynamic bearings and naturally the designer would like to have a quick estimate on what kind of magnification these terms induce on the rotor at resonance. Rao [70] addressed this issue and his analysis is briefly presented here.

The rotor is first considered on rigid bearings and a Jeffcott rotor model in Figure 14.2 is first deduced. The bearings are then accounted for eight coefficient bearings whose stiffness and damping properties are evaluated. Then we have the equations of motion

$$\begin{aligned} M\ddot{z} + K_1{}^*z + K_{12}{}^*y &= Ma\omega^2\cos\omega t \\ M\ddot{y} + K_2{}^*y + K_{21}{}^*z &= Ma\omega^2\sin\omega t \end{aligned} \tag{14.26}$$

where

$$K_1{}^* = \frac{K^*[2K_{zz}{}^*(2K_{yy}{}^* + K^*) - 4K_{zy}{}^*K_{yz}{}^*]}{(2K_{zz}{}^* + K^*)(2K_{yy}{}^* + K^*) - 4K_{zy}{}^*K_{yz}{}^*}$$

$$K_2{}^* = \frac{K^*[2K_{yy}{}^*(2K_{zz}{}^* + K^*) - 4K_{zy}{}^*K_{yz}{}^*]}{(2K_{zz}{}^* + K^*)(2K_{yy}{}^* + K^*) - 4K_{zy}{}^*K_{yz}{}^*}$$

$$K_{12}{}^{*} = \frac{2K_{zy}{}^{*}K^{*2}}{(2K_{zz}{}^{*} + K^{*})(2K_{yy}{}^{*} + K^{*}) - 4K_{zy}{}^{*}K_{yz}{}^{*}}$$

$$K_{21}{}^{*} = \frac{2K_{yz}{}^{*}K^{*2}}{(2K_{zz}{}^{*} + K^{*})(2K_{yy}{}^{*} + K^{*}) - 4K_{zy}{}^{*}K_{yz}{}^{*}}$$

$$K^{*} = K + i\omega C; \quad K_{zz}{}^{*} = K_{zz} + i\omega C_{zz}$$

$$\ldots \tag{14.27}$$

The unbalance response can be obtained as

$$R = \frac{1}{2}\omega^2 \frac{\{(\omega_1{}^{*2} + \omega_2{}^{*2} - 2\omega^2) - i(\mu_2{}^{*}\omega_2{}^{*2} - \mu_1{}^{*}\omega_1{}^{*2})\}}{(\omega_1{}^{*2} - \omega^2)(\omega_2{}^{*2} - \omega^2) - \mu_1{}^{*}\omega_1{}^{*2}\mu_2{}^{*}\omega_2{}^{*2}} e^{i\omega t}$$
$$- \frac{1}{2}\omega^2 \frac{\{(\omega_1{}^{*2} - \omega_2{}^{*2}) + i(\mu_2{}^{*}\omega_2{}^{*2} + \mu_1{}^{*}\omega_1{}^{*2})\}}{(\omega_1{}^{*2} - \omega^2)(\omega_2{}^{*2} - \omega^2) - \mu_1{}^{*}\omega_1{}^{*2}\mu_2{}^{*}\omega_2{}^{*2}} e^{-i\omega t} \tag{14.28}$$

where

$$\omega_1{}^{*2} = \frac{K_1{}^{*}}{M}, \quad \omega_2{}^{*2} = \frac{K_2{}^{*}}{M}$$
$$\mu_1{}^{*} = \frac{K_{12}{}^{*}}{K_1{}^{*}}, \quad \mu_2{}^{*} = \frac{K_{21}{}^{*}}{K_2{}^{*}} \tag{14.29}$$

The unbalance response given by equation (14.28) can be determined as a function of speed and where the peak occurs; we can obtain the quality factor. To get an idea of the quality factor, Rao [70] assumed the cross-coupled stiffness and damping influence can be neglected and retained the direct stiffness and damping terms. In that case, the response can be simplified. Then equations in (14.26) get decoupled:

$$M\ddot{z} + \bar{C}_z\dot{z} + \bar{K}_1 z = Ma\omega^2 \cos\omega t$$
$$M\ddot{y} + \bar{C}_y\dot{y} + \bar{K}_2 y = Ma\omega^2 \sin\omega t \tag{14.30}$$

where

$$\bar{C}_z = \frac{2(2CK_{zz}{}^2 + 2C\omega^2C_{zz}{}^2 + \omega^2C^2C_{zz} + K^2C_{zz})}{(2K_{zz} + K)^2 + \omega^2(2C_{zz} + C)^2}$$

$$\bar{C}_y = \frac{2(2CK_{yy}{}^2 + 2C\omega^2C_{yy}{}^2 + \omega^2C^2C_{yy} + K^2C_{yy})}{(2K_{yy} + K)^2 + \omega^2(2C_{yy} + C)^2}$$

$$\bar{K}_z = \frac{2(KK_{zz} - \omega^2CC_{zz})(2K_{zz} + K) + 2\omega^2(CK_{zz} + KC_{zz})(2C_{zz} + C)}{(2K_{zz} + K)^2 + \omega^2(2C_{zz} + C)^2}$$

$$K_y = \frac{2(KK_{yy} - \omega^2CC_{yy})(2K_{yy} + K) + 2\omega^2(CK_{yy} + KC_{yy})(2C_{yy} + C)}{(2K_{yy} + K)^2 + \omega^2(2C_{yy} + C)^2} \tag{14.31}$$

For (14.30), the damping ratios in z and y directions are

$$\bar{\xi}_z = \frac{\bar{C}_z}{2\sqrt{\bar{K}_z M}}$$

$$\bar{\xi}_y = \frac{\bar{C}_y}{2\sqrt{\bar{K}_y M}} \tag{14.32}$$

If the shaft damping is assumed to be negligible compared to oil film damping, (14.31) becomes simplified as

$$\tilde{C}_z = \frac{2K^2 C_{zz}}{(2K_{zz}+K)^2 + \omega^2(2C_{zz})^2}$$

$$\tilde{C}_y = \frac{2K^2 C_{yy}}{(2K_{yy}+K)^2 + \omega^2(2C_{yy})^2}$$

$$\tilde{K}_z = \frac{K[2K_{zz}(2K_{zz}+K)+\omega^2(2C_{zz})^2]}{(2K_{zz}+K)^2+\omega^2(2C_{zz})^2}$$

$$\tilde{K}_y = \frac{K[2K_{yy}(2K_{yy}+K)+\omega^2(2C_{yy})^2]}{(2K_{yy}+K)^2+\omega^2(2C_{yy})^2} \tag{14.31a}$$

The corresponding damping ratios are

$$\tilde{\xi}_z = \frac{\hat{C}_z}{2\sqrt{\hat{K}_z M}}, \quad \tilde{\xi}_y = \frac{\hat{C}_y}{2\sqrt{\hat{K}_y M}} \tag{14.32a}$$

Case 1: Lightly Damped Rotor

Consider Gunter's rotor in Section 14.3. From (14.31a) $\tilde{C}_z = 1955.2$ Ns/m; $\tilde{K}_z = 1.25 \times 10^7$ N/m; $\tilde{C}_y = 637.9$ Ns/m; $\tilde{K}_y = 0.824 \times 10^7$ N/m.

In z direction: $p_z = 479.212$ rad/s or 4576.14 RPM; $C_{cz} = 52000$ Ns/m, $\xi_z = 0.0375$; $Q_z = 13.3333$.

It may be noted that the quality factor is at undamped resonance and the peak value may be away from resonance condition.

The peak value occurs at

$$r = \frac{1}{\sqrt{1-2\xi^2}}$$

The response of the uncoupled equation is

$$\frac{Z}{a} = \frac{r^2}{\sqrt{(1-r^2)^2 + (2\xi_z r)^2}}$$

The amplification factor at peak response is therefore

$$A_z = \frac{1}{2\xi_z\sqrt{1 - 2\xi_z{}^2}} = 13.3427$$

at 4582.55 RPM.

In y direction: $p_y = 339.078$ rad/s or 3715.42 RPM; $C_{cy} = 42355$ Ns/m, $\xi_y = 0.015$; $Q_z = 33.3333$.

The amplification factor at peak response is therefore $A_y = 33.337$ at 3716.257 RPM.

Case 2: Heavily Damped Rotor
Consider Morton's rotor in Section 14.3: $\tilde{C}_z = 135000$ Ns/m; $\tilde{K}_z = 3.13 \times 10^8$ N/m; $\tilde{C}_y = 133000$ Ns/m; $\tilde{K}_y = 2.86 \times 10^8$ N/m.

In z direction: $p_z = 830.6836$ rad/s or 7932.444 RPM; $C_{cz} = 753520$ Ns/m, $\xi_z = 0.179$; $Q_z = 2.839$.

The peak amplitude occurs at 8199.5 RPM, $A_z = 2.839$.

In y direction: $p_y = 794.1$ rad/s or 7583 RPM; $C_{cy} = 720408$ Ns/m, $\xi_y = 0.185$; $Q_z = 2.7$.

The peak amplitude occurs at 7856.24 RPM, $A_y = 2.750$.

14.6 Gyroscopic Effects

Rayleigh [76] identified the effect of rotary inertia of a disk mounted on a stationary shaft. Stodola [82] considered this effect of rotary inertia and identified it as a gyroscopic effect that gives rise to split natural frequencies and backward whirl between them. Green [28] extended these gyroscopic studies of the critical speeds of flexible rotors. Den Hartog [11] and Timoshenko [90] discussed gyroscopic effects on synchronous and non-synchronous whirls. Carnegie [6] used energy methods to determine gyroscopic effects.

Consider a lumped mass model as shown in Figure 14.16. We find in the II Mode there is no Disk motion and if the shaft is massless, there is no kinetic energy. Therefore we cannot predict the II Mode. If the mass is a disk with transverse inertia, the picture is altogether different as in Figure 14.17 since in the II Mode, there is kinetic energy due to rotation. The disk undergoes a rotation given by an angle equal to the slope $(\partial v/\partial x)$ – the derivative of that gives angular velocity and therefore, the expression for kinetic energy in rotation is

$$\frac{1}{2} I_T \left\{ \frac{d}{dt} \left(\frac{\partial v}{\partial x} \right) \right\}^2$$

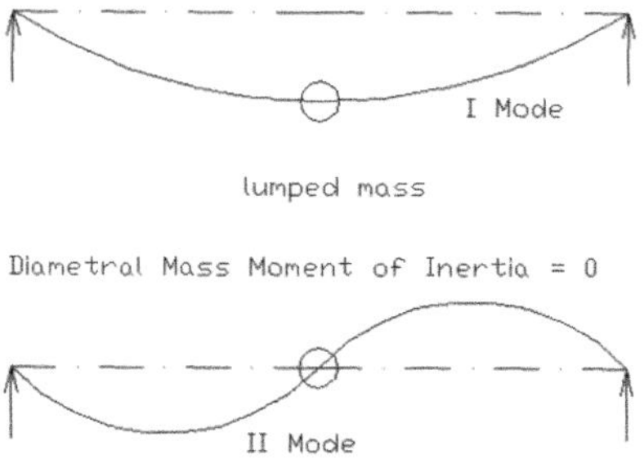

Fig. 14.16 Lumped mass model

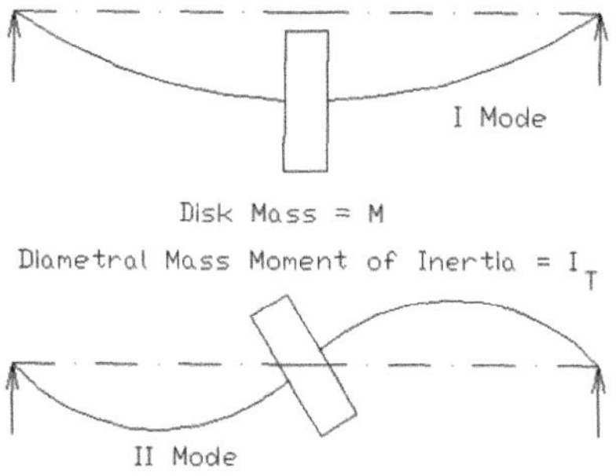

Fig. 14.17 Disk with transverse inertia I_T

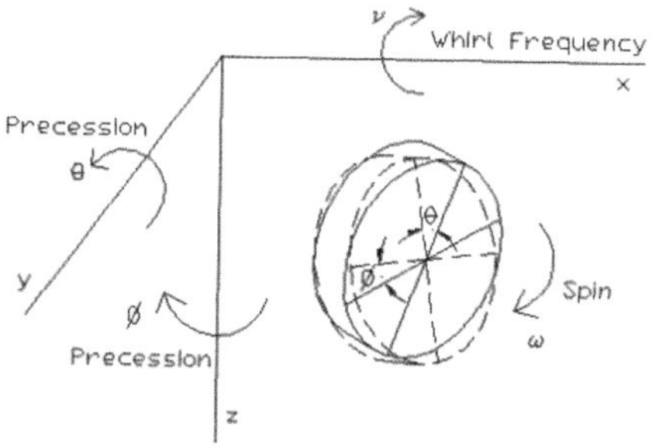

Fig. 14.18 A freely spinning disk

For a stationary beam, this was called by Rayleigh as rotary inertia. When the shaft rotates, the disk is just like a spin top and gyroscope. This has significant effect in rotor dynamics.

A spinning disk in three dimensions is shown in Figure 14.18, whirling about the x axis. Because of whirling, precessional motions are introduced corresponding to the shaft slopes at the disk center. These precessional motions make the disk's motion similar to a spin top and they introduce the gyroscopic effect.

(a)

(b)

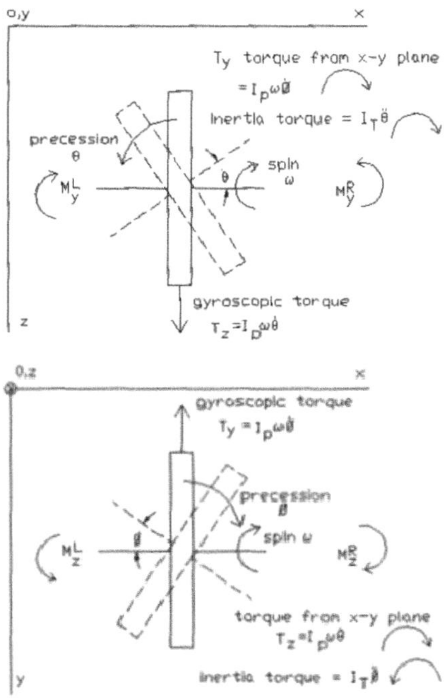

Fig. 14.19 (a) Precession in xz plane. (b) Precession in yz plane

A disk spinning and precessing gives a gyroscopic couple; it is given as the product of polar mass moment of inertia, spin angular velocity and precessional angular velocity and the resulting couple is in the direction following the right-hand cork screw rule. The two precessional motions give rise to gyroscopic couples in the two orthogonal planes of bending – these are depicted in Figure 14.19.

The inertia and gyroscopic torques make the bending moment jump across the disk element as shown above. We can write these moment relations across the disk I in xz and yz planes:

$$M_{yi}^R = M_{yi}^L + I_p\omega\dot{\phi}_i + I_T\ddot{\theta}_i$$

$$M_{zi}^R = M_{zi}^L - I_p\omega\dot{\theta}_i + I_T\ddot{\phi}_i \qquad (14.33)$$

We can use a Myklestad–Thomson type of formulation in transfer form to derive the governing equations. Consider the case of a mass less shaft with an overhung disk in Figure 14.20. The two bearings are taken very close to each other, so that we have a cantilevered rotor, The shaft is spinning at an angular velocity ω and simultaneously whirling at the same angular velocity in the same direction.

Synchronous whirl is similar to the moon spinning about its axis and going around the earth. In the first position, the shaft center is at S_1 and the outer most point in line with OS is at P_1. The shaft whirls about the bearing centerline O to

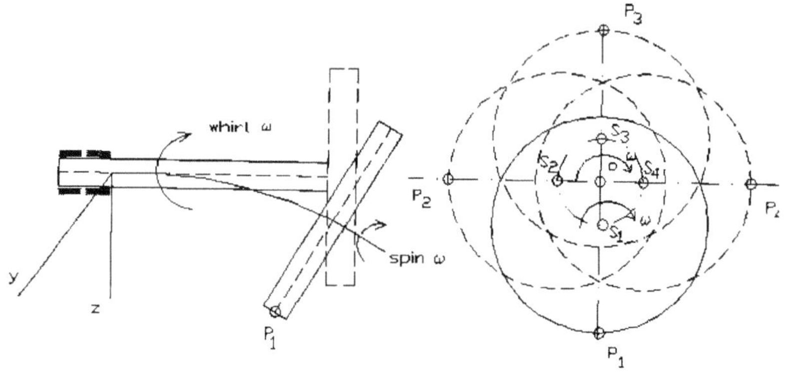

Fig. 14.20 Overhung shaft with synchronous whirl

a position S_2 from S_1 in a clockwise direction, while it has rotated simultaneously with the outermost point P occupying the position P_2.

This process repeats every complete rotation and we will always find point P remaining at the outer most position. If one observes the disk which is spinning at angular velocity ω while sitting at O and whirling at angular velocity ω in the same direction, point P remains at the same location.

For the overhung rotor, one can derive a relation which gives the natural frequency. Here m is disk mass, EI is flexural rigidity, l is the length and I_T is the transverse mass moment of inertia of the disk.

$$p^4 I_T + p^2 \frac{12EI}{ml^3}\left(\frac{1}{3}ml^2 - I_T\right) - \frac{12E^2I^2}{ml^4} = 0 \tag{14.34}$$

With lumped mass model, the above gives

$$p^2 \frac{12EI}{ml^3}\left(\frac{1}{3}ml^2\right) - \frac{12E^2I^2}{ml^4} = 0$$

$$p_{\text{lumped}} = \sqrt{\frac{3EI}{ml^3}} \text{ rad/s} \tag{14.35}$$

With infinite transverse inertia, equation (14.34) reduces to

$$I_T \to \infty, \quad p^4 + p^2 \frac{12EI}{ml^3} = 0$$

$$p_{\text{disk}} = \sqrt{\frac{12EI}{ml^3}} \text{ rad/s} \tag{14.36}$$

In general, the frequency defined by non-dimensional frequency parameter $\lambda = p\sqrt{ml^3/EI}$ is obtained as a function of disk parameter $\delta = I_T/ml^2$

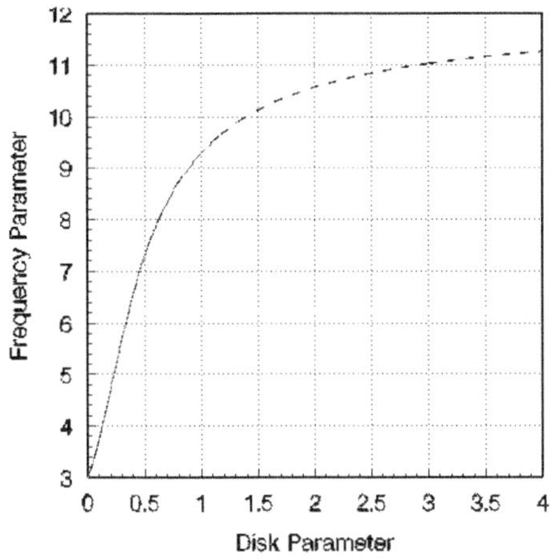

Fig. 14.21 Synchronous whirl of a cantilevered disk

$$\lambda_{1,2}^2 = \left(6 - \frac{2}{\delta}\right) \pm \sqrt{\left(6 - \frac{2}{\delta}\right)^2 + \frac{12}{\delta}} \tag{14.37}$$

The above relation is plotted in Figure 14.21.

We deduce from here that in a whirl that is synchronous and in the same direction (forward synchronous whirl) the natural frequency increases because of disk effect from lumped model to disk with infinite radius having the same mass. This is a significant effect and must be considered in all rotor dynamic analyses.

We now notice a significant fact – when a rotor is spinning and disturbed from its steady condition, the response depends on how the disk is distributed. This response is a whirl condition whose frequency will be the same as the natural frequency shown here.

In reality we must consider the case of non-synchronous whirl, i.e., while the rotor spins at angular velocity ω the whirling takes place at an angular velocity ν either in forward direction to spin or backward direction to spin. Without going through derivations here, a relation for the natural frequencies, whirling frequency parameter $\lambda = \nu\sqrt{ml^3/3EI}$ of spinning rotors with non-dimensional rotational parameter $\Omega = \omega\sqrt{ml^3/3EI}$ is given as a function of the disk parameter $\delta = 3I_T/ml^2$ in equation (14.38). This relation is also plotted in Figure 14.22.

$$\nu^4 - 2\omega\nu^3 - EI\left(\frac{12}{ml^3} + \frac{4}{I_T l}\right)\nu^2 + 24\omega\frac{EI}{ml^3}\nu + \frac{12E^2I^2}{I_T ml^4} = 0$$

$$\lambda^4 - 2\Omega\lambda^3 - 4\left(\frac{\delta + 1}{\delta}\right)\lambda^2 + 8\Omega\lambda + \frac{4}{\delta} = 0 \tag{14.38}$$

Negative roots of λ for positive rotational speeds ω indicate that the shaft whirls in an opposite direction to spin – backward whirl. These backward whirl roots are

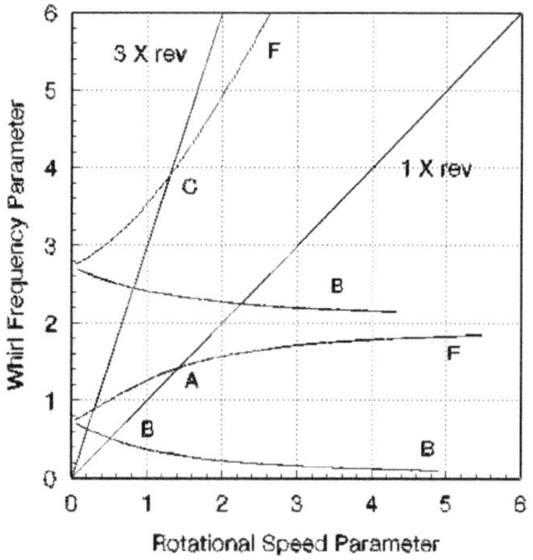

Fig. 14.22 Non-synchronous whirl of a cantilevered rotor

plotted as positive values denoted by B. Forward whirl roots, i.e., positive roots of λ are marked F.

Each frequency of a stationary shaft is split into two components. Per rev excitations are also given, making this a Campbell diagram. When the shaft is stationary, if we do a bump test, it responds with the structural natural frequencies.

As we increase the shaft speed, first it crosses at $1B$ with a resonance – this is usually not observed as there is no excitation in the backward direction; then it crosses $1A$, with peak response whirl, this is first critical speed. Notice the difference in first critical speed because of the disk effect.

One thing should be noted is that the gyroscopic effect introduces the effect of rotation on the natural frequencies; but any variation with the speed is not connected directly with the centrifugal field. It is merely introduced by the disk gyroscopic effect. Conventionally the rotor dynamics models are beam models and therefore there is no centrifugal force even though there is rotor eccentricity. All this changed in the 21st century with the introduction of solid rotor model analysis that brings in stress stiffening and spin softening and our understanding of forward whirl and backward critical speeds has considerably changed. We will discuss this subsequently.

14.7 Internal Friction, Hysteresis

Internal friction, also called Hysteresis causes instabilities. A loosely mounted part on the rotating shaft can also have relative motion in rotating coordinates and is equivalent to internal friction.

Newkirk [55] observed this phenomenon first experimentally and Kimball [38] discussed this analytically.

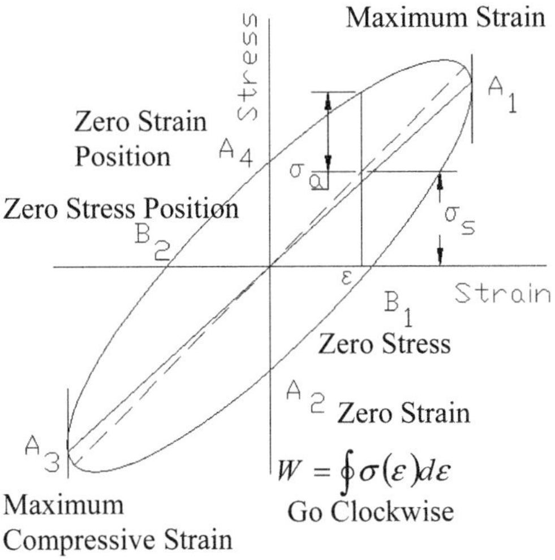

Fig. 14.23 Hysteresis loop

Figure 14.23 shows a typical hysteresis loop, the work lost in one cycle is given by the area of the loop. Conventionally, we go clockwise from maximum strain A_1 to zero stress position B_1 then to A_2 at zero strain; at A_3 we have maximum compressive strain, then zero stress at B_2 and again to zero strain at A_4 before completing the full cycle.

At any point with a strain ε, the stress is made of a steady part σ_s and an alternating component σ_a. The work lost in each cycle is equivalent to damping in rotating coordinates and to determine this, we express the material property as viscoelastic, instead of purely elastic. Here D is damping modulus Ns/m^2.

$$\sigma = E\varepsilon + D\dot{\varepsilon} \tag{14.39}$$

The work of a hysteresis loop is compared with work in deforming the body over a quarter period, which gives relatively the damping in the system, also called specific damping capacity. For convenience, the work of hysteresis is taken over one radian, rather than one cycle, that gives the well-known Loss Factor.

Work done per unit volume over one period is

$$W = \oint \sigma(\varepsilon) d\varepsilon \tag{14.40}$$

Work done per unit volume in deformation during one quarter period is

$$U = \frac{1}{2}\hat{\sigma}\hat{\varepsilon} \tag{14.41}$$

Relative damping or Specific damping capacity is then given by

$$\beta = \frac{W}{U} = \frac{\oint \sigma(\varepsilon) d\varepsilon}{\frac{1}{2}\hat{\sigma}\hat{\varepsilon}} \tag{14.42}$$

$$\therefore \text{Loss Factor} \eta = \frac{\beta}{2\pi} \tag{14.43}$$

For harmonic strain, the strain rate is $\dot{\varepsilon} = \hat{\varepsilon}\omega \cos \omega t$. Using

$$\cos \omega t = \sqrt{1 - \sin^2 \omega t} = \sqrt{1 - \left(\frac{\varepsilon}{\hat{\varepsilon}}\right)^2}$$

equation (14.39) is

$$\sigma = E\varepsilon + D\omega\hat{\varepsilon}\sqrt{1 - \left(\frac{\varepsilon}{\hat{\varepsilon}}\right)^2}$$

$$= E\varepsilon + D^* \sqrt{\hat{\varepsilon}^2 - \varepsilon^2} = \sigma_s + \sigma_a \tag{14.44}$$

where $D^* = D\omega$. The harmonic component $\sigma_a = D^* \sqrt{\hat{\varepsilon}^2 - \varepsilon^2}$ varies periodically, zero at peak strain points. The viscous effect is proportional to strain rate with its coefficient called the Damping Modulus. We can use this alternating stress for a harmonic strain and determine the loss factor. The internal damping is calculated, using a viscoelastic model, as follows now:

$$W = \oint \sigma(\varepsilon) d\varepsilon = \pi D^* \hat{\varepsilon}^2 \tag{14.40a}$$

$$U = \frac{1}{2}\hat{\sigma}\hat{\varepsilon} = \frac{1}{2}E\hat{\varepsilon}^2 \tag{14.41a}$$

$$\beta = \frac{W}{U} = 2\pi\frac{D^*}{E} \tag{14.42a}$$

$$\eta = \frac{D^*}{E} \tag{14.43a}$$

Let the equivalent internal damping factor be h. Then

$$W_d = \pi h \omega X^2 \tag{14.40b}$$

$$U = \frac{1}{2}kX^2 \tag{14.41b}$$

$$\beta = \frac{W_d}{U} = 2\pi\frac{h\omega}{k} \tag{14.42b}$$

Or h is given by

$$h = \frac{1}{2\pi}\frac{\beta k}{\omega} = \frac{\eta k}{\omega} \tag{14.44a}$$

Substituting for loss factor η the equivalent viscous damping ratio is

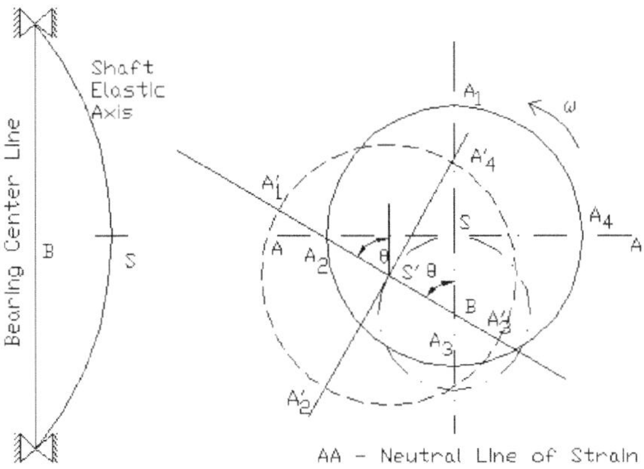

Fig. 14.24 Shaft in synchronous whirl

$$\xi_v = \frac{hp}{2k} = \frac{\eta p}{2\omega} \tag{14.45}$$

The loss factor for a given material is determined experimentally, so the internal damping coefficient h or equivalent internal damping can be determined as above.

Shaft in Synchronous whirl: Consider the disk at shaft center S in Figure 14.24, which whirls about bearing centerline B through angle $\theta = \omega t$. In synchronous whirl, the shaft spins about its center S through the same angle $\theta = \omega t$, point A_1 in maximum tension position it remains the same as A_1', etc. Note that A_2, A_4 are neutral strain points, not neutral stress points.

Shaft in Subsynchronous whirl $\nu < \omega$: Let us now consider the case of whirl frequency ν lower than spin speed ω. Now, the shaft center S whirls about bearing centerline B through angle $\theta = \nu t$. The shaft spins about its center S through an angle $> \theta$ by ωt. A_1 goes beyond A_1' of the synchronous whirl upto A_1'' under subsynchronous whirl, i.e., it tries to catch up with A_2'. Therefore, we have the zero stress position B_1, between A_1 and A_2.

The component of P in the horizontal direction has a moment about S aiding the whirl, thus making the shaft unstable.

A shaft spinning at ω when disturbed, whirls with ν = natural frequency p. Therefore this instability can occur only when $\omega > p$.

Shaft in supersynchronous whirl $\nu > \omega$: Here shaft center S whirls about bearing centerline B through angle $\theta = \nu t$, see Figure 14.25. The shaft spins about its center S through an angle $< \theta$ by ωt. A shaft spinning at ω when disturbed whirls with ν = natural frequency p. Therefore this instability can never occur when $\omega < p$.

Tondl [91] provided an analysis of a Jeffcott type rotor with internal damping. Figure 14.26a shows the disk in rotating coordinates $\eta\xi$. Equilibrium conditions are defined in rotating coordinates. Use accelerations in rotating coordinates for inertia forces and velocities in rotating coordinates for external damping.

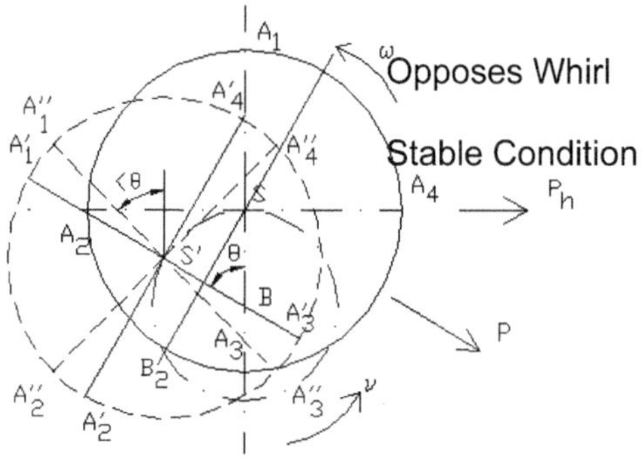

Fig. 14.25 Supersynchronous whirl

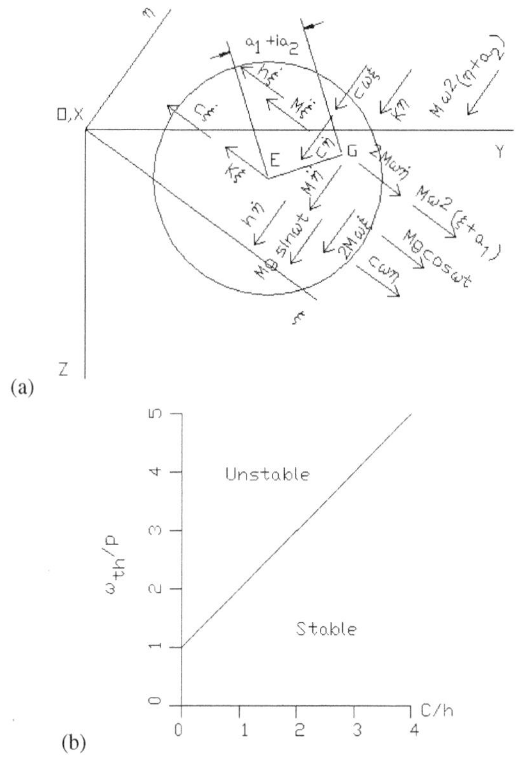

Fig. 14.26 (a) Shaft with internal damping. (b) Stability chart of a rotor with internal damping or loosely mounted part

The shaft center in stationary coordinates is $r = z + iy$ and in rotating coordinates it is $\zeta = \xi + i\eta$. The velocities and accelerations of G are

$$
\begin{aligned}
v_\xi &= \dot{\xi} - \omega(\eta + a_2) \\
v_\eta &= \dot{\eta} + \omega(\xi + a_1) \\
a_\xi &= \ddot{\xi} - 2\omega\dot{\eta} - \omega^2(\xi + a_1) \\
a_\eta &= \ddot{\eta} + 2\omega\dot{\xi} - \omega^2(\eta + a_2)
\end{aligned} \tag{14.46}
$$

The governing equation in rotating coordinates is

$$M(\ddot{\xi} - 2\omega\dot{\eta} - \omega^2\xi) + C(\dot{\xi} - \omega\eta) + h\dot{\xi} + K\xi = Ma_1\omega^2 + Mg\cos\omega t$$

$$M(\ddot{\eta} + 2\omega\dot{\xi} - \omega^2\eta) + C(\dot{\eta} + \omega\xi) + h\dot{\eta} + K\eta = Ma_2\omega^2 + Mg\sin\omega t$$

$$M(\ddot{\zeta} + 2\omega\dot{\zeta} - \omega^2\zeta) + C(\dot{\zeta} + i\omega\zeta) + h\dot{\zeta} + K\zeta = Ma\omega^2 + Mg\exp(-i\omega t)$$

$$\frac{C}{M} = 2\delta_v, \quad \frac{h}{M} = 2\delta_h$$

$$\ddot{\zeta} + 2\omega\dot{\zeta} - \omega^2\zeta + 2\delta_v(\dot{\zeta} + i\omega\zeta) + 2\delta_h\dot{\zeta} + p^2\zeta = a\omega^2 + g\exp(-i\omega t) \tag{14.47}$$

Convert the above to stationary coordinates and solve. Stability depends on the two exponentially decaying or growing type terms.

$$
\begin{aligned}
&\zeta = r\exp(-i\omega t) \\
&\ddot{r} + 2\delta_v\dot{r} + 2\delta_h(\dot{r} - i\omega r) + p^2 r = a\omega^2\exp(i\omega t) + g \\
&r = \exp(i\lambda t) \\
&\lambda^2 - 2i\lambda(\delta_v + \delta_h) + 2i\omega\delta_h - p^2 = 0 \\
&r = A_1\exp\left[-\left(\delta_v + \delta_h + \frac{\delta_h\omega}{p}\right)t\right]\exp(-ipt) \\
&\qquad + A_2\exp\left[-\left(\delta_v + \delta_h - \frac{\delta_h\omega}{p}\right)t\right]\exp(ipt)
\end{aligned} \tag{14.48}
$$

A stability chart for a rotor with internal friction is given in Figure 14.26b. In practice, a loosely mounted disk, gear, flywheel, etc., cause friction between the two surfaces, the rotor and the mounted part. Unlike friction between a stationary part and moving part, here the friction is with respect to the rotor, i.e., in rotating coordinates. The friction coefficient between two such surfaces is h. For large h, or low external damping C, the instability can set in just after the first critical speed. We need external friction to fight the instability due to a loosely mounted part. If C is, say, $3h$, then the instability takes place at a speed four times the critical speed. This instability can never occur at speeds below the first critical speed.

Ehrich [15] also considered hysteretic whirl and contributed to the *Shock and Vibration Handbook* by Harris and Crede [32].

14.8 Shafts with Gravity and Variable Elasticity

Stodola [82–84] recognized the influence of gravity and variable elasticity causing minor phenomena, though we know today that they play a significant role. In magnitude they are infinitesimals of the second order; but although they must be described as "secondary", they may still cause trouble in operation. He explains the gravity phenomenon as follows.

The motion of a single disk consists of a fairly uniform rotation with the angular velocity ω about the center of gravity S. The latter itself describes during the stationary motion (Jeffcott's analysis is published in 1919 [37]) a circular path with angular velocity ω and an elastic vibration with circular frequency ω_k which may be rectilinear, elliptical or circular and in the latter case may proceed in the same direction (*forward whirl*) or in the opposite direction (*backward whirl*) of ωThe relative motion of S with respect to space rotation with ω may be a motion with velocity $\omega_k - \omega$, which should be regarded as a combination of two harmonic vibrations. Its weight G is converted in the space ω into a rotating force with components $G \sin \omega t$ and $G \cos \omega t$ and which gets into resonance with the relative vibration in case of frequency $\omega_k - \omega$ and ω coincide, i.e.,

$$\omega_k - \omega = \omega$$

$$\omega = \frac{1}{2}\omega_k \tag{14.49}$$

According to Stodola, the Siemens–Schuckert Works reported such a vibration as far back as 1910. Stodola [82, 85] discussed the variable elasticity effect of the shaft material that necessitates taking the bending moments in two perpendicular directions which amounts to the same thing as though the shaft, for instance, had two principal axes of inertia as in the case of a shaft having a groove all the way in longitudinal direction (or as a generator rotor and equivalent rectangular cross-section shaft). Stodola [82] correctly identified the two critical speeds due to variable elasticity (or stiffness) and also he states that for the intermediate speeds the equilibrium is not a stable one. We know this to be true only for vertical rotors but horizontal rotors with gravity effect; the response between these two critical speeds is stable. What is interesting here to note is that Stodola says that Prandtl [63], a famous name in Fluid Mechanics, has shown that a new critical speed becomes noticeable at about half the value of the usual critical speed. So it seems Prandtl is the first one to work on Gravity Critical.

Gümbel [29, 30] and Föppl [21] discussed the gravity effect. Taylor [87], Timoshenko [90] and Den Hartog [11] discussed the physics of the shafts with variable stiffness using a rectangular shaft model carrying a central disk. Mathieu [47], also see [51], studied stability of equations governed by variable stiffness. Whittaker [95] provided general solutions of Mathieu's equation; see also [35, 69]. The vertical shaft configuration (e.g., generator rotor in a hydraulic turbine generator system) is shown in Figure 14.27. The response of a vertical shaft without gravity

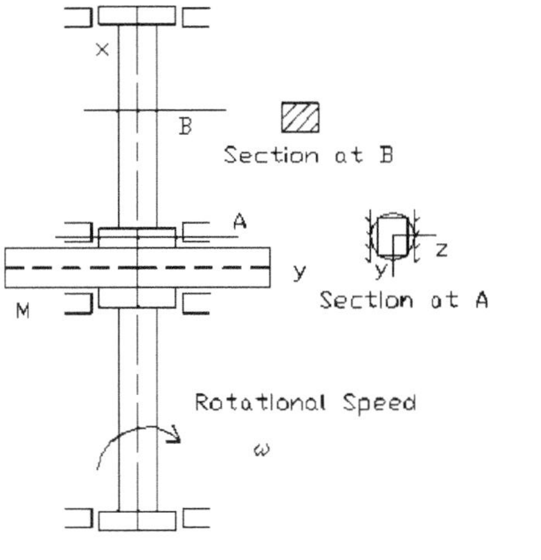

Fig. 14.27 A vertical rectangular shaft carrying a central mass

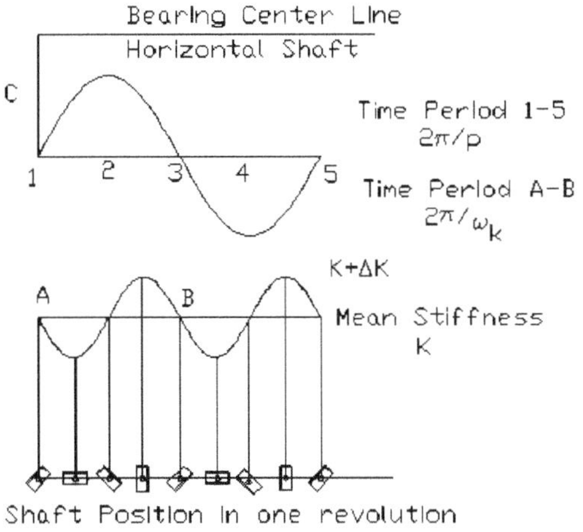

Fig. 14.28 Response of a horizontal shaft at critical speed

and the response of a horizontal shaft, whose disk is displaced by C, is shown in Figure 14.28.

The mean stiffness is taken as K and the variation in stiffness by ΔK. As can be seen the stiffness varies twice in one revolution. Let the rotor spin speed be ω and the stiffness variation be at angular velocity $\omega_k = 2\omega$. Let us first consider the horizontal shaft as shown in Figure 14.28 running at critical speed $\omega = p$. The shaft stiffness varies at $\omega_k = 2\omega = 2p$. When the shaft center moves away from the bearing centerline between points 2 and 4, the stiffness varies once, with the mean at K. Therefore there is no energy input to the system and the horizontal shaft is stable at critical speed.

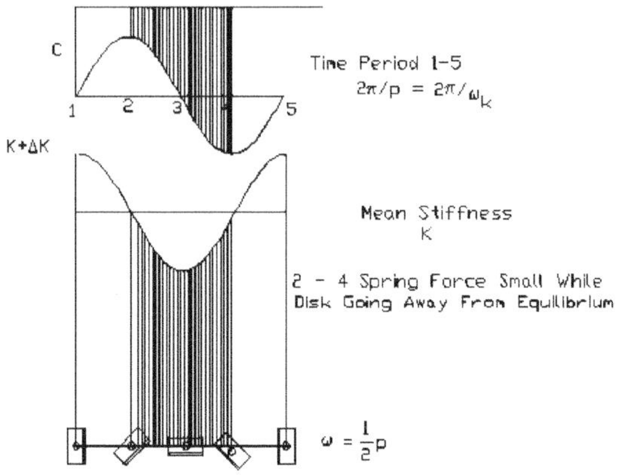

Fig. 14.29 Horizontal shaft at half critical speed

For a vertical rotor, the response has pure sinusoidal variation with a frequency p corresponding to the natural frequency, and when the rotor is horizontal, we have to take this variation down by the gravity deflection C.

For vertical rotors (hydro machines), at critical speed, between 1–2 and 3–4, when the shaft is going away from mean position, the stiffness is lower, therefore it requires less energy to push the shaft away, thus it is unstable. Besides unbalance forced vibration, a vertical shaft is unstable as well.

If the shaft is horizontal, it goes away from mean equilibrium, between 2–4 and the stiffness here averages out, therefore there is no energy input and the system is stable. Horizontal shafts are not unstable; they are subjected to conventional forced vibration alone. At half critical speed, the horizontal shaft behaves in a different manner. At this speed, the stiffness is varying at twice the half critical speed, i.e., the stiffness variation is at the same frequency as the natural frequency. When disturbed, the shaft responds at critical speed and the stiffness is also varying at the same frequency as shown in Figure 14.29. For a horizontal shaft, between 2–4, the stiffness is lower; therefore, it is unstable as shown.

The response is due to conventional unbalance which is low at this speed; however, the instability at this speed causes large whirls. Initially, this speed was thought to be some kind of critical, when the instability was not clearly understood and therefore, this speed is called gravity critical.

If however, the shaft runs at half critical speed as in Figure 14.29, the situation is entirely different. Between 2 and 4 when the shaft center is going away from the bearing centerline, the stiffness is below mean value and therefore offers less resistance. This leads to energy input into the system and a self excited vibration takes place.

Timoshenko [90] gave an elegant analysis of an asymmetric shaft by solving a single degree of freedom system with constant stiffness $K + (\Delta K/K)$ for half of stiffness variation period and $K - (\Delta K/K)$ for the other half and applying appropriate boundary conditions. We will however go through the classical solution

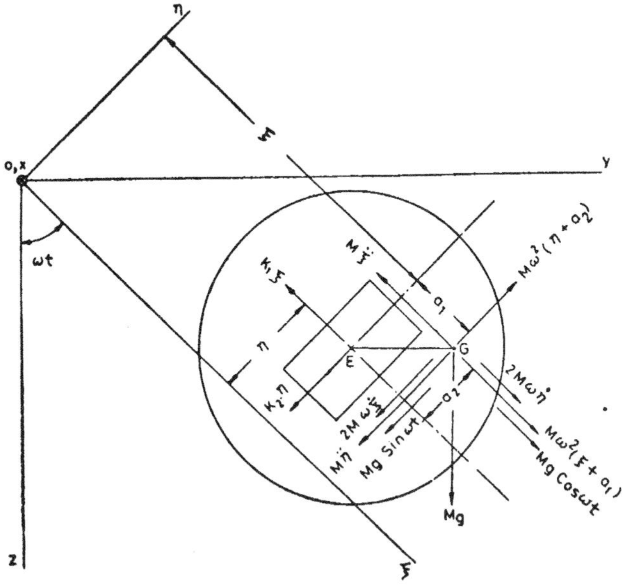

Fig. 14.30 Whirling of a rotor with a rectangular cross-section shaft

provided by Tondl [91] here. Figure 14.30 gives the rectangular cross-section shaft in rotating coordinates.

Similar to equations (14.47), the governing equations for the rectangular shaft are

$$M[\ddot{\xi} - 2\omega\dot{\eta} - \omega^2(\xi + a_1)] + K_1\xi = Mg\cos\omega t$$
$$M\ddot{\eta} + 2\omega\dot{\xi} - \omega^2(\eta + a_2) + K_2\eta = Mg\sin\omega t \tag{14.50}$$

where

$$K_1 = K - \Delta K$$
$$K_2 = K + \Delta K \tag{14.51}$$

Equations (14.50) are simplified with no unbalance as

$$\ddot{\xi} - 2\omega\dot{\eta}(p_1^2 - \omega^2)\xi = g\cos\omega t$$
$$\ddot{\eta} + 2\omega\dot{\xi}(p_2^2 - \omega^2)\eta = -g\sin\omega t \tag{14.52}$$

where

$$p_1^2 = \frac{K_1}{M} = p^2\left(1 - \frac{\Delta K}{K}\right)$$

$$p_2^2 = \frac{K_2}{M} = p^2\left(1 + \frac{\Delta K}{K}\right) \tag{14.53}$$

Vertical Shaft – No gravity: The free vibration solution of (14.52) is

$$\xi = \xi_0 e^{i\lambda_0 t}$$

$$\eta = \eta_0 e^{i\lambda_0 t} \tag{14.54}$$

where λ_0 is the frequency of vibration in rotating coordinates. Substituting (14.54) in (14.52), we get the frequency equation

$$\lambda_0{}^4 - (p_1{}^2 + p_2{}^2 + 2\omega^2)\lambda_0{}^2 + (p_1{}^2 - \omega^2)(p_2{}^2 - \omega^2) = 0 \tag{14.55}$$

The roots of the above equation expressed in stationary coordinates system $\lambda_i = \lambda_{0i} + \omega$ are

$$\frac{\lambda}{p} = \frac{\omega}{p} \pm \sqrt{1 + \left(\frac{\omega}{p}\right)^2 \pm \sqrt{\left[1 + \left(\frac{\omega}{p}\right)^2\right]^2 - \left[1 - \frac{\Delta K}{K} - \left(\frac{\omega}{p}\right)^2\right]\left[1 + \frac{\Delta K}{K} - \left(\frac{\omega}{p}\right)^2\right]}} \tag{14.56}$$

For $\Delta K/K = 0.4$, the above is plotted in Figure 14.31. There are four real roots for all λ/p, for all ω/p except when

$$1 - \frac{\Delta K}{K} < \left(\frac{\omega}{p}\right)^2 < 1 + \frac{\Delta K}{K} \tag{14.57}$$

For $\omega/p = 0$, the four roots are $\lambda/p = \pm 1.183$ and ± 0.77. The stability diagram is shown in Figure 14.31. From this we can find that a vertical asymmetric shaft is unstable in the critical speed region.

Effect of Disk Unbalance and No Gravity: Equation (14.50) reduces for this case to

$$\ddot{\xi} - 2\omega\dot{\eta}(p_1^2 - \omega^2)\xi = a_1\omega^2$$

$$\ddot{\eta} + 2\omega\dot{\xi}(p_2^2 - \omega^2)\eta = a_2\omega^2 \tag{14.58}$$

Setting

$$\zeta = \xi + i\eta$$

$$\bar{\zeta} = \xi - i\eta \tag{14.59}$$

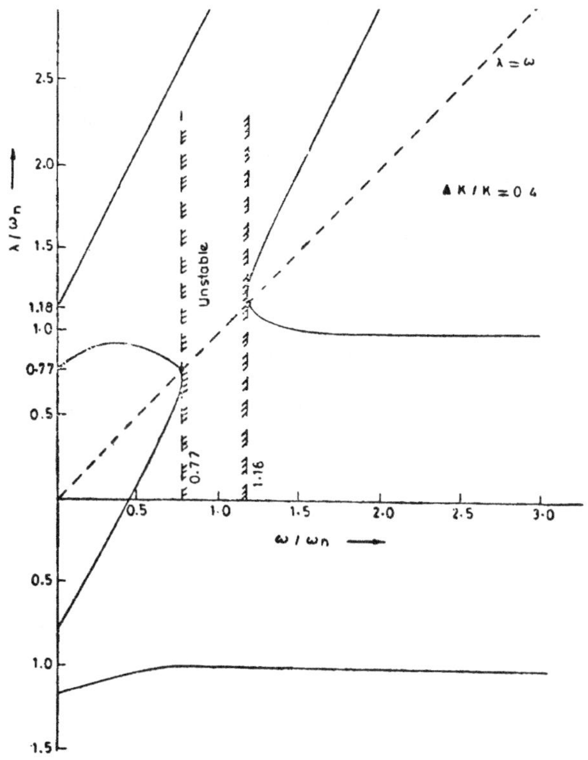

Fig. 14.31 Stability of a balanced vertical rotor on a rectangular shaft

equations (14.58) can be combined to give

$$\ddot{\zeta} - 2i\omega\dot{\zeta}(p^2 - \omega^2) - \frac{\Delta K}{K}p^2\bar{\zeta} = \omega^2(a_1 + a_2) \tag{14.60}$$

For the steady-state solution, we write

$$\begin{aligned} \zeta_u &= A + iB \\ \bar{\zeta}_u &= A - iB \end{aligned} \tag{14.61}$$

where A and B are constants and reconstruct equation (14.60) to give

$$(p^2 - \omega^2)(A + iB) - \frac{\Delta K}{K}p^2(A - iB) = \omega^2(a_1 + a_2) \tag{14.60a}$$

Separating real and imaginary parts

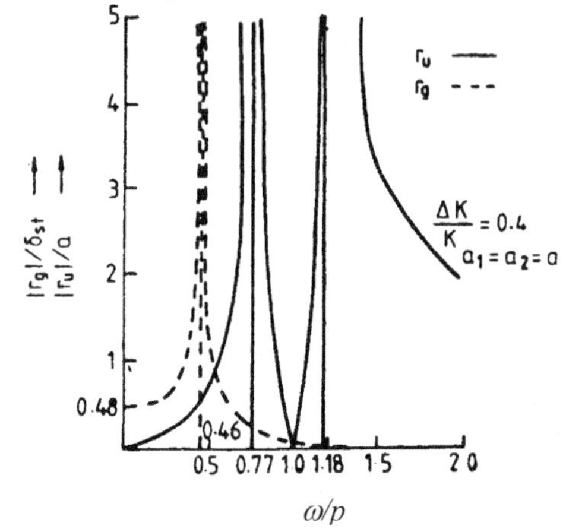

Fig. 14.32 Unbalance and gravity response of a rectangular shaft with disk

$$A = \frac{a_1 \frac{\omega^2}{p^2}}{1 - \frac{\omega^2}{p^2} - \frac{\Delta K}{K}}, \quad B = \frac{a_2 \frac{\omega^2}{p^2}}{1 - \frac{\omega^2}{p^2} + \frac{\Delta K}{K}} \tag{14.62}$$

The whirl radius due to unbalance is $r + u = \sqrt{A^2 + B^2}$

$$r_u = \frac{\omega^2}{p^2} \frac{\sqrt{a_1{}^2[1 - \frac{\omega^2}{p^2} + \frac{\Delta K}{K}]^2 + a_2{}^2[1 - \frac{\omega^2}{p^2} - \frac{\Delta K}{K}]^2}}{(1 - \frac{\omega^2}{p^2})^2 - (\frac{\Delta K}{K})^2} \tag{14.63}$$

The two critical speeds are

$$\frac{\omega^2}{p^2} = 1 - \frac{\Delta K}{K} \text{or} 1 + \frac{\Delta K}{K} \tag{14.64}$$

For a vertical rotor the response is unstable between the critical speeds, for a horizontal rotor, gravity removes this instability and gives rise to pure unbalance response with two peaks (see Figure 14.32).

Effect of Gravity on a Balanced Disk: Since the mean equilibrium keeps changing as the stiffness varies, gravity plays a significant role. Equation (14.50) for this case is

$$\ddot{\zeta} - 2i\omega\dot{\zeta}(p^2 - \omega^2) - \frac{\Delta K}{K} p^2 \bar{\zeta} = g \exp(-i\omega t) \tag{14.65}$$

Writing the solution due to gravity

$$\begin{aligned} \zeta_g &= Ae^{i\omega t} + Be^{-i\omega t} \\ \bar{\zeta}_g &= \bar{A}e^{-i\omega t} + \bar{B}e^{i\omega t} \end{aligned} \tag{14.66}$$

where A and B are now complex constants $\bar{A}$ and $\bar{B}$ are their conjugates, we can show that

$$\frac{A}{\delta_{st}} = \frac{\frac{\Delta K}{K}}{1 - 4r^2 - (\frac{\Delta K}{K})^2}, \quad \frac{B}{\delta_{st}} = \frac{1 - 4r^2}{1 - 4r^2 - (\frac{\Delta K}{K})^2} \tag{14.67}$$

where $\delta_{st} = Mg/K$. The solution in the stationary coordinates is

$$r_g = \zeta_g e^{i\omega t} = Ae^{2i\omega t} + B \tag{14.68}$$

and

$$\frac{r_g}{\delta_{st}} = \frac{1 - 4r^2}{1 - 4r^2 - (\frac{\Delta K}{K})^2} + \frac{\frac{\Delta K}{K}}{1 - 4r^2 - (\frac{\Delta K}{K})^2} e^{2i\omega t} \tag{14.69}$$

The above response is twice the rotational speed and resonance takes place when

$$4r^2 = 1 - \left(\frac{\Delta K}{K}\right)^2 \quad \text{or when} \quad \omega = \frac{1}{2} p \sqrt{1 - \left(\frac{\Delta K}{K}\right)^2}$$

The response from (14.69) is also plotted in Figure 14.32. Because of this resonance at $\omega = 0.46p$, we are somewhat justified in saying that this critical condition is due to gravity and therefore it is a gravity critical. However, the solution here is not stable. To obtain the stability condition of the second harmonic, we modify the free vibration system equation with ξ and η coordinates rotating at 2ω, by virtue of the forced vibration of the horizontal shaft that can occur at 2ω frequency. Then equation (14.52) becomes modified

$$\begin{aligned} \ddot{\xi} - 4\omega\dot{\eta}(p_1^2 - 4\omega^2)\xi &= 0 \\ \ddot{\eta} + 4\omega\dot{\xi}(p_2^2 - 4\omega^2)\eta &= 0 \end{aligned} \tag{14.52a}$$

Using (14.54), the frequency equation is

$$\lambda_0{}^4 - (p_1{}^2 + p_2{}^2 + 8\omega^2)\lambda_0{}^2 + (p_1{}^2 - 4\omega^2)(p_2{}^2 - 4\omega^2) = 0 \tag{14.54a}$$

Also, since $\lambda = \lambda_0 + 2\omega$ we get equation (14.56) for this case,

$$\frac{\lambda}{p} = \frac{2\omega}{p} \pm$$

$$\sqrt{1 + 4\left(\frac{\omega}{p}\right)^2 \pm \sqrt{\left[1 + 4\left(\frac{\omega}{p}\right)^2\right]^2 - \left[1 - \frac{\Delta K}{K} - 4\left(\frac{\omega}{p}\right)^2\right]\left[1 + \frac{\Delta K}{K} - 4\left(\frac{\omega}{p}\right)^2\right]}} \tag{14.56a}$$

In general there are four real roots for, except when

$$1 - \frac{\Delta K}{K} < 4\left(\frac{\omega}{p}\right)^2 < 1 + \frac{\Delta K}{K} \tag{14.57a}$$

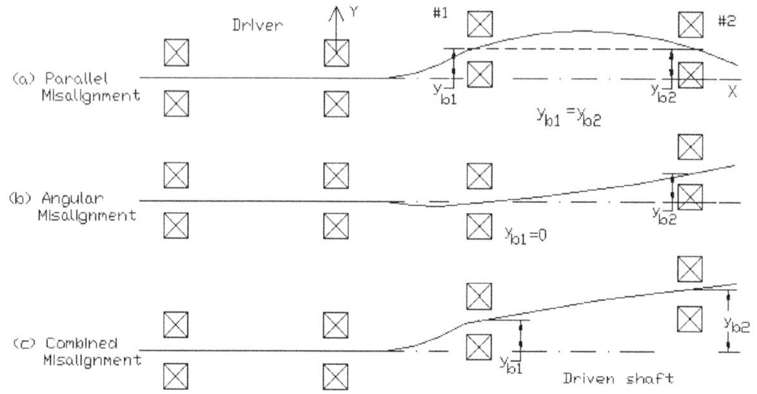

Fig. 14.33 Different misalignments

which is $p_1/2 < \omega < p_2/2$. Under these conditions there are two real roots and two complex roots which give rise to oscillatory solutions. However the real parts being positive, the solutions are unstable.

14.9 Misalignment

Stodola [82] records that due to periodic variation of the tangential force or intermediacy of a flexible coupling with inaccurate alignment of the shafts, variable impulse act on the shaft. According to Föppl [22] if the variation of the deformation causes ν impulses per revolution, a new critical disturbance with the velocity

$$\omega_1 = \frac{\omega_k}{\nu + 1} \tag{14.70}$$

comes into evidence. The complete solution shows also the presence of disturbance

$$\omega_2 = \frac{\omega_k}{\nu - 1} \tag{14.71}$$

Misalignment is a very common problem that occurs in machinery after some amount of running. Like residual unbalance, misalignment is always present to some extent and if it deteriorates, the machine runs rough and its life is affected. There are three types of misalignment, Parallel misalignment is indicated by $y_{b2} = y_{b1}$ and $z_{b2} = z_{b1}$ in Figure 14.33.

The angular misalignment is indicated by y_{b2} with $y_{b1} = 0$ and z_{b2} with $z_{b1} = 0$. Combined misalignment is indicated by $y_{b2} - y_{b1}$ and $z_{b2} - z_{b1}$.

A 3-D view of the driver and driven shaft axes along with the deformed coupling element of length l_c is shown in Figure 14.34 giving parallel and combined parallel

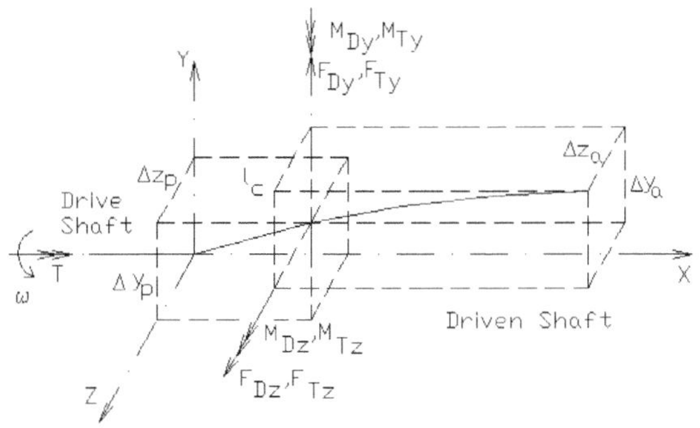

Fig. 14.34 Forces and moments due to misalignment

and angular misalignment values. Due to the deformation of the coupling element and the drive shaft torque, forces and moments are induced as indicated.

Reaction forces and moments due to transmitted torque are

$$\begin{aligned}
M_{Ty} &= T \sin\phi_a \cos\phi_p + T \cos\phi_a \sin\phi_p \\
M_{Tz} &= -T \sin\theta_a \cos\theta_p - T \cos\theta_a \sin\theta_p \\
F_{Ty} &= \frac{M_{Tz}}{l_c} \\
\sin\theta_a &= \frac{\Delta z_a}{l_b}, \quad \sin\theta_p = \frac{\Delta z_p}{l_c} \\
\sin\phi_a &= \frac{\Delta y_a}{l_b}, \quad \sin\phi_p = \frac{\Delta y_p}{l_c}
\end{aligned} \tag{14.72}$$

Reaction forces and moments due to coupling element deformation are

$$\begin{aligned}
F_{Dy} &= \frac{12EI_c}{l_c^3}(\Delta y_p - \frac{1}{2}l_c\phi_a) \\
F_{Dz} &= \frac{12EI_c}{l_c^3}(\Delta z_p - \frac{1}{2}l_c\theta_a) \\
M_{Dy} &= \frac{6EI_c}{l_c^2}(\Delta y_p - \frac{2}{3}l_c\phi_a) \\
M_{Dz} &= \frac{6EI_c}{l_c^2}(\Delta z_p - \frac{2}{3}l_c\theta_a)
\end{aligned} \tag{14.73}$$

The axial force is

$$F_{Dx} = \frac{EA_c}{l_c}\Delta x \tag{14.74}$$

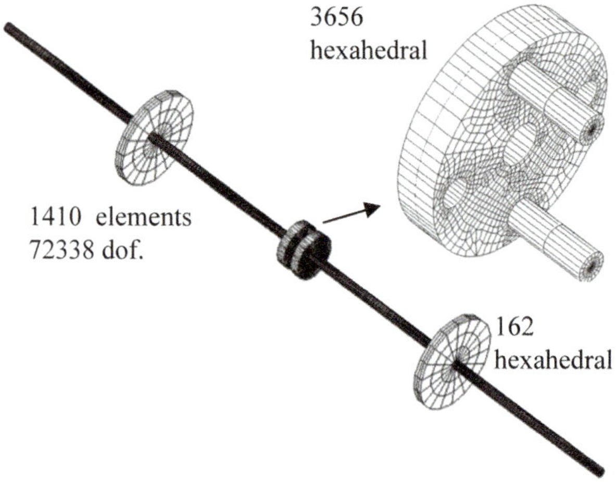

Fig. 14.35 FE model of a misaligned shaft system

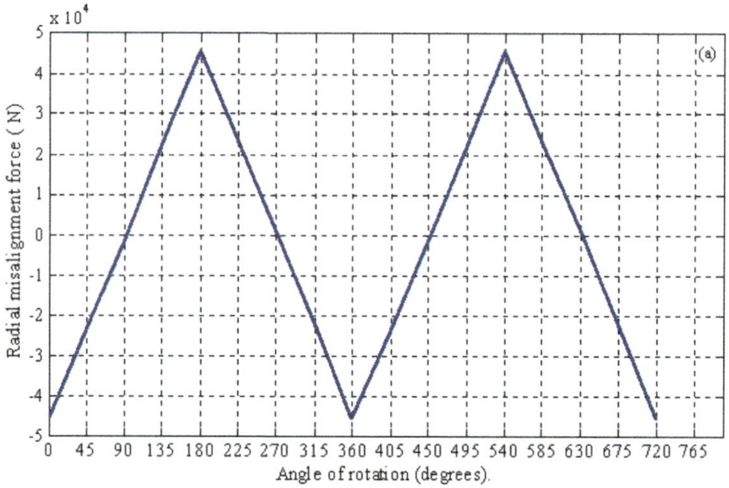

Fig. 14.36 Radial misalignment force due to parallel misalignment

According to Ehrich [17] preloads due to misalignment vary at the frequency of the shaft speed. Lee [42] modeled and performed vibration analysis of misaligned rotors on ball bearing supports.

To understand the influence of rotation, Rao and Sreenivas [74] made a FE model of two Jeffcott rotors coupled by a two pin element as shown in Figure 14.35. Revolving this system, we can determine the forces and stresses as a function of time and their peak values.

The peak values obtained from FE analysis are the same as strength of material values derived by Gibbons [25], see Figure 14.36. Also, due to parallel misalignment

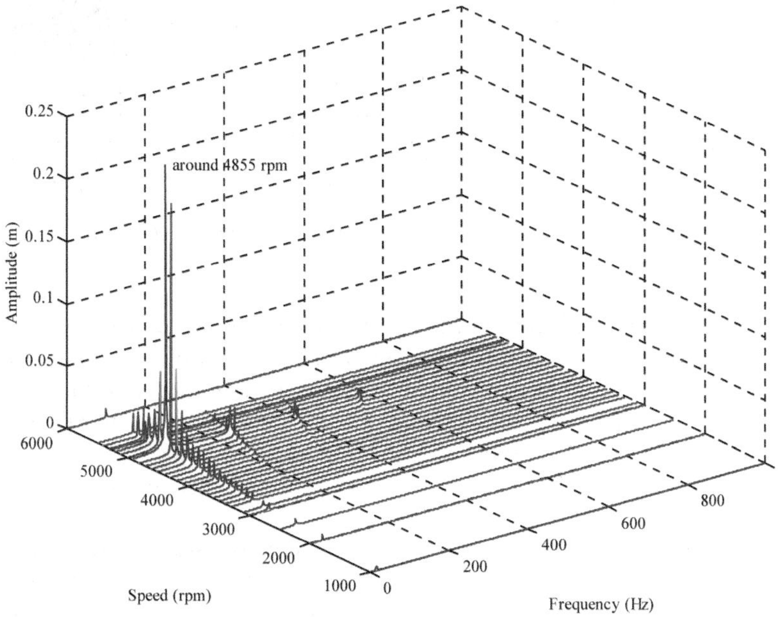

Fig. 14.37 Transient response due to parallel misalignment

of 1 mm, the radial force obtained is found to vary linearly, once per revolution, as observed by Ehrich [17], such a triangular wave has several harmonics given by $1\times$, $3\times$, $5\times$ ... These components are responsible for higher harmonic responses in the misaligned shaft system. Similarly other misalignments produced a response with one per revolution variation.

The transient response obtained with these misalignment values is given in Figure 14.37. One per rev component increases significantly while approaching the critical speed and the magnitude of higher harmonics become significant with more misalignment as shown.

An experimental study was conducted as shown in Figure 14.38.

Figure 14.39 shows a signal captured during the coasting up of the rotor system to a steady state running frequency of 36 Hz. The response in the radial direction is high at 1/3 and 1/2 critical speeds.

The axial vibration of the driving shaft disc was also picked up using a proximity probe. The coasting up signal of the axial vibration is shown in Figure 14.40. It was found that the axial vibration undergoes a significant change around the one-half natural frequency region.

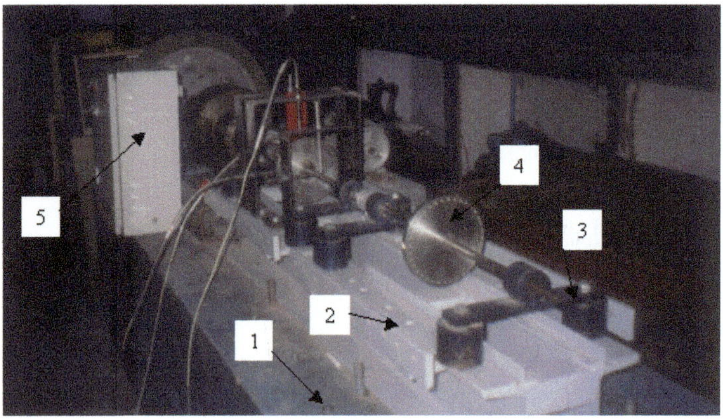

Fig. 14.38 Misalignment experimental studies [74]

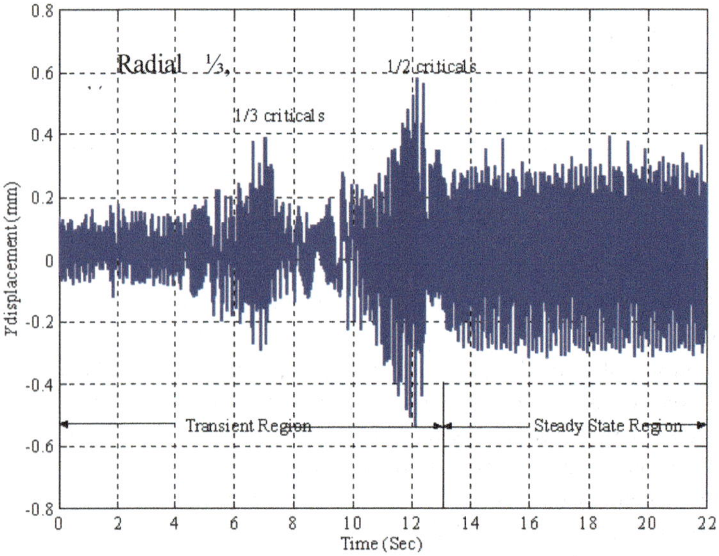

Fig. 14.39 Response in radial direction shows 1/3 and 1/2 transient criticals

14.10 Bowed Rotors

Well-balanced rotors are sometimes subjected to deformations while running or under stationary conditions. Typically they arise from thermal stresses when the rotor is not properly stabilized. Such a rotor is called Bowed Rotor or a Rotor with Permanent Bow. A bowed rotor can be considered as the Jeffcott model with a static deflection from the bearing centerline with a magnitude r_0 at a phase angle a_0 (see Figure 14.41) and we can write the governing equation as

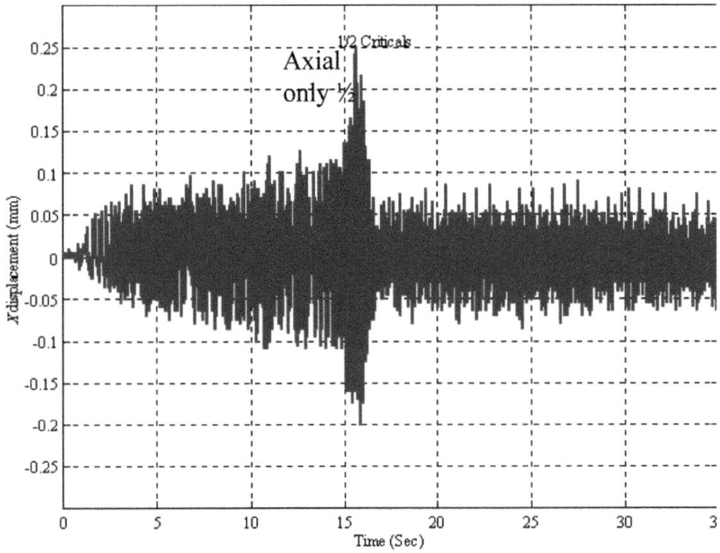

Fig. 14.40 Response in axial direction shows only 1/2 critical transient

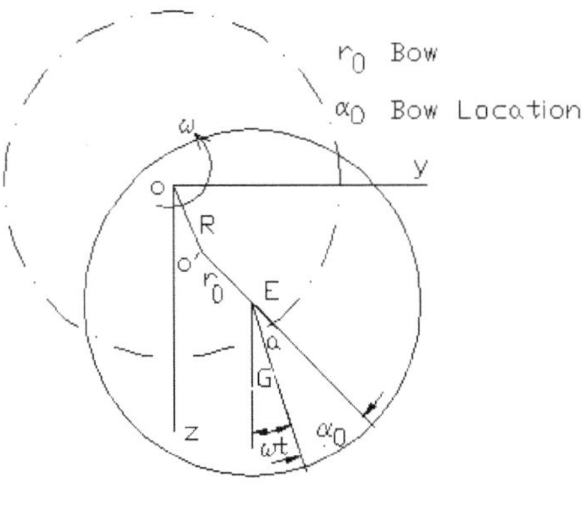

Fig. 14.41 Rotor on a bowed shaft

$$M\frac{d^2r}{dt^2} + C\frac{dr}{dt} + Kr = Ma\omega^2 e^{i\omega t} + Kr_0 e^{i(\omega t + \alpha_0)} \tag{14.75}$$

The solution consists of two parts:

$$\bar{R} = \frac{R}{a} = \frac{\Omega^2}{\sqrt{(1-\Omega^2)^2 + (2\xi\Omega)^2}} e^{i(\omega t - \phi)} + \frac{R_0 e^{i\alpha_0}}{\sqrt{(1-\Omega^2)^2 + (2\xi\Omega)^2}} e^{i(\omega t - \phi_b)} \tag{14.76}$$

where $R_0 = r_0/a$ is the bow factor; $\phi = 2\xi\Omega/1 - \Omega^2$ is the unbalance phase angle; and $\phi_b = \phi + \alpha_0$ is the bow phase angle.

Rotors get statically bent or bowed or warped due to sudden thermal loads, or leaving the rotor unattended for long periods without barring. Sometimes, the rotor is balanced in a tunnel and left for long periods in a crate without adequate support to avoid gravity sag. In all these cases, the rotor comes to a halt at the heavy spot, with the rotor sag and eccentricity in one line. If the rotor is dropped from a height, the bow location angle is 180°, which is in a direction opposite to the mass center.

The response is rewritten as

$$\begin{aligned}\bar{R} &= [Ae^{-i\phi} + Be^{-i(\phi+\alpha_0)}]e^{i\omega t} \\ &= [A + Be^{-i\alpha_0}]e^{i(\omega t-\psi)}\end{aligned} \tag{14.77}$$

where A is the response due to conventional unbalance and B due to the bow unbalance.

$$A = \frac{\Omega^2}{\sqrt{(1-\Omega^2)^2 + (2\xi\Omega)^2}}$$

$$B = \frac{R_0 e^{i\alpha_0}}{\sqrt{(1-\Omega^2)^2 + (2\xi\Omega)^2}}$$

$$\psi = \tan^{-1}\left(\frac{A\sin\phi + B\sin\phi_b}{A\cos\phi + B\cos\phi_b}\right)$$

Usually, the bow gives $\alpha_0 = 0$, then these two responses get added, i.e., the unbalance increases.

$$\alpha_0 = 0°, \quad \bar{R} = \frac{\Omega^2 + R_0}{\sqrt{(1-\Omega^2)^2 + (2\xi\Omega)^2}} \tag{14.78}$$

For a dropped rotor the bow usually is $\alpha_0 = 180°$, then these two responses are opposed.

$$\alpha_0 = 180°, \quad \bar{R} = \frac{\Omega^2 - R_0}{\sqrt{(1-\Omega^2)^2 + (2\xi\Omega)^2}} \tag{14.79}$$

This condition leads to a self balancing speed, Ω_s, the speed at which the response becomes zero.

$$\Omega_s = \sqrt{R_0} \tag{14.80}$$

The response with bow phase 180° for a bow $R_0 = 0.5$ is given in Figure 14.42. The self balancing speed $\Omega_s = \sqrt{R_0} = \sqrt{0.5} = 0.707$, is shown in Figure 14.42, where a phase change of 180° takes place.

The self balancing and phase difference predicted was tested in the laboratory by Rao and Sharma [73]. A field observation of the bode plot of start up of a 220 MW machine in Raichur, Karnataka, India is shown in Figure 14.43. This bode plot clearly shows that the rotor is bent.

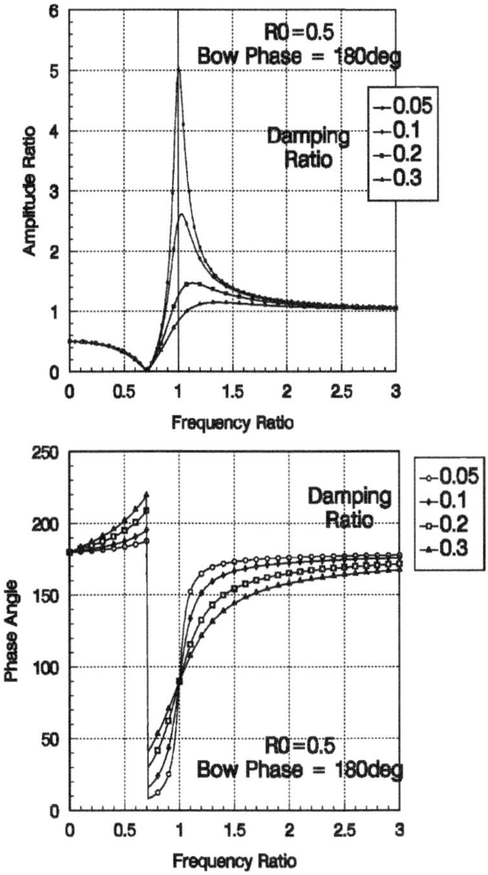

Fig. 14.42 Response with bow phase 1800 and self balancing speed

14.11 Variable Inertia

In reciprocating machinery, Goldsborough [26] has shown that torsional vibrations of large amplitudes occur within a series of ranges of instability, hence increasing the probability of crankshaft failure in fatigue. He later conducted theoretical investigations and tests of a single cylinder engine [27]. Draminsky [13] and Archer [2] in their analysis of failures of marine diesel engine crankshafts have clearly established the need to calculate vibration stresses that accounts for variable inertia effects. Carnegie et al. [7] used Runge–Kutta method to predict the instability regimes. Their model is derived here. For simplicity the piston displacement is assumed to be harmonic $x = a\cos\theta$. Then the potential and kinetic energies are written

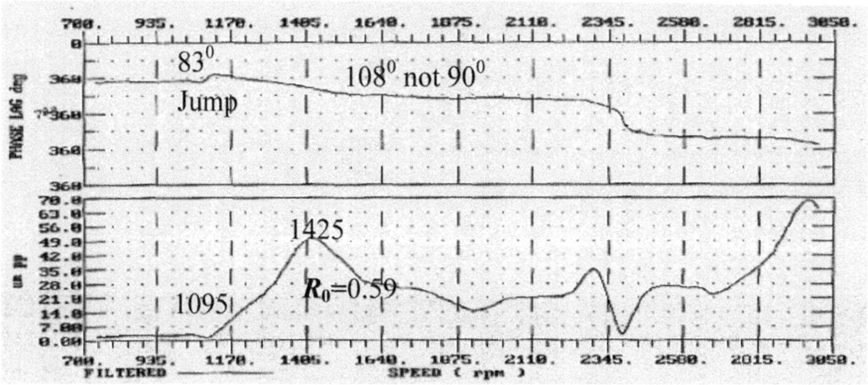

Fig. 14.43 Field observation of a bowed rotor

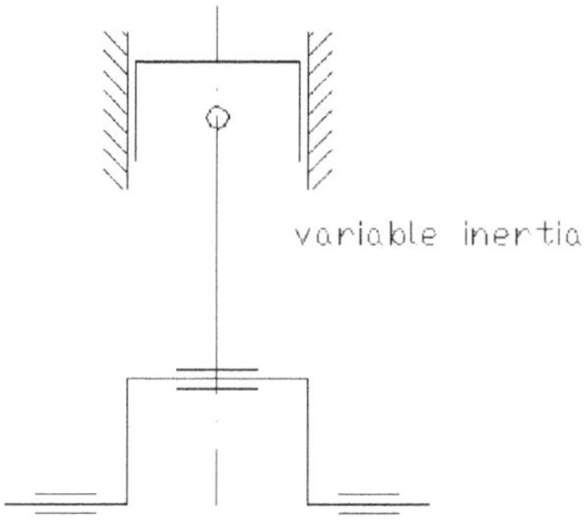

Fig. 14.44 Variable inertia system

$$T = \frac{1}{2}I\dot{\theta}^2 + \frac{1}{2}M\frac{1}{2}a^2\dot{\theta}^2(1 - \cos 2\theta) + \frac{1}{2}I_A\omega^2$$
$$V = \frac{1}{2}K(\theta - \theta_1)^2 \quad (14.81)$$

Neglecting higher order terms, the equation of motion is derived as

$$\left(I + \frac{1}{2}Ma^2 - \frac{1}{2}Ma^2\cos 2\theta\right)\ddot{\theta} + \frac{1}{2}Ma^2\dot{\theta}^2\sin 2\theta + K(\theta - \theta_1) = 0 \quad (14.82)$$

The equation of motion is non-dimensionalized as

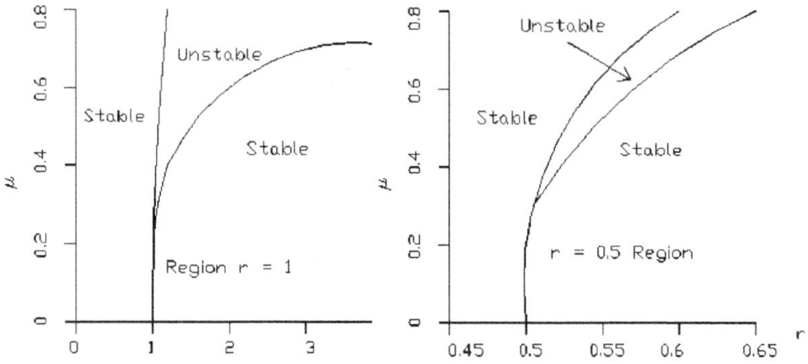

Fig. 14.45 Unstable regions of a variable inertia system

$$\theta = \omega t + \gamma, \quad \tau = \omega t$$

$$p^2 = \frac{K}{I + \frac{1}{2}Ma^2}, \quad \mu = \frac{\frac{1}{2}Ma^2}{I + \frac{1}{2}Ma^2}$$

$$(1 - \mu\cos 2\tau)\frac{d^2\gamma}{d\tau^2} + 2\mu\sin 2\tau\frac{d\gamma}{d\tau} + \left(\frac{1}{r^2} + 2\mu\cos 2\tau\right)\gamma = -\mu\sin 2\tau \tag{14.83}$$

Goldsborough solved for the complementary function of (14.83) using the Floquet theory [19], see [51].

$$\gamma = a_1 e^{c\tau}\phi(\tau) + a_2 e^{-c\tau}\phi(-\tau) \tag{14.84}$$

The exponent c and function ϕ are developed using variable inertia parameter μ

$$c = \mu C_1 + \mu^2 C_2 + \cdots$$
$$\phi = \cos\kappa\tau + A\sin\kappa\tau + \mu x_1 + \mu^2 x_2 + \cdots \tag{14.85}$$

where $\kappa = 1/r$. Then a solution is sought in the neighborhood of integer values of the non-dimensional parameter $\kappa^2 = \rho^2 + \mu\alpha + \cdots$ where $\rho = \pm1, \pm2, \ldots$ After a lengthy calculation [69] one gets analytical expressions for unknown coefficients c_1, c_2 and A and corresponding bounds of stability.

Oravsky and Rao [60] discussed another system representing an autonomous basic aggregate and showed that it too is governed by the same equations. A numerical solution by the Runge–Kutta method [41, 79] was obtained by Carnegie et al. [7]. The unstable regions around $r = 1$ and 1/2 are shown in Figure 14.45.

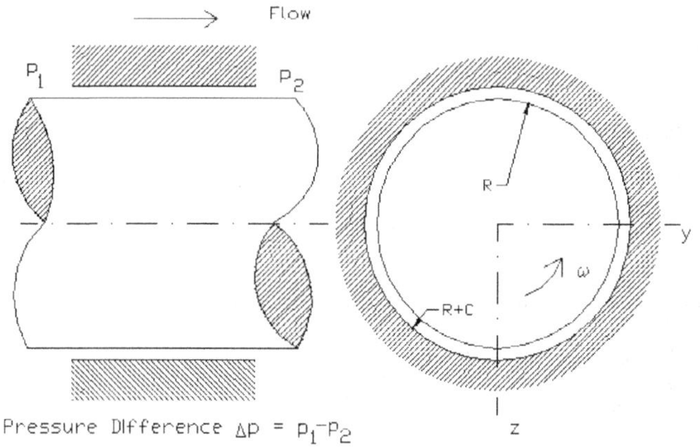

Fig. 14.46 Simple smooth seal

14.12 Seals and Instabilities

Seals are used to prevent leakage of a high pressure fluid medium into low pressure sections or into an atmosphere from one section of the rotor to another along the axial length. They are usually located between the shaft and the housing. Sometimes they are located between a rotating impeller shroud and the housing; they are then called neck ring seals. If they are located between the hub and the housing, they form an inter stage seal. They do not form a support to the rotor, however their characteristics are similar to those of fluid film bearings and therefore we study the properties of seals in this chapter.

Seals can be classified into two general categories, viz., smooth and labyrinth seals. A simple seal is shown in Figure 14.46. If the shaft is stationary and concentric with the seal there will be an axial flow due to the pressure difference across the seal. Due to any eccentricity, an unsymmetrical velocity and pressure distribution around the circumference takes place and the resulting force acting on the shaft is directly in opposition to the shaft displacement. This gives rise to a direct stiffness coefficient which was first determined by Lomakin [43]. Once the shaft rotates, the flow in the circumferential direction gets diverted, resulting in a transverse component to the direction of shaft displacement. This gives rise to the cross-coupled stiffness term. The displacement is usually accompanied by a velocity and it results in damping terms. The stiffness and damping forces for seals can be represented in a general form similar to equation (14.86).

$$-\begin{Bmatrix} F_x \\ F_y \end{Bmatrix} = \begin{bmatrix} M & 0 \\ 0 & M \end{bmatrix} \begin{Bmatrix} \ddot{x} \\ \ddot{y} \end{Bmatrix} + \begin{bmatrix} C_d & C_c \\ -C_c & C_d \end{bmatrix} \begin{Bmatrix} \dot{x} \\ \dot{y} \end{Bmatrix} + \begin{bmatrix} K_d & K_c \\ -K_c & K_d \end{bmatrix} \begin{Bmatrix} x \\ y \end{Bmatrix} \quad (14.86)$$

Lomakin [43, 44] provided a method of calculation of critical speeds and dynamic stability of hydraulic high pressure machines with reference to the forces acting in the gap seals. Black [4] gave the procedure for calculation of this stiffness and damping coefficients, this procedure is outlined by Rao [69] and Krämer [40]. Brown [5] adopted the same procedure for calculating the stiffness and damping coefficients in his analysis.

Consider a plain seal with length L, radius R and clearance C, with pressure difference across given by Δp. Let the average flow velocity in the seal be V, as given here.

$$V = \sqrt{\frac{2\Delta p}{\rho(1+\xi+2\sigma)}} \tag{14.87}$$

Resistance to flow is represented by σ with the resistance coefficient λ which depends on its Reynolds number.

$$\begin{aligned}
\sigma &= \lambda\frac{L}{C}\\
\lambda &= 0.066R_a^{-0.25}\left(1+\frac{1}{4b^2}\right)^{0.375}\\
R_a &= V\frac{C}{\nu}\\
b &= \frac{V}{R\omega}\\
\xi &\approx 0.5
\end{aligned} \tag{14.88}$$

Here, ρ is density, ν is kinematic viscosity and ξ is entrance loss coefficient, which can be taken as 0.5. The coefficient 0.066 and the exponents −0.25 and 0.375 are Hir's coefficients. The rotational speed is ω rad/s. In terms of characteristic time for the seal $T = L/V$, the stiffness and damping coefficients are

$$\begin{aligned}
K_d &= \left(a_0 - \frac{1}{4}a_2\omega^2T^2\right)K^*\\
K_c &= \frac{1}{2}a_1\omega T K^*\\
C_d &= a_1C^*\\
C_c &= a_2\omega T C^*
\end{aligned} \tag{14.89}$$

where

$$\begin{aligned}
K^* &= \Delta p\frac{LR}{c}\\
C^* &= K^*T
\end{aligned} \tag{14.90}$$

The coefficients a_0, a_1, and a_2 are given by

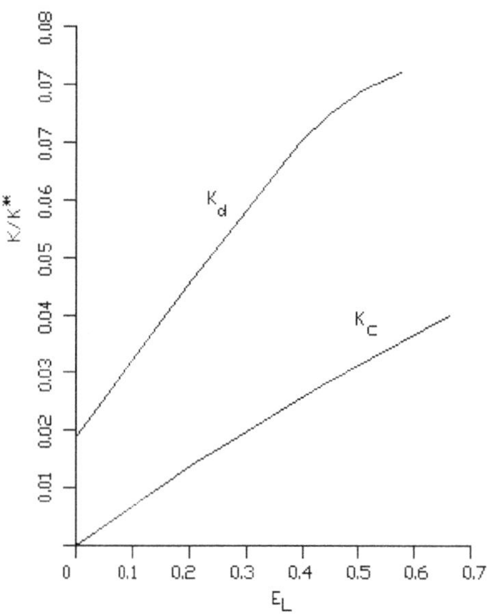

Fig. 14.47 Stiffness and damping coefficients of seals

$$a_0 = 2.5AE$$

$$a_1 = 24\left\{\frac{E}{\sigma} + \frac{1}{2}B\left(E + \frac{1}{6}\right)\right\}$$

$$a_2 = \frac{A}{\sigma}\left(E + \frac{1}{6}\right) \tag{14.91}$$

where A, B and E are obtained from

$$A = \frac{\pi\omega}{1 + \xi + 2\sigma}$$

$$B = \frac{1 + 7b^2}{1 + 4b^2}$$

$$E = \frac{1 + \xi}{2(1 + \xi + B\sigma)} \tag{14.92}$$

For labyrinth seals a coefficient E_L is first determined

$$E_L = \frac{1}{2}\bar{\rho}\omega^2 R^2 \frac{1}{\Delta p + \frac{1}{2}\rho_0 u_0^2} \tag{14.93}$$

where ρ_0 is the inlet density; $\bar{\rho}$ is the average density; and u_0 is the inlet axial velocity. Then the direct and cross-coupled stiffness are given by Figure 14.47.

The stiffness and damping coefficients of seals being important in stability analysis of high speed pumps, such as cryogenic pumps used in space applications [75], several attempts have been made to design test rigs for this purpose, see for example [9, 10, 36].

Childs and his team in Texas have an extensive facility for testing seals and have conducted several tests to determine stiffness and damping coefficients, e.g., many results are reported in [8]. Nordmann has also conducted extensive tests on seals, e.g., Nordmann and Massman [58] and Nordmann and Dietzen [57] gave a finite difference analysis to determine stiffness and damping coefficients, whereas Rao and Saravana [72] used a CFD code to determine these coefficients.

To assess stability of rotors due to seals, we can consider a Jeffcott rotor with mass $M = 50$ kg, $K = 5.0 \times 10^6$ N/m, rigid bearing critical speed is 3020 RPM. The seal dimensions and properties are:

$$R = 0.05 \text{ m}$$

$$L = 0.05 \text{ m}$$

$$C = 0.0002 \text{ m}$$

$$\Delta p = 8 \text{ MPa}$$

$$\rho = 5 \text{ kg/m}^3$$

$$\xi = 0.5$$

$$\nu = 2$$

At 3000 RPM:

$$K_d = 3.64 \times 10^6 \text{ N/m}$$

$$K_c = 1.99 \times 10^6 \text{ N/m}$$

The governing equations without damping are

$$\begin{bmatrix} M & \\ & M \end{bmatrix} \begin{Bmatrix} \ddot{z} \\ \ddot{y} \end{Bmatrix} + \begin{bmatrix} K + K_d & K_C \\ -K_C & K + K_d \end{bmatrix} \begin{Bmatrix} z \\ y \end{Bmatrix} = 0$$

The eigen-values are

$$\lambda^4 + 0.346 \times 10^6 \lambda^2 + 0.031 \times 10^{12} = 0$$

$$\lambda_{1,2}^2 = \frac{1}{2}\left[-0.346 \times 10^6 \pm \sqrt{0.12 \times 10^{12} - 0.126 \times 10^{12}}\right]$$

$$= -0.173 \times 10^6 \pm 0.078 \times 10^6 i$$

$$\lambda_{1,2} = 91.57 \pm 425.89 i$$

The response is

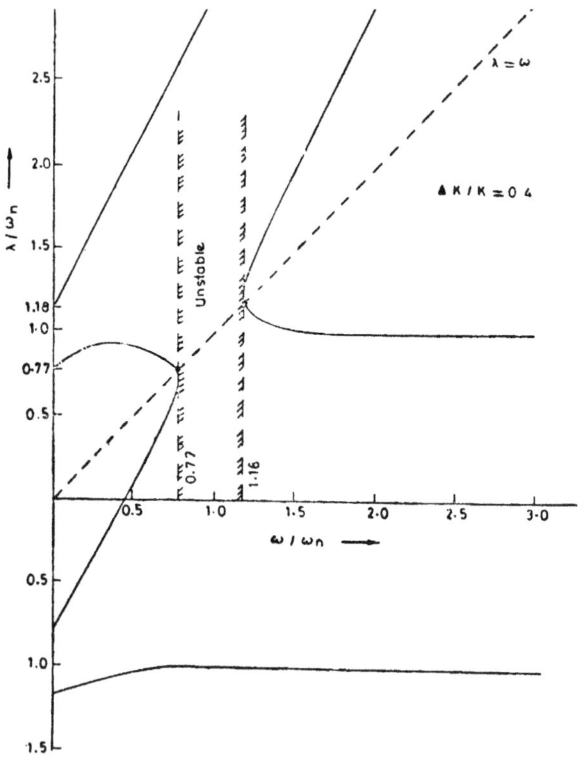

Fig. 14.48 Steam whirl

$$z = e^{91.57t}(Ae^{i425.89t} + Be^{-i425.89t})$$

The real part is positive and therefore the solution is unstable and the response occurs at 3000 RPM with a frequency of 425.89 rad/s = 4067 RPM.

14.13 Steam Whirl

The earliest publication addressing steam whirl is from Thomas [88]. Due to shaft whirling, the clearance does not remain constant and therefore a resultant couple acts arising from the steam forces, see Figure 14.48. Alford [1] proposed the same explanation for the gas turbines in the US. In the US, the net destabilizing force is called the Alford force, in Europe it is often referred to as the Thomas–Alford force. Urlichs [93] and Wohlrab [96] gave extensive experimental results.

Let P be the stage power, then the tangential force F_u (N) is

$$F_u = \frac{P}{\omega R}$$

$$f_u = \frac{F_u}{2\pi R} = \frac{P}{2\pi\omega R^2} \tag{14.94}$$

where f_u is force per unit circumferential length (N/m). The clearance c at an angle ϕ for a displacement in z direction is

$$c_\phi = c + \Delta c = c - z\cos\phi \tag{14.95}$$

Under ideal circumstances with clearance $c = 0$, $f_{u0} = [f_u]_{c=0}$. With a clearance, force per unit circumferential length is

$$f_u = f_{u0}\left(1 - \gamma\frac{c}{l}\right) \tag{14.96}$$

where l is blade length and c is the clearance factor. We can now write down the tangential force at an angle ϕ

$$\begin{aligned} f_{u\phi} &= f_u + \Delta f_u = f_u + \frac{\partial f_u}{\partial c}\Delta c \\ &= f_u\left(1 + \frac{\gamma}{l}z\cos\phi\right) \end{aligned} \tag{14.97}$$

We can sum up the forces around the periphery by integration and obtain the stiffnesses for displacement in the z direction

$$\begin{aligned} F_z &= \int_0^{2\pi} f_{u\phi} R\sin\phi d\phi = 0 \\ F_y &= \int_0^{2\pi} f_{u\phi} R\cos\phi d\phi = \left(\pi\frac{\gamma}{l}Rf_u\right)z = k_s z \\ k_s &= \pi\frac{\gamma}{l}Rf_u = \frac{\gamma}{l}\frac{P}{2\omega R}\ \text{N/m} \end{aligned} \tag{14.98}$$

We notice that the steam whirl gives no direct stiffness and only cross-coupled stiffness, the resulting force relation is

$$\begin{aligned} F_z &= -k_s y \\ F_y &= 0 \\ \begin{Bmatrix} F_z \\ F_y \end{Bmatrix} &= \begin{bmatrix} 0 & -k_s \\ k_s & 0 \end{bmatrix} \begin{Bmatrix} z \\ y \end{Bmatrix} \end{aligned} \tag{14.99}$$

The Thomas clearance factor γ in (14.96) is $2K_2$ where

$$K_2 \approx 2 - 0.4\left(\psi - \frac{3}{2}\right) \pm 0.2 \quad \text{50\% Reaction Stage}$$

$$K_2 \approx 3.2 - 0.27\left(\psi - \frac{3}{2}\right) \pm 0.2 \quad \text{Weak Reaction Stage}$$

$$\text{Blade Pressure Factor } \psi = 2\frac{\Delta h_s}{u^2}$$

As an example consider a 25 MW stage operating at 3000 RPM. The rotor mass is 5000 kg and the shaft stiffness is 200 MN/m. The rigid bearing critical speed is 200 rad/s.

Let the blade length l be 6 cm and disk radius 40 cm. γ is taken as 5. Then the steam whirl stiffness coefficient is

$$\begin{aligned} k_s &= \frac{\gamma}{l}\frac{P}{2\omega R} \\ &= \frac{5}{0.06}\frac{25 \times 10^6}{2 \times \frac{3000\pi}{30} \times 0.4} \\ &= 8.3 \text{ MN/m} \end{aligned}$$

The equations of motion and eigen-values are obtained as

$$\begin{bmatrix} M & \\ & M \end{bmatrix}\begin{Bmatrix} \ddot{z} \\ \ddot{y} \end{Bmatrix} + \begin{bmatrix} K & K_s \\ -K_s & K \end{bmatrix}\begin{Bmatrix} z \\ y \end{Bmatrix} = 0$$

$$\lambda^4 + 0.08 \times 10^6\lambda^2 + 0.001603 \times 10^{12} = 0$$

$$\begin{aligned} \lambda_{1,2}^2 &= \frac{1}{2}\left[-0.08 \times 10^6 \pm \sqrt{0.0064 \times 10^{12} - 0.006411 \times 10^{12}}\right] \\ &= -0.04 \times 10^6 \pm 0.0033 \times 10^6 i \end{aligned}$$

$$\lambda_{1,2} = 8.24 \pm 200i$$

The response is

$$z = e^{8.24t}(Ae^{i200t} + Be^{-i200t})$$

The real parts of the eigen-values are positive and therefore the system is unstable at this operation. The response occurs at 200 rad/s = 1910 RPM.

14.14 Cracked Shafts

It is commonly observed that high speed and heavy duty rotor shafts develop transverse cross-sectional cracks due to fatigue at some time during their life period.

Usually, the crack is initiated at a local defect or at a location where there is a stress raiser. For crack initiation, the stress required is usually large. In the case of a fatigue crack the combination of steady and dynamic stresses play a role. Since the stresses required are large, sudden transient loads are generally responsible for crack initiation. Once a crack is initiated it propagates and the stress required for propagation is generally much smaller than that required for crack initiation. The operating stresses may be sufficient to propagate the crack. The crack propagation takes place over a certain length when it is sufficient to create unstable conditions and fracture takes place. When the failure occurs it could be quite sudden without much notice. Hence it becomes important to understand the behavior of a cracked rotor.

Crack propagation is a slow process; it takes several cycles of rotation for the crack to advance a small distance over few microns. Therefore the shaft can be considered with a transverse crack of a given constant depth for the purpose of dynamic analysis. A Jeffcott rotor with a transverse crack is shown in Figure 14.49. The shaft is considered symmetric and the bearings rigid. Let K be the stiffness of the uncracked shaft; the corresponding deflection due to the gravity of the rotor is given in Figure 14.49.

When the cracked shaft is rotated, the rotor deflection varies due to change in the stiffness of the shaft as a function of the location of the crack. When the crack is occupying the lowest position, say 0°, the shaft offers minimum stiffness. Here, the crack is located amongst the shaft fibers which are in tension, consequently it is wide open. As the shaft rotates in a counter clockwise direction, to the 90° position, a portion of the crack is in compression and hence it tends to close as shown by the double hatched area in Figure 14.49. Therefore the shaft stiffness increases and the rotor deflection decreases. When the shaft is in the 180 deg. position, it can be seen that the crack is located amongst the fibers which are in compression and hence it is completely closed. The stiffness here corresponds to the uncracked shaft. The crack begins to open again as the shaft is rotated further, until it is wide open at the 360° position.

The stiffness of the shaft is significantly diminished only when the crack is of sufficient depth. For shallow cracks the stiffness decreases in a limited area surrounding the crack. This aspect makes the detection of a crack at an early stage to be difficult.

Mayes and Davies [48–50]), were amongst the earlier ones to have modeled a transverse crack. The periodic closing and opening of the crack is called breathing action [23]. Figure 14.50 shows the shaft with a crack in stationary as well as rotating coordinates. For convenience the η axis is located parallel to the crack. We can write

$$\begin{Bmatrix} F_\xi \\ F_\eta \end{Bmatrix} = \begin{bmatrix} K - \Delta K_c & 0 \\ 0 & K \end{bmatrix} \begin{Bmatrix} \xi \\ \eta \end{Bmatrix} \tag{14.100}$$

where the reduction in stiffness due to crack, ΔK_c is taken as

$$\begin{aligned} \Delta K_c &\neq 0, \quad \xi > 0 \\ \Delta K_c &= 0, \quad \xi < 0 \end{aligned} \tag{14.101}$$

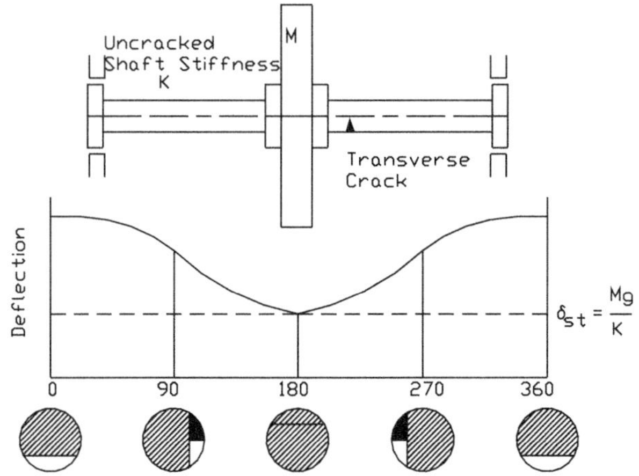

Fig. 14.49 Cracked shaft deflection under gravity

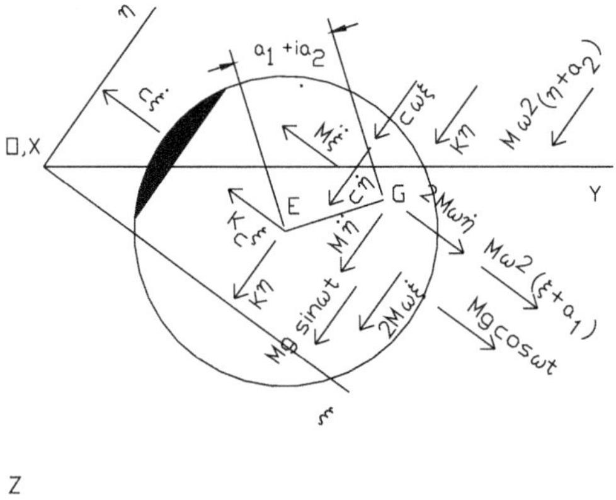

Fig. 14.50 Cracked shaft in rotating coordinates

The above breathing model can be represented by a hinge which closes under compression and opens under tension as shown in Figure 14.51. The governing equations can be written to give

$$M(\ddot{\xi} - 2\omega\dot{\eta} - \omega^2\xi) + C(\dot{\xi} - \omega\eta) + K_c\xi = Ma_1\omega^2 + Mg\cos\omega t$$
$$M(\ddot{\eta} + 2\omega\dot{\xi} - \omega^2\eta) + C(\dot{\eta} + \omega\xi) + K\eta = Ma_2\omega^2 + Mg\sin\omega t \quad (14.102)$$

where

$$K_c = K - \Delta K_c, \quad \xi > 0$$
$$= 0, \quad \xi < 0 \quad (14.103)$$

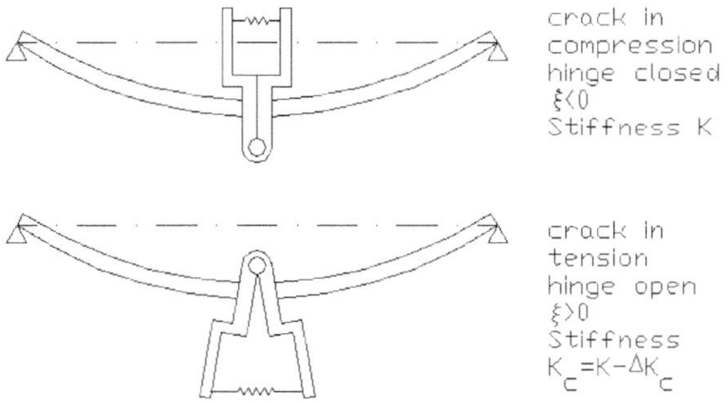

Fig. 14.51 Breathing crack model

The above equations can be non-dimensionalized and written as

$$[I]\begin{Bmatrix} \bar{\xi}'' \\ \bar{\eta}'' \end{Bmatrix} + \begin{bmatrix} 2\zeta & -2r \\ 2r & 2\zeta \end{bmatrix}\begin{Bmatrix} \bar{\xi}' \\ \bar{\eta}' \end{Bmatrix} + \begin{bmatrix} \frac{r^2}{r_c{}^2}(1-r_c{}^2) & -2\zeta r \\ 2\zeta r & (1-r^2) \end{bmatrix}\begin{Bmatrix} \bar{\xi} \\ \bar{\eta} \end{Bmatrix}$$

$$= \begin{Bmatrix} \cos r\tau \\ -\sin r\tau \end{Bmatrix} + \begin{Bmatrix} \frac{a_1 r^2}{\eta_{st}} \\ \frac{a_2 r^2}{\eta_{st}} \end{Bmatrix} \tag{14.104}$$

where

$$p = \sqrt{\frac{K}{M}}; \quad p_c = \sqrt{\frac{K_c}{M}}$$
$$r = \frac{\omega}{p}; \quad r_c = \frac{\omega}{p_c} \tag{14.105}$$

$$\eta_{st} = \frac{Mg}{K}; \quad \zeta = \frac{C}{2Mp}$$
$$\bar{\xi} = \frac{\xi}{\eta_{st}}; \quad \bar{\eta} = \frac{\eta}{\eta_{st}} \tag{14.106}$$

$$\tau = pt$$
$$' = \frac{d}{d\tau} \tag{14.107}$$

Equations (14.104) are nonlinear by virtue of (14.103). Gasch and Pfutzner [24] obtained analog computer solutions for a crack with $r/r_c = 0.9$ when the hinge is open and a damping ratio 0.005. Figure 14.52 shows the response due to gravity alone with no unbalance, i.e. $a = 0$. The response peak at $r = 0.5$ (gravity critical) is

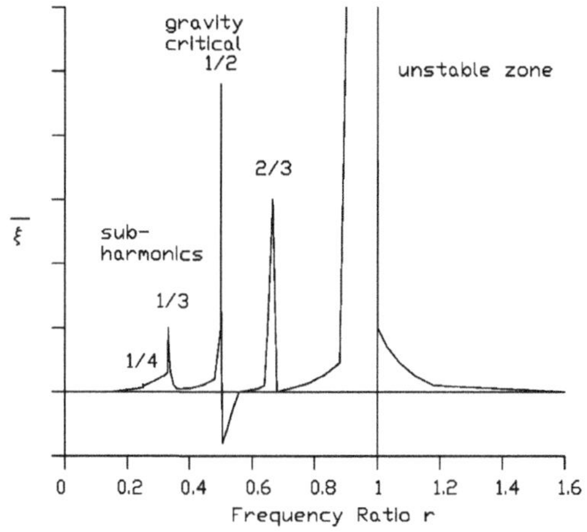

Fig. 14.52 Response of cracked shaft under gravity

clearly visible. There is an unstable zone in the range $0.9 < r < 1.0$. There are also other secondary resonances at $r = 1/3$, 1/4, … and 2/3, which are subharmonics.

The breathing action of the crack can be modeled more accurately by considering the stiffness in both directions as a function of time rather than as a step function in (14.101). Schmied and Krämer [80] calculated the stiffnesses in two directions z and y as a ratio of the uncracked shaft stiffness for a simply supported beam as a function of crack depth. These stiffnesses vary with the shaft rotation similar to Figure 14.49. They are expressed as

$$K_1(t) = K_{m1} + \Delta K_1 \cos\omega t$$
$$K_2(t) = K_{m2} + \Delta K_2 \cos\omega t \tag{14.108}$$

where

$$K_{m1} = \frac{1}{2}(K_0 + K_1), \quad \Delta K_1(K_0 - K_1)$$
$$K_{m2} = \frac{1}{2}(K_0 + K_2), \quad \Delta K_2(K_0 - K_2) \tag{14.109}$$

The variation of the above stiffnesses with respect to the coordinates 1, 2 fixed on the rotor are shown in Figure 14.53. They can be expressed in stationary coordinates by the following stiffness matrix

$$[K(t)] = \begin{bmatrix} K_m + \Delta K \cos 2\omega t & \Delta K \sin 2\omega t \\ \Delta K \sin 2\omega t & K_m - \Delta K \cos 2\omega t \end{bmatrix} \tag{14.110}$$

where

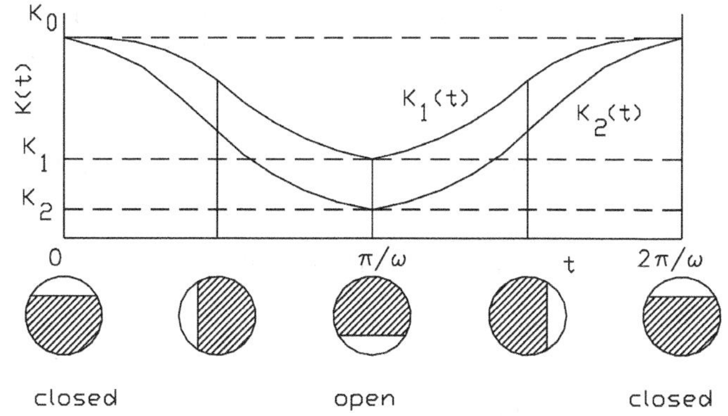

Fig. 14.53 Stiffness variation for a breathing crack

$$K_m = \frac{1}{2}[K_1(t) + K_2(t)]$$

$$\Delta K = \frac{1}{2}[K_1(t) - K_2(t)] \tag{14.111}$$

Using (14.108), (14.109) and (14.111), equation (14.110) can be recast as

$$[K(t)] = [K_c] + [K_t(t)]$$
$$= \begin{bmatrix} K'_m & 0 \\ 0 & K'_m \end{bmatrix} + \Delta K' \begin{bmatrix} f_1(t) & f_3(t) \\ f_3(t) & f_2(t) \end{bmatrix} \tag{14.112}$$

where

$$K'_m = \frac{1}{4}K_0(2 + \kappa_1 + \kappa_2)$$

$$\Delta K' = \frac{1}{8}K_0(\kappa_1 - \kappa_2)$$

$$\kappa_1 = \frac{K_1}{K_0}; \quad \kappa_2 = \frac{K_2}{K_0} \tag{14.113}$$

and

$$f_1(t) = C' \cos \omega t + 2 \cos 2\omega t - \cos 3\omega t$$

$$f_2(t) = (C' + 2) \cos \omega t - 2 \cos 2\omega t + \cos 3\omega t$$

$$f_3(t) = - \sin \omega t + 2 \sin 2\omega t - \sin 3\omega t$$

$$C' = \frac{4 - 3\kappa_1 - \kappa_2}{\kappa_1 - \kappa_2} \tag{14.114}$$

The stiffness matrix of the cracked shaft therefore contains harmonic terms with 1 per rev, 2 per rev and 3 per rev components. One can therefore expect the response to contain half and one-third subharmonics. For a general rotor, we have several natural frequencies and therefore on the Campbell diagram there will be several critical speeds with 1 per rev excitation. For each of these critical speeds we can imagine a Jeffcott rotor model. Hence for a cracked shaft, resonances occur at half and one-third values of all the critical speeds. Of specific importance are those concerned with the first critical speed. There are several other contributions for cracked shafts [12,54].

This chapter essentially shows how the subject of Rotor Dynamics developed once Jeffcott gave a clear understanding on the unbalance and response when the stationary shaft rotates. This is like a single degree of freedom modal model that one can derive from a multidegree of freedom stationary system. Once we get the natural frequencies of a stationary rotor by using, e.g., Prohl's method, i.e., rigid bearing rotor critical speed, we can construct the Jeffcott rotor and from there include the rotor dynamic specific aspects, viz., oil film bearings, asymmetric shaft, seals, etc. This approach even today is the best, since determining rigid bearing critical speed is the simplest and easily accomplishable.

While Myklestad and Prohl provided numerical methods around World War II, the initial computer era exploited transfer matrix methods before finite element methods for rotors were developed. However, the 20th century by and large was restricted to beam models only.

References

1. Alford, J. (1965) Protecting Turbomachinery from Self Excited Rotor Whirl, *J. Engng. Power*, vol. 87, No. 4, p. 333.
2. Archer, S. (1964) Some Factors influencing the Life of Marine Crankshafts, *Trans. Instn. Marine Engineers*.
3. Bhat, R.B., Rao, J.S. and Sankar, T.S. (1982) Optimum Journal Bearing Parameters for Minimum Unbalance Response in Synchronous Whirl, *J. Mech. Des., ASME*, vol. 104, p. 339.
4. Black, H.F. (1977) The Effect of Inlet Flow Swirl on the Dynamic Coefficients of High Pressure Annular Clearance Seals, Univ. of Virginia, Charlottesville.
5. Brown, R.D. (1992) Dynamic Characteristics of Long Annular Seals in Centrifugal Pumps, in *Proceedings 5th International I Mech E Conf on Vibrations in Rotating Machinery*, p. 467.
6. Carnegie, W. (1964) Rotary Inertia and Gyroscopic Effects in Overhung Shaft Systems, *Bull. Mech. Engng. Educ.*, vol. 3, p. 191.
7. Carnegie, W., Rao, J.S. and Pasricha, M.S. (1971) A Theoretical Study of the Effects of Variable Inertia on the Torsional Vibrations of a Single Cylinder Engine System, Presented at Vibration Section of Institution of Marine Engineers, 17th November.
8. Childs, D.W. (1993) *Turbomachinery Rotordynamics: Phenomena, Modeling, and Analysis*, Wiley Inter Science.
9. Childs, D.W. and Dressman, J.B. (1982) Testing of Turbulent Seals for Rotordynamic Coefficients, NASA CP 2250, p. 157.
10. Childs, D.W. and Wade, J. (2004) Rotordynamic-Coefficient and Leakage Characteristics for Hole-Pattern-Stator Annular Gas Seals – Measurements Versus Predictions, *Journal of Tribology, Transactions of the ASME*, vol. 126, p. 326.

11. Den Hartog, J.P. (1940, 1956) *Mechanical Vibration*, McGraw-Hill Book Co.
12. Dimarogonas, A.D. and Papadopoulos, C.A. (1983) Vibration of Cracked Shafts in Bending, *Journal of Sound and Vibration*, vol. 91, no. 4, p. 583.
13. Draminsky, P. (1961) *Secondary Resonance and Subharmonics in Torsional Vibration*, Acta Polytechnica, Scandinavia, No. 10, Copenhagen.
14. Dunkerley, S. (1894) On the Whirling of Vibration of Shafts, *Philos. Trans. Roy. Soc., Series A*, vol. 185, p. 279.
15. Ehrich, F.F. (1964) Shaft Whirl Induced by Rotor Internal Damping, *Journal of Applied Mechanics*, vol. 23, no. 1, p. 109.
16. Ehrich, F.F. (1987) *Self Excited Vibration, Shock and Vibration Handbook*, McGraw-Hill.
17. Ehrich, F.F. (1992) *Handbook of Rotor Dynamics*, McGraw-Hill.
18. Ehrich, F.F. (1995) Nonlinear Phenomena in Dynamic Response of Rotors in Anisotropic Mounting Systems, in *ASME Special 50th Anniversary Design Issue*, vol. 117, p. 154.
19. Floquet, G. (1883) Sur les équations différentielles linéaires à coefficients périodiques, *Ann. École Norm. Sup.*, vol. 12, p. 47.
20. Föppl, O. (1895) Das Problem der Lavalschen Turbinenwelle, *Der Civilingenieur*, vol. 4, p. 335.
21. Föppl, O. (1918) Z. f. gesamte Turb., *Wesen*, nos. 18 and 19.
22. Föppl, O. (1919) *Zeitschrift der VDI*, p. 866.
23. Gasch, R. (1976) Dynamic Behavior of a Simple Rotor with a Cross-Sectional Crack, in *Proc. I Mech. E Conf. Vibrations in Rotating Machinery*, Cambridge, C178/76, p. 123.
24. Gasch, R. and Pfutzner, H. (1975) *Rotordynamik*, Springer Verlag.
25. Gibbons, C.B. (1976) Coupling Misalignment Forces, in *Proceedings Fifth Turbomachinery Symposium*, Gas Turbine Laboratories, Texas A&M University, pp. 111–116.
26. Goldsborough, G.R. (1925) Torsional Vibration in Reciprocating Engine Shafts, *Proc. Roy. Soc.*, vol. 109, p. 99.
27. Goldsborough, G.R. (1926) The Properties of Torsional Vibration in Reciprocating Engine Shafts, *Proc. Roy. Soc.*, vol. 113, p. 259.
28. Green, R. (1948) Gyroscopic Effects of the Critical Speeds of Flexible Rotors, *Journal Applied Mechanics*, vol. 15, p. 369.
29. Gümbel (1917) *Dinglers Polytechnic Journal*, p. 235.
30. Gümbel (1918) *Dinglers Polytechnic Journal*, p. 71.
31. Gunter, E.J. (1966) Dynamic Stability of Rotor Bearing System, NASA report SP-113.
32. Harris, C.M. and Crede, C.E. (1981) *Shock and Vibration Handbook*, McGraw-Hill.
33. Holzer, H. (1921) *Die Berechnung der Drehschwingungen*, Springer Verlag.
34. Holzer, H. (1922) Tabular Method for Torsional Vibration Analysis of Multiple-Rotor ShSft systems, *Machine Design*, May, p. 141.
35. Iwatsubo, T. (1971) Vibration of Asymmetric Shaft, *JSME*, vol. 37. p. 1503.
36. Iwatsubo, T., Sheng, B.C. and Matsumoto, T. (1988) An Experimental Study on the Static and Dynamic Characteristics of Pump Annular Seals, NASA CP 3026, p. 229.
37. Jeffcott, H.H. (1919) The Lateral Vibration of Loaded Shafts in the Neighborhood of a Whirling Speed – The Effect of Want of Balance, *Philos. Mag., Series 6*, vol. 37, p. 304.
38. Kimball, A.L. (1924) Internal Friction Theory of Shaft Whipping, *General Electric Review*, vol. 27, p. 244.
39. Klompas, N. (2001) Nature of Vibratory Waves in Bladed Disks, 2001-GT-0291.
40. Krämer, E. (1993) *Dynamics of Rotors and Foundations*, Springer-Verlag, Berlin.
41. Kutta, W. (1900) Beiträge zur näherungsweisen Integration totaler Differentialgleichungen, Ph.D. Thesis, University of Munich.
42. Lee, Y.S. (1998) Modeling and Vibration Analysis of Misaligned Rotor-Ball Bearing Systems, Ph.D. Thesis, KAIST, Korea.
43. Lomakin, A.A. (1955) Feed Pumps of the SWP-220-280 Type with Ultra-High Operating Data, *Energomashinotroenie*, vol. 2.
44. Lomakin, A.A. (1958) Calculation of Critical Speeds and Securing of the Dynamic Stability of Hydraulic High-Pressure Machines with Reference of the Forces Arising in the Gap Seals, *Energomashinotroenie*, vol. 4, no. 1.

45. Lund, J.W. (1965) *Rotor Bearing Dynamic Design Technology, Part III: Design Handbook for Fluid Film Bearings* and *Part V: Computer Program for Unbalance Response and Stability*, Mechanical Technology Inc., AFAPL-Tr-65-45.
46. Lund, J.W. (1974) Stability and Damped Critical Speeds of a Flexible Rotor in Fluid Film Bearings, *Journal of Engineering for Industry, Trans. ASME*, vol. 92, p. 509.
47. Mathieu, E. (1868) Mémoire sur le Mouvement Vibratoire d'une Membrane de forme Elliptique, *Journal des Mathématique Pures et Appliquées*, p. 137.
48. Mayes, I.W. and Davies W.G.R. (1976) The Vibrational Behavior of A Rotating Shaft System Containing A Transverse Crack, in *Proceedings I Mech. E Conf. Vibrations in Rotating Machinery*, Cambridge, C168/76, p. 53.
49. Mayes, I.W. and Davies, W.G.R. (1980) A Method of Calculating Vibrational Behavior of Coupled Rotating Shafts Containing a Transverse Crack, in *Proceedings I Mech. E Conf. Vibrations in Rotating Machinery*, Cambridge, 1980, C254/80, p. 18.
50. Mayes, I.W. and Davies, W.G.R. (1984) Analysis of the Response of a Multi-Rotor-Bearing System Containing a Transverse Crack in a Rotor, *Journal Vib. Acoust. Stress and Rel. in Des.*, vol. 106, p. 139.
51. McLachlan, N.W. (1962) *Theory and Application of Mathieu Functions*, Dover.
52. Morrison, D. and Peterson, A.N. (1964-65) Criteria for Unstable Oil Whirl of Flexible Rotor, *Proc. I Mech E*, vol. 179, part 3J, p. 45.
53. Morton, P.G. (1967) Influence of Coupled Asymmetric Bearings on the Motion of a Massive Flexible Rotor, *Proc. Inst. Mech. Engrs.*, vol. 182, part 1, no. 13, p. 255.
54. Nataraj, C. and Nelson, H.D. (1986) The Dynamics of a Rotor System with a Cracked Shaft, *Journal Vib. Acoustic. Stress and Rel. in Des.*, vol. 108, p. 189.
55. Newkirk, B.L. (1924) Shaft Whipping, *General Electric Review*, vol. 27, p. 169.
56. Newkirk, B.L. and Taylor, H.D. (1925) Oil Film Whirl – An Investigation of Disturbances on Oil Film in Journal Bearings, *General Electric Review*, vol. 28, p. 559.
57. Nordmann, R. and Dietzen F.J. (1988) Finite-Difference Analysis of Rotor dynamic Seal Coefficients for An Eccentric Shaft Position, in *Proceedings Vibrations in Rotating Machinery Conf. I. Mech. E.*, p. 379.
58. Nordmann, R. and Massman, H. (1984)Identification of Dynamic Coefficients of Annular Turbulent Seals, NASA 2338, p. 295.
59. Ocvirk, F.W. (1952) Short Bearing Approximation for Full Journal Bearings, NASA TN 2808.
60. Oravsky, V. and Rao, J.S. (1996) Dynamic Characteristics of Two Different Systems with Variable Inertia controlled by Same Equations, in *Proceedings I Mech. E Conference Transactions, Sixth Intl. Conf. on Vibrations in Rotating Machinery*, Oxford, p. 609.
61. Petroff, N. (1883) Reibung in Maschinen und Wirkung des Schmiermittels, Original Russisch, Neue Theorie der Reibung, Leipzig 1887.
62. Pinkus, O. and Sternlicht, B. (1961) *Theory of Hydrodynamic Lubrication*, McGraw-Hill.
63. Prandtl, L. (1918) Beitrage zur Frage der Kritischen Drehzahlen, *Dinglers Polytechnic Journal*, p. 179.
64. Rankine, W.J.M. (1869) On the Centrifugal Force of Rotating Shafts, *Engineer*, vol. 27, p. 249.
65. Rao, J.S. (1982) Synchronous Whirl of a Flexible Rotor in Hydrodynamic Bearings, *Mechanism and Machine Theory*, vol. 17, no. 2, p. 143.
66. Rao, J.S. (1983) *Rotor Dynamics*, John Wiley & Sons, New Age International, 1996.
67. Rao, J.S. (1983) Instability of Rotors in Fluid Film Bearings, *ASME J. Vib. Acoustic. Stress Rel. Des.*, vol. 105, p. 274.
68. Rao, J.S. (1985) Instability of Rotors Mounted on Fluid Film Bearings with a Negative Cross-Coupled Stiffness Coefficient, *Mechanism and Machine Theory*, vol. 20, No. 3, p. 181.
69. Rao, J.S. (1992) *Advanced Theory of Vibration*, John Wiley & Sons.
70. Rao, J.S. (1993) A Note on Quality Factor of Rotor with Hydrodynamic Bearings, *Journal of Engineering for Gas Turbines and Power, Trans. ASME*, vol. 115, p. 261.
71. Rao, J.S., Raju, R.J. and Reddy, K.B.V. (1970) Experimental Investigation on Oil Whip of Flexible Rotors, *Tribology*, p. 100.

72. Rao, J.S. and Saravana, M. (2006) Numerical Simulation of Seal Flow and Determination of Stiffness and Damping Coefficients, in *Proceedings 7th IFToMM-Conference on Rotor Dynamics*, Vienna, Austria, pp. 25–28.
73. Rao, J.S. and Sharma, M. (2003) Dynamic Analysis of Bowed Rotors, Advances in Vibration Engineering, *Journal of Vibration Institute of India*, vol. 2, no. 2, p. 128.
74. Rao, J.S. and Sreenivas, R. (2003) Dynamic Analysis of Misaligned Rotor Systems, Advances in Vibration Engineering, *Journal of Vib Institute of India*, vol. 2, no. 1, p. 1.
75. Rao, J.S., Sreenivas, R. and George, P. (2004) Dynamics of High Speed Cryo Pump Rotors, in *Proceedings 8th International I Mech E Conference on Vibrations in Rotating Machinery*, C623/103/2004, p. 467.
76. Rayleigh, J.W.S. (1877) *Theory of Sound*, Macmillan, London.
77. Reynolds, O. (1886) On the Theory of Lubrication and its Application to Mr. Beauchamp Tower's Experiments, *Philos. Trans. Royal Society*, Part 1, p. 177.
78. Robertson, D. (1933) Whirling of Journal in Sleeve Bearings, *Philos. Mag.*, vol. 15, p. 113.
79. Scarborough, J.B. (1950) *Numerical Mathematical Analysis*, Johns Hopkins Press.
80. Schmied, J. and Krämer, E. (1984) Vibrational Behavior of a Rotor with a Cross-sectional Crack, in *Proc. I Mech. E Conf. Vibrations in Rotating Machinery*, Edinburgh, C279/84, p. 183.
81. Sommerfeld, O.Z. (1904) *Math. Phys*, vol. 50, p. 97.
82. Stodola, A. (1910) *Dampf- und Gasturbinen*, Springer, Berlin. Translation (1927) *Steam and Gas Turbines*, McGraw-Hill.
83. Stodola, A. (1917) *Schweiz Bauztg.*, vol. 69, pp. 93, 229.
84. Stodola, A. (1918a) *Dinglers Polytechnic Journal*, pp. 1, 17, 117 and 135.
85. Stodola, A. (1918b) *Dinglers Polytechnic Journal*, p. 182.
86. Subbaiah, R., Bhat, R.B., Sankar, T.S. and Rao, J.S. (1985) Backward Whirl in a Simple Rotor Supported on Hydrodynamic Bearings, NASA Conf. Publication 2409, p. 145.
87. Taylor, H.D. (1945) Critical Speed Behavior of Unsymmetrical Shafts, *ASME Journal of Applied Mechanics*, vol. 84, p. A77.
88. Thomas, H. (1958) Instabile Eigenschwingungen von Turbinenläufern angefacht durch die Spaltströmungen Stopfbuschen und Beschauflungen, *Bulletin de l'AIM*, vol. 71, p. 1039.
89. Thomson, W.T. (1949) Matrix Solution of Vibration of Non-Uniform Beams, ASME Paper 49 A-11.
90. Timoshenko, S.P. (1955) *Vibration Problems in Engineering*, D. Van Nostrand Co. Inc.
91. Tondl, A. (1965) *Some Problems of Rotordynamics*, Chapman and Hall.
92. Tower, B. (1883) First Report on Friction Experiments (Friction of Lubricated Bearings), *Proc. Instn. of Mech. Engrs.*, p. 632.
93. Urlichs, K. (1976) Leakage Flow in Thermal Turbomachines as the Origin of Vibration Exciting Lateral Forces, NASA TT F-17409.
94. Vanderplaats, G.N. (1973) Structural Optimization by Methods of Feasible Directions, *Computers and Structures*, vol. 3, p. 739.
95. Whittaker, E.T. (1914) On the General Solution of Mathieu's Equation, *Proc. Edinburgh Math. Soc.*, vol. 32, p.75.
96. Wohlrab, R. (1975) Experimentelle Ermittlung Spaltsströmungsbedingter Kräfte an Turbinenstufen und Deren Einfluss auf die Laufstabilität Einfacher Rotoren, Doctoral Thesis, Technical University, Munich.

Chapter 15
Transfer Matrix Methods

We have seen in Sections 12.6 and 12.7 the Holzer and Myklestad–Prohl methods in transfer matrix form for stationary shafts. We will illustrate here transfer matrix methods for rotors.

15.1 Torsional Vibration due to Short Circuit of Generators

Linke [6] measured the main current and exciting current after a short circuit. The main current rose to about 32 times the normal current. Brown Boverie Co. [1] reported that for a 3000 RPM 8800 kva machine, the maximum short circuit current is about 10–20 times that at full load. The current diminishes rapidly, but the steady state value is reached only after several seconds.

Stodola [11] investigated whether there is a possibility for the period of the alternating current to be very near to the period of the natural vibration of the shaft. He used a two-degree freedom torsional system, with inertia for generator rotor Θ_0 and another one Θ of the turbine. The polar moment of the shaft is J_p and G is the coefficient of rigidity. The periodic torque of electrical forces is $M = M_0 \cos \omega_e t$. The twists of turbine and generator rotors are ϕ_1 and ϕ_2. The twist of the shaft produces a twisting moment

$$M_t = \frac{J_p G}{L} (\phi_1 - \phi_2) \tag{15.1}$$

The equations of motion are

$$\begin{aligned} \Theta \ddot{\phi}_1 &= -M_t \\ \Theta_0 \ddot{\phi}_2 &= M_t - M \end{aligned} \tag{15.2}$$

The above two equations are combined to give

$$\ddot{\phi}_1 - \ddot{\phi}_2 = -\frac{J_p G}{L} \left(\frac{1}{\Theta} + \frac{1}{\Theta_0} \right) (\phi_1 - \phi_2) - \frac{M_0}{\Theta_0} \cos \omega_e t \tag{15.3}$$

Let

$$\varepsilon^2 = \frac{J_p G}{L}\left(\frac{1}{\Theta} + \frac{1}{\Theta_0}\right)$$

be the natural frequency of the turbine-generator system in torsional vibration, and let $\phi_1 - \phi_2 = \phi$ and $M_0/\Theta_0 = a$, equation (15.3) is simplified to

$$\ddot{\phi} = -\varepsilon^2\phi - a\cos\omega_e t \tag{15.3a}$$

The excitation from electrical forces is assumed to be constant for a small period instead of decaying after some time; then for forced vibration, equation (15.3a) gives

$$\phi_0 = A\cos\omega_0 t = \frac{a}{\varepsilon^2 - \omega^2}\cos\omega_0 t \tag{15.4}$$

The complete solution is

$$\phi = \frac{a}{\varepsilon^2 - \omega^2}\cos\omega_0 t + B\cos\varepsilon t + C\sin\varepsilon t \tag{15.5}$$

For $\phi = 0, \dot{\phi} = 0$ at $t = 0$, the above yields

$$\phi = \frac{a}{\varepsilon^2 - \omega^2}(\cos\omega_0 t - \cos\varepsilon t) \tag{15.6}$$

The maximum value of the response is

$$\phi_{\max} = \frac{2a}{\varepsilon^2 - \omega^2} = \frac{2M_0}{\Theta_0\left(\varepsilon^2 - \omega^2\right)} \tag{15.7}$$

Present day technologies would allow us to calculate the response more accurately. A sudden short circuit at generator terminals sets up large torques in the rotor and induces severe stresses of the order of 4–5 times the normal value [9]. A short circuit in a modern generator terminals occurs for a very short period, just about 0.2 seconds and is removed by automatic controls. During this time, the short circuit can be divided into two parts, the first of which lasts less than 0.05 seconds and is known as a subtransient. The current across the terminals raises up to 10–15 times the rated value. In the second part, the transient, the current fluctuates between 3–5 times. The torque consists of two components:

- Air gap torque T_a that decays depending on machine constant and more predominant during the subtransient. Let T_n be the normal torque, then the air gap torque is given by [2]

$$T_a = T_n m_a e^{-\alpha t}\sin\omega t \tag{15.8}$$

  where α is subtransient saliency constant m_a is the oscillating moment and ω is the rotational speed (line frequency).
- Armature torque T_s which is predominant in the transient period is given by

$$T_s = T_n m_s e^{-\beta t}\sin 2\omega t \tag{15.9}$$

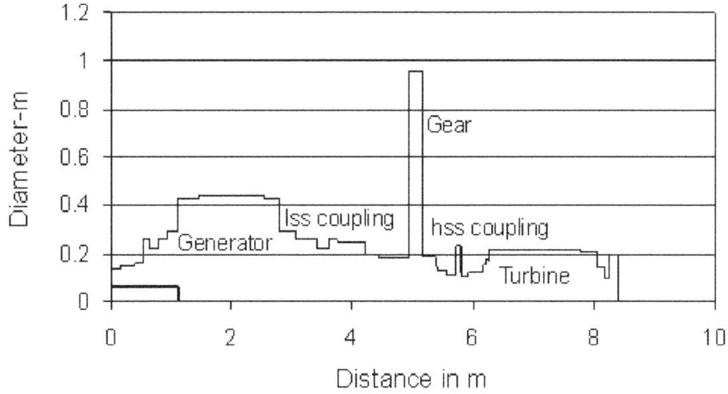

Fig. 15.1 6 MW turbo-generator set geometry

where m_s and β are transient saliency constants. The armature torque occurs at twice rotational speed.

The torsional analysis of such systems is discussed by Hammons [5], Rao et al. [8] and Schwibinger et al. [10]. Usually, the couplings are designed to act as a fuse between the turbine and generator so that the rotor system is protected from any damage. The transfer matrix method discussed in Section 12.6 is of advantage in torsional analysis as one can keep long elements of shafting in comparison to the finite element modeling [3].

The generator considered is 7508 kW operating at 1500 RPM (rated torque 47800 Nm). The speed ratio 5.5416 of the gear box is taken into account while modeling the system. Drawings of the turbine and generator are used to obtain the geometry, see Figure 15.1. The rotors are divided into the required number of elements using any change in the shaft cross-section diameter as the beginning and end of the elements, as shown in Table 15.1, 59 elements are used here. The mounted parts, e.g., blades, are taken into account as additional discrete polar mass moments of inertia, shear modulus is 0.800×10^5 N/mm^2 and density is 0.785×10^{-5} kg/mm^3.

Using the transfer matrix method, the first three natural frequencies obtained are 1,318.03, 4,325.55 and 6,173.07 RPM. The mode shapes of the TG set are given in Figure 15.2. The first mode is at 1318 RPM with the node falling at the low speed side (lss) coupling and the turbine suffers a large displacement relative to the generator. The second mode is at 4325 RPM with the two nodes at the hollow end of the generator shaft and the high speed side (hss) coupling. In this mode only the generator is responding. The third mode is 6173 RPM with the first node very close to the generator shaft end, the second node before the lss coupling and the third node just beyond the hss coupling. The gearbox suffers maximum displacement in the third mode.

An equivalent discrete model is made to identify the generator to which the excitation torque is applied, the gearbox and the turbine rotor. The approximate CG of each disk location is chosen at stations 6, 12, 26 and 47 as given in Table 15.1.

Table 15.1 Data of turbo-generator set

Sectn.	Cplng Locn 1	Length mm	Outer Dia mm	Inner Dia mm	Polar Inertia kg-mm^2	Tr. Ratio Ref. Gen.	CG Locn Approx
1	0	0.140E+03	140.00	60.00	0.300E+08	1.00	
2	0	0.243E+03	140.00	60.00	0.000E+00	1.00	
3	0	0.155E+03	150.00	60.00	0.160E+07	1.00	
4	0	0.100E+03	160.00	60.00	0.000E+00	1.00	
5	0	0.132E+03	260.00	60.00	0.000E+00	1.00	
6	0	0.156E+03	220.00	60.00	0.000E+00	1.00	1
7	0	0.200E+03	260.00	60.00	0.000E+00	1.00	
8	0	0.343E+03	290.00	60.00	0.000E+00	1.00	
9	0	0.275E+03	430.00	0.00	0.520E+08	1.00	
10	0	0.250E+03	440.00	0.00	0.116E+09	1.00	
11	0	0.250E+03	440.00	0.00	0.116E+09	1.00	
12	0	0.440E+02	440.00	0.00	0.200E+08	1.00	1
13	0	0.250E+03	440.00	0.00	0.116E+09	1.00	
14	0	0.250E+03	440.00	0.00	0.116E+09	1.00	
15	0	0.275E+03	430.00	0.00	0.520E+08	1.00	
16	0	0.343E+03	290.00	0.00	0.000E+00	1.00	
17	0	0.202E+03	260.00	0.00	0.000E+00	1.00	
18	0	0.152E+03	220.00	0.00	0.000E+00	1.00	
19	0	0.174E+03	260.00	0.00	0.000E+00	1.00	
20	0	0.260E+03	250.00	0.00	0.000E+00	1.00	
21	1	0.157E+08	0.00	0.00	0.382E+07	1.00	
22	0	0.245E+03	200.00	0.00	0.000E+00	1.00	
23	0	0.130E+03	180.00	0.00	0.000E+00	1.00	
24	0	0.980E+02	180.00	0.00	0.000E+00	1.00	
25	0	0.277E+03	180.00	0.00	0.000E+00	1.00	
26	0	0.220E+03	958.00	0.00	0.000E+00	1.00	1
27	0	0.220E+03	189.00	0.00	0.000E+00	5.50	
28	0	0.275E+02	148.00	0.00	0.000E+00	5.50	
29	0	0.165E+03	130.00	0.00	0.000E+00	5.50	
30	0	0.164E+03	110.00	0.00	0.000E+00	5.50	
31	0	0.400E+02	230.00	0.00	0.000E+00	5.50	
32	1	0.495E+07	0.00	0.00	0.569E+06	5.50	
33	0	0.635E+01	126.00	0.00	0.000E+00	5.50	
34	0	0.349E+02	230.00	0.00	0.000E+00	5.50	
35	0	0.111E+03	102.00	0.00	0.000E+00	5.50	
36	0	0.508E+02	120.34	0.00	0.000E+00	5.50	
37	0	0.476E+02	126.00	0.00	0.000E+00	5.50	
38	0	0.476E+02	126.00	0.00	0.000E+00	5.50	
39	0	0.905E+02	126.69	0.00	0.000E+00	5.50	
40	0	0.539E+02	155.42	0.00	0.000E+00	5.50	
41	0	0.333E+02	174.50	0.00	0.000E+00	5.50	
42	0	0.123E+02	171.30	0.00	0.000E+00	5.50	
43	0	0.592E+02	212.60	0.00	0.520E+06	5.50	
44	0	0.305E+03	212.70	0.00	0.520E+06	5.50	
45	0	0.191E+03	212.60	0.00	0.600E+06	5.50	
46	0	0.815E+02	212.60	0.00	0.736E+06	5.50	
47	0	0.780E+02	212.60	0.00	0.613E+06	5.50	1
48	0	0.380E+03	212.60	0.00	0.600E+06	5.50	
49	0	0.700E+02	212.60	0.00	0.550E+06	5.50	
50	0	0.700E+02	212.60	0.00	0.630E+06	5.50	
51	0	0.700E+02	212.60	0.00	0.626E+06	5.50	
52	0	0.700E+02	212.60	0.00	0.626E+06	5.50	
53	0	0.111E+03	212.60	0.00	0.626E+06	5.50	
54	0	0.296E+03	209.40	0.00	0.000E+00	5.50	
55	0	0.117E+03	145.80	0.00	0.000E+00	5.50	
56	0	0.476E+02	101.00	0.00	0.000E+00	5.50	
57	0	0.476E+02	101.00	0.00	0.000E+00	5.50	
58	0	0.313E+02	195.26	0.00	0.000E+00	5.50	
59	0	0.106E+03	88.24	0.00	0.000E+00	5.50	

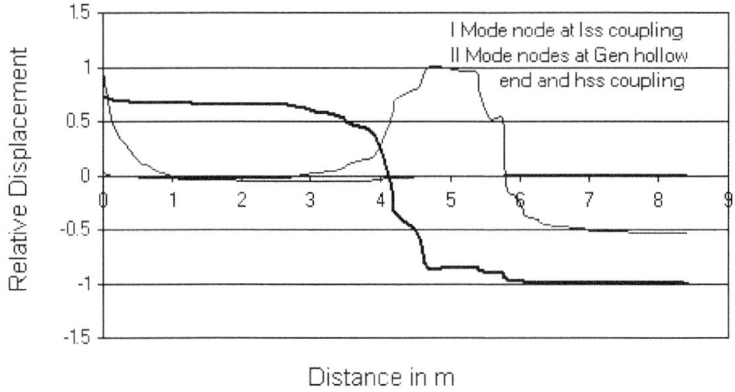

Fig. 15.2 First two modes

Table 15.2 Equivalent model

Mass No.	CG Locn.	Mass Moment kgm^2	Actual Stiffness Nm/rad	Approx. Stiffness Nm/rad
1	6	30.4	1.197	6.832
2	12	636.33	0.592	0.554
3	26	182.23	2.05	5.253
4	47	291.07	–	–

The model obtained is given in Table 15.2. The natural frequencies obtained with the above model are 1317.812, 4325.754 and 6173.256 RPM. This four-rotor model is used to determine the response of the system under a short circuit.

In equations (15.7) and (15.8), $m_a = 6.1538, m_s = -3.0769, \alpha = 12$ and $\beta = 4$. The transient response obtained is given in Figure 15.3. The response in section 3–4 (right on the gearbox) reaches a peak value 2.6 times the nominal torque at 0.09 s after the shock.

During a possible gearbox failure scenario and the resulting distress situation, the system passes through a resonance at 1318 RPM while coasting down until the gear box fails completely. Therefore, a resonance analysis with per rev component is also considered. The resulting stresses are given in Figure 15.4 from which it can be observed that the response continually increases. In section 2–3 (lss coupling) a magnification of about 4.5 is reached at 0.275 s. At a speed close to I mode, 1317 RPM, this magnification can even be higher during this period.

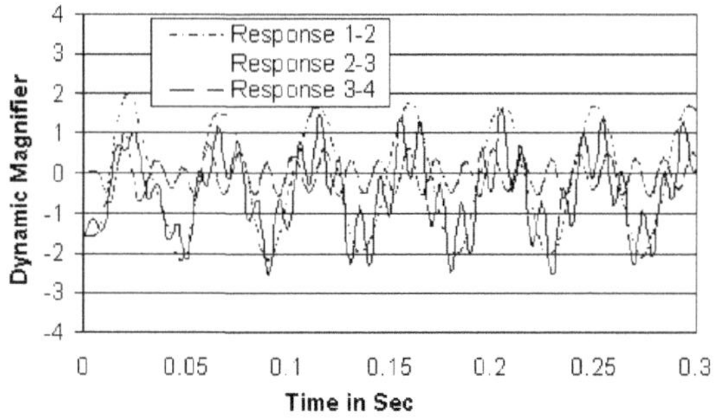

Fig. 15.3 Torque magnification due to ground fault

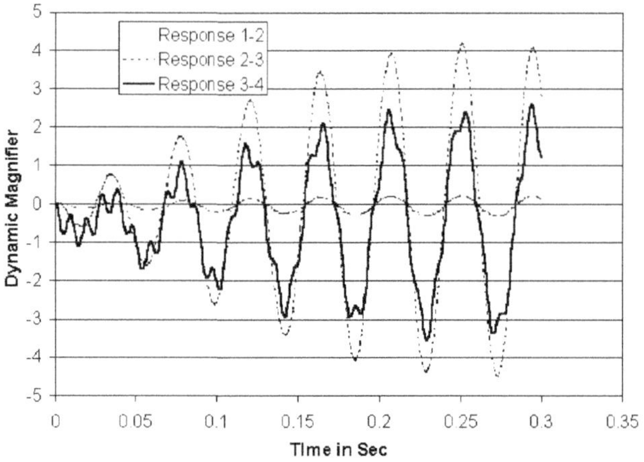

Fig. 15.4 Torque magnification at resonance

15.2 Transfer Matrix Method for Lateral Vibrations of Rotors

In Section 12.7 we discussed Myklestad and Prohl's methods in transfer matrix form for stationary structures. We will now discuss the Out-of-Balance Response of Rotors using the Transfer Matrix Method [7].

Figure 15.5 shows a shaft element with bending in both planes. The state quantities are shown in both planes in Figure 15.6. The transfer matrices are built from one station to another station using the equilibrium relations. The state vector is written in both the x–z plane of bending and the x–y plane of bending as

$$\{S\} = \begin{Bmatrix} S_z \\ S_y \end{Bmatrix} \tag{15.10}$$

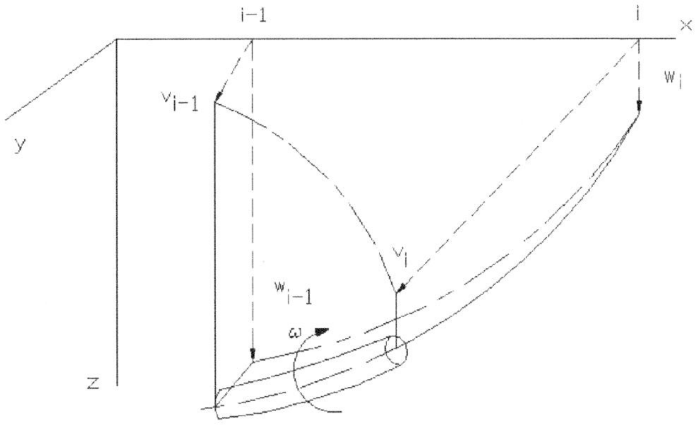

Fig. 15.5 Rotor shaft element

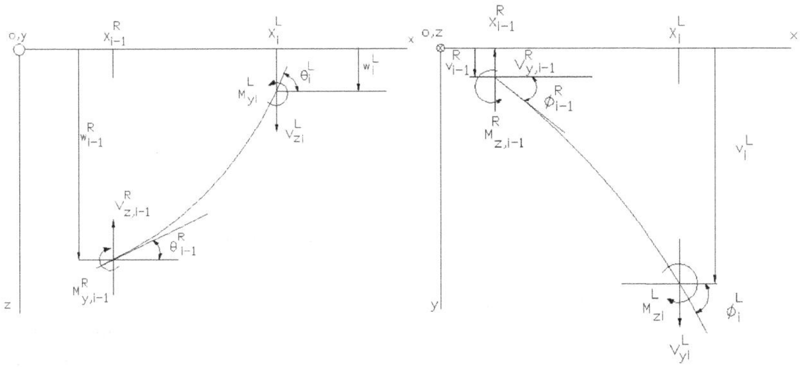

Fig. 15.6 State quantities for transfer matrices

$$\{S_z\} = \begin{Bmatrix} -w \\ \theta \\ M_y \\ V_z \end{Bmatrix}; \quad \{S_y\} = \begin{Bmatrix} v \\ \phi \\ M_z \\ -V_y \end{Bmatrix} \tag{15.11}$$

In the x–z plane, the field matrix is given by (12.68). In the x–y plane this is

$$\begin{Bmatrix} v \\ \phi \\ M_z \\ -V_y \end{Bmatrix}_i^L = \begin{bmatrix} 1 & l & \frac{l^2}{2EI} & \frac{l^3}{6EI} \\ 0 & 1 & \frac{l}{EI} & \frac{l^2}{2EI} \\ 0 & 0 & 1 & l \\ 0 & 0 & 0 & 1 \end{bmatrix} \begin{Bmatrix} v \\ \phi \\ M_z \\ -V_y \end{Bmatrix}_{i-1}^R$$

$$\{S_y\}_i^L = [F]_i \{S_y\}_{i-1}^R \tag{15.12}$$

Combining (12.68) and (15.12), we get the ith field matrix

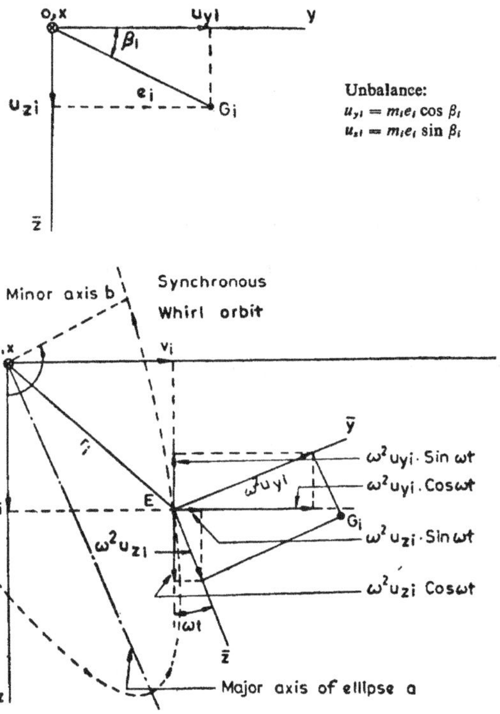

(a)

(b)

Fig. 15.7 (a) The ith unbalanced mass in rotating coordinates. (b) Equilibrium relations of ith unbalanced mass

$$\begin{Bmatrix} \{S_z\} \\ \{S_y\} \end{Bmatrix}_i^L = \begin{bmatrix} [F] & 0 \\ 0 & [F] \end{bmatrix}_i \begin{Bmatrix} \{S_z\} \\ \{S_y\} \end{Bmatrix}_{i-1}^R \tag{15.13}$$

The point matrix is given by (12.69).

In general a response in the y direction will be different from that in the z direction. Further from Figure 15.7b the response will consist of cosine and sine components. Hence

$$\{S_z\} = \begin{Bmatrix} \{S_{zc}\} \\ \{S_{zs}\} \end{Bmatrix}; \quad \{S_y\} = \begin{Bmatrix} \{S_{yc}\} \\ \{S_{ys}\} \end{Bmatrix} \tag{15.14}$$

$$\{S_{zc}\} = \begin{Bmatrix} -w_c \\ \theta_c \\ M_{yc} \\ V_{zc} \end{Bmatrix}; \quad \{S_{zs}\} = \begin{Bmatrix} -w_s \\ \theta_s \\ M_{ys} \\ V_{zs} \end{Bmatrix} \tag{15.15}$$

$$\{S_{yc}\} = \begin{Bmatrix} v_c \\ \phi_c \\ M_{zc} \\ -V_{yc} \end{Bmatrix}; \quad \{S_{ys}\} = \begin{Bmatrix} v_s \\ \phi_s \\ M_{zs} \\ -V_{ys} \end{Bmatrix} \tag{15.16}$$

For the purpose of response calculations, we add one more column and row using an Identity matrix and write a general transfer relation as

$$\begin{Bmatrix} S_{zc} \\ S_{zs} \\ S_{yc} \\ S_{ys} \\ 1 \end{Bmatrix}_i^L = \begin{bmatrix} [F] & [0] & [0] & [0] & \{0\} \\ [0] & [F] & [0] & [0] & \{0\} \\ [0] & [0] & [F] & [0] & \{0\} \\ [0] & [0] & [0] & [F] & \{0\} \\ \{0\}^T & \{0\}^T & \{0\}^T & \{0\}^T & 1 \end{bmatrix} \begin{Bmatrix} S_{zc} \\ S_{zs} \\ S_{yc} \\ S_{ys} \\ 1 \end{Bmatrix}_{i-1}^R$$

$$\{\bar{S}\}_i^L = [\bar{F}]_i \{\bar{S}\}_{i-1}^R \tag{15.17}$$

We can write the general point matrix relation in a similar manner to Figure 15.7b

$$\begin{Bmatrix} S_{zc} \\ S_{zs} \\ S_{yc} \\ S_{ys} \\ 1 \end{Bmatrix}_i^R = \begin{bmatrix} [P] & [0] & [0] & [0] & \{m_{zc}\} \\ [0] & [P] & [0] & [0] & \{m_{zs}\} \\ [0] & [0] & [P] & [0] & \{m_{yc}\} \\ [0] & [0] & [0] & [P] & \{m_{ys}\} \\ \{0\}^T & \{0\}^T & \{0\}^T & \{0\}^T & 1 \end{bmatrix} \begin{Bmatrix} S_{zc} \\ S_{zs} \\ S_{yc} \\ S_{ys} \\ 1 \end{Bmatrix}_i^L$$

$$\{\bar{S}\}_i^R = [\bar{P}]_i \{\bar{S}\}_i^L \tag{15.18}$$

where

$$\{m_{zc}\} = \begin{Bmatrix} 0 \\ 0 \\ 0 \\ -u_z\omega^2 \end{Bmatrix}_i; \quad \{m_{zs}\} = \begin{Bmatrix} 0 \\ 0 \\ 0 \\ u_y\omega^2 \end{Bmatrix}_i$$

$$\{m_{yc}\} = \begin{Bmatrix} 0 \\ 0 \\ 0 \\ u_y\omega^2 \end{Bmatrix}_i; \quad \{m_{zs}\} = \begin{Bmatrix} 0 \\ 0 \\ 0 \\ u_z\omega^2 \end{Bmatrix}_i \tag{15.19}$$

We can now form the following non-homogenous equations for the unbalance response for a given set of boundary conditions, e.g., a rotor on rigid bearing end supports:

$$\{\bar{S}\}_{n+1} = [\bar{F}]_{n+1}[\bar{P}]_n[\bar{F}]_n[\bar{P}]_{n-1}\cdots[\bar{F}]_1\{\bar{S}\}_0$$

$$\{\bar{S}\}_{n+1} = [\bar{U}]\{\bar{S}\}_0 \tag{15.20}$$

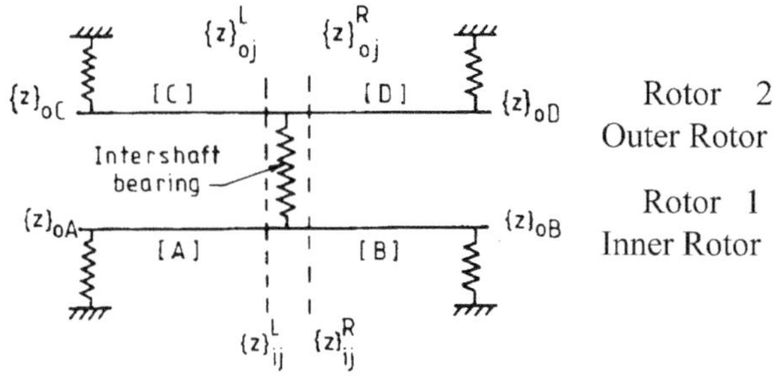

Fig. 15.8 Two-spool rotor configuration

$$\begin{bmatrix} u_{1,2} & u_{1,4} & u_{1,6} & u_{1,8} & u_{1,10} & u_{1,12} & u_{1,14} & u_{1,16} \\ u_{3,2} & u_{3,4} & & & & & & u_{3,16} \\ u_{5,2} & & & & & & & \\ u_{7,2} & & & & & & & \\ u_{9,2} & & & & & & & \\ u_{11,2} & & & & & & & \\ u_{13,2} & & & & & & & \\ u_{15,2} & & & & & & & u_{15,16} \end{bmatrix} \begin{Bmatrix} \theta_c \\ V_{zc} \\ \theta_s \\ V_{zs} \\ \phi_c \\ -V_{yc} \\ \phi_s \\ -V_{ys} \end{Bmatrix}_0 = - \begin{Bmatrix} u_{1,17} \\ u_{3,17} \\ - \\ - \\ - \\ - \\ - \\ u_{15,17} \end{Bmatrix} \tag{15.21}$$

At any station, the whirl orbit can be constructed from

$$r(t) = e^{i\omega t}\left[\frac{1}{2}(w_c + v_s) + \frac{1}{2}i(v_c - w_s)\right] + e^{-i\omega t}\left[\frac{1}{2}(w_c - v_s) + \frac{1}{2}i(v_c + w_s)\right] \tag{15.22}$$

15.3 Twin Spool Rotor Analysis

The analysis in Section 15.2 is extended for two-spool rotor analysis by Gupta et al. [4] wherein an aircraft engine is modeled with its LP rotor as rotor 1 (inner rotor) over which the HP rotor, rotor 2 (outer rotor) is mounted through an intershaft bearing, see Figure 15.8.

The inner and outer rotors have different speeds, say, ω_m inner rotor speed ω_n outer rotor speed. The response then is

$$w = w_{cm}\cos\omega_m t + w_{sm}\sin\omega_m t + w_{cn}\cos\omega_n t + w_{sn}\sin\omega_n t$$

$$\theta = \theta_{cm}\cos\omega_m t + \theta_{sm}\sin\omega_m t + \theta_{cn}\cos\omega_n t + \theta_{sn}\sin\omega_n t$$

$$M_y = M_{ycm}\cos\omega_m t + M_{ysm}\sin\omega_m t + M_{ycn}\cos\omega_n t + M_{ysn}\sin\omega_n t$$

$$V_z = V_{zcm}\cos\omega_m t + V_{zsm}\sin\omega_m t + V_{zcn}\cos\omega_n t + V_{zsn}\sin\omega_n t$$

$$v = v_{cm} \cos \omega_m t + v_{sm} \sin \omega_m t + v_{cn} \cos \omega_n t + v_{sn} \sin \omega_n t$$

$$\phi = \phi_{cm} \cos \omega_m t + \phi_{sm} \sin \omega_m t + \phi_{cn} \cos \omega_n t + \phi_{sn} \sin \omega_n t$$

$$M_z = M_{zcm} \cos \omega_m t + M_{zsm} \sin \omega_m t + M_{zcn} \cos \omega_n t + M_{zsn} \sin \omega_n t$$

$$V_y = V_{ycm} \cos \omega_m t + V_{ysm} \sin \omega_m t + V_{ycn} \cos \omega_n t + V_{ysn} \sin \omega_n t \quad (15.23)$$

The state vector has now 33 quantities; the field and point matrices are also 33 × 33. The bearing matrix can be shown to be

$$[B]_{33\times 33} = \begin{bmatrix} [B_{mz1}] & 0 & [B_{mz2}] & 0 & 0 \\ 0 & [B_{nz1}] & 0 & [B_{nz2}] & 0 \\ [B_{my2}] & 0 & [B_{my1}] & 0 & 0 \\ 0 & [B_{ny2}] & 0 & [B_{ny1}] & 0 \\ 0 & 0 & 0 & 0 & 1 \end{bmatrix} \quad (15.24)$$

$$\begin{aligned}
&[B_{mz1}](i,i) = 1, \quad i = 1, 2, \ldots, 8 \\
&[B_{mz1}](4,1) = [B_{mz1}](8,5) = -K_{zz} + m\omega_m{}^2 \\
&[B_{mz1}](4,5) = -C_{zz}\omega_m \\
&[B_{mz1}](8,1) = C_{zz}\omega_m \\
&[B_{mz2}](4,1) = [B_{mz2}](8,5) = K_{zy} \\
&[B_{mz2}](4,5) = C_{zy}\omega_m \\
&[B_{mz2}](8,1) = -C_{zy}\omega_m \\
&\text{Rest}[B_{mz1}] = [B_{mz2}] = 0 \qquad (15.25)
\end{aligned}$$

Matrices $[B_{nz1}]$ and $[B_{nz2}]$ are obtained from the above by replacing ω_m with ω_n. Matrices $[B_{my1}]$ and $[B_{my2}]$ are obtained from the above by replacing subscripts z with y and y with z. Similarly by replacing ω_n with ω_m we can obtain $[B_{ny1}]$ and $[B_{ny2}]$.

Similarly we can derive the following point matrix for the ith unbalance on the inner rotor:

$$\begin{Bmatrix} \{S_{zm}\} \\ \{S_{zn}\} \\ \{S_{ym}\} \\ \{S_{yn}\} \\ 1 \end{Bmatrix}_i^R = \begin{bmatrix} [P_m] & 0 & 0 & 0 & \{m_{zm}\} \\ 0 & [P_n] & 0 & 0 & 0 \\ 0 & 0 & [P_m] & 0 & \{m_{zn}\} \\ 0 & 0 & 0 & [P_n] & 0 \\ 0 & 0 & 0 & 0 & 1 \end{bmatrix} \begin{Bmatrix} \{S_{zm}\} \\ \{S_{zn}\} \\ \{S_{ym}\} \\ \{S_{yn}\} \\ 1 \end{Bmatrix}_i^L \quad (15.26)$$

where

$$\{m_{zm}\} = \begin{Bmatrix} 0 \\ 0 \\ 0 \\ -u_z\omega_m{}^2 \\ 0 \\ 0 \\ 0 \\ u_y\omega_m{}^2 \end{Bmatrix}, \quad \{m_{ym}\} = \begin{Bmatrix} 0 \\ 0 \\ 0 \\ u_y\omega_m{}^2 \\ 0 \\ 0 \\ 0 \\ u_z\omega_m{}^2 \end{Bmatrix} \tag{15.27}$$

We can write similarly for the outer rotor using ω_n. We need to write junction conditions where two shafts are connected through intershaft bearing. At junction station, j we have two shafts running at ω_m (inner) and ω_n (outer) and let us reach this junction through transfer matrices from a starting point on the left side on the inner rotor $[A]$, right side on the inner rotor $[B]$, left side of the outer rotor $[C]$, and right side of the outer rotor through transfer matrix $[D]$. Then

$$\begin{aligned} \{S\}_{mj}^L &= [A]\{S\}_{0A} \\ \{S\}_{0B} &= [B]\{S\}_{mj}^R \\ \{S\}_{nj}^L &= [C]\{S\}_{0C} \\ \{S\}_{0D} &= [B]\{S\}_{nj}^R \end{aligned} \tag{15.28}$$

At the junction of this inner rotor

$$\begin{aligned} \{S\}_{mj}^R &= [I]\{S\}_{mj}^L + [B']\{S\}_{mj}^L - [B']\{S\}_{nj}^L \\ &= [[I] + [B']]\{S\}_{mj}^L - [B']\{S\}_{nj}^L \end{aligned} \tag{15.29}$$

Similarly at the junction of the outer rotor

$$\{S\}_{nj}^R = [[I] + [B']]\{S\}_{nj}^L - [B']\{S\}_{mj}^L \tag{15.30}$$

Using the above two equations in the second and fourth equations of (15.28) we get

$$\begin{aligned} \{S\}_{OB} &= [B][[I] + [B']]\{S\}_{mj}^L - [B][B']\{S\}_{nj}^L \\ \{S\}_{OD} &= [D][[I] + [B']]\{S\}_{nj}^L - [D][B']\{S\}_{mj}^L \end{aligned} \tag{15.31}$$

With the help of the first and third equations in (15.28), the above becomes

$$\begin{aligned} \{S\}_{OB} &= [B][[I] + [B']][A]\{S\}_{OA} - [B][B'][C]\{S\}_{OC} \\ \{S\}_{OD} &= [D][[I] + [B']][C]\{S\}_{OC} - [D][B'][A]\{S\}_{OA} \end{aligned} \tag{15.32}$$

Premultiplying the first equation by $[B]^{-1}$ and the second by $[D]^{-1}$, the above equations are written in a matrix equation as

$$\begin{bmatrix} [A_m] & [B_m] & [C_m] & 0 \\ [A_n] & 0 & [C_n] & [D_n] \end{bmatrix}_{66\times 132} \begin{Bmatrix} \{S\}_{OA} \\ \{S\}_{OB} \\ \{S\}_{OC} \\ \{S\}_{OD} \end{Bmatrix}_{132\times 1} = 0 \tag{15.33}$$

where

$$\begin{aligned} [A_m] &= [[I] + [B']][A] \\ [C_n] &= [[I] + [B']][C] \\ [B_m] &= -[B]^{-1} \\ [D_n] &= -[D]^{-1} \\ [C_m] &= -[B'][C] \\ [A_n] &= -[B'][A] \end{aligned} \tag{15.34}$$

Consider free ends of all the shaft sections A, B, C and D for the initial state quantities, then all M_y, V_z, M_z, V_y in $\{S\}_0$ vectors are zero. The unknown quantities are w, θ, v and ϕ at the four starting stations. For each starting station, we have 16 unknowns, $-w_{cm}, \theta_{cm}, -w_{sm}, \theta_{sm}, -w_{cn}, \theta_{cn}, -w_{sn}, \theta_{sn}, v_{cm}, \phi_{cm}, v_{sm}, \phi_{sm}, v_{cn}, \phi_{cn}, v_{sn}, \phi_{sn}$, in $\{S_{OA}\}$; similarly in $\{S_{OB}\}$ $\{S_{OC}\}$ and $\{S_{OD}\}$. Since M_y, V_z, etc., in (15.33) are zero, we can drop 3, 4, 7, 8, 11, 12, etc.; 31, 32, 36, 37, etc.; 64, 65, 69, 70, etc.; 97, 98, 102, 103, etc.; 130, 131 columns in the 66 × 132 matrix in (15.33). We also note that the 33rd, 66th, 99th and 132nd elements of the vector are unity and hence when multiplied with the corresponding column quantity in the matrix, give rise to a constant quantity independent of any state quantity. These four constants can be added and taken to the right hand side of the corresponding equation. We can also delete the 33rd and 66th rows in (15.33) as they simply give identity equations. Then equation (15.33) reduces to

$$[\bar{A}]_{64\times 64}\{S_0\}_{64\times 1} = \{\bar{B}\}_{64\times 1} \tag{15.35}$$

where

$$\{S_0\} = \begin{Bmatrix} \{S_{OA}\} \\ \{S_{OB}\} \\ \{S_{OC}\} \\ \{S_{OD}\} \end{Bmatrix}$$

and

$$\bar{B}_i = A_{mi,33} + B_{mi,33} + C_{mi,33} + D_{mi,33}$$

The linear system of equations (15.35) can be solved to determine $\{S_0\}$. From there the state vector quantities at any station can be determined using the appropriate transfer matrices. We need to write a computer program to do this job.

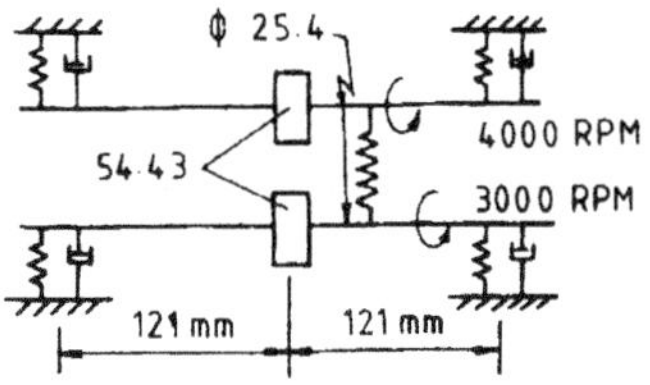

Fig. 15.9 A simple dual rotor system

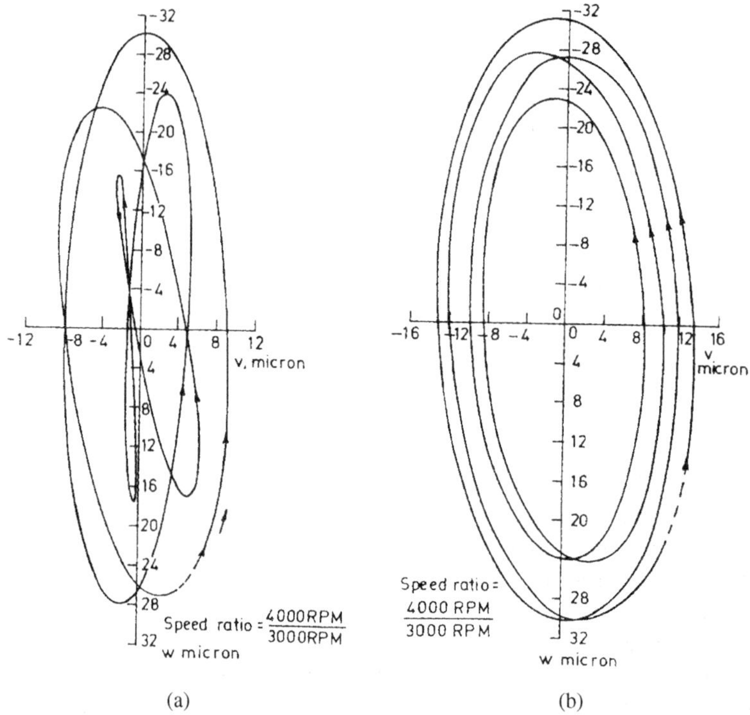

Fig. 15.10 (a) Whirl orbit of inner rotor. (b) Whirl orbit of outer rotor

As an example, consider a simple dual rotor carrying 54.43 kg each as shown in Figure 15.9.

The support and intershaft bearings are considered as follows: K_{yy} = 17.5 MN/m; K_{yz} = 10.0 MN/m; K_{zy} = 10.0 MN/m; K_{zz} = 17.5 MN/m; C_{yy} = 700 KN-s/m; C_{yz} = 400 KN-s/m; C_{zy} = 700 KN-s/m; C_{zz} = 400 KN-s/m.

The response is shown in Figures 15.10a (on inner rotor) and 15.10b (on outer rotor) for one full period. They exhibit elliptical loops. The time period T is

$$T = \frac{2\pi}{\frac{\omega_n - \omega_m}{p-q}}$$

where p and q are the smallest integers such that $p/q = \omega_n/\omega_m$. The number of loops in one period is given by

$$n = \{\max[\omega_n, \omega_m]\}\left\{\frac{p-q}{\omega_n - \omega_m}\right\}$$

Transfer matrix methods played an important role in the early computer era when machine capacities were low. They needed 4×4 matrices for simple bending problems to store the final or overall transfer matrices and as we have seen in this section, to a maximum of 133×133 size. This storage is irrespective of number of stations adopted in the analysis. But they suffered from making beam models of actual complex rotor systems, which takes considerable skilled hours and in some cases writing special programs. Finite element methods removed this limitation for common applications and became popular after the 1980s. The finite element methods also paved the way quickly to solid rotor models, thus removing several simplifications and approximations in beam models.

References

1. Brown Boveri & Cie. (1920) Report Publications, p. 32.
2. Concordia, C. (1956) *Synchronous Machine Theory and Performance*, John Wiley & Sons.
3. Eshleman, R.L. (1974) Torsional Response of Internal Combustion Engines, *Journal of Engineering for Industry, Trans. ASME*, p. 441.
4. Gupta, K.D., Gupta, K., Rao, J.S. and Bhaskara Sarma, K.V. (1988) Dynamic Response of a Dual Rotor System by Extended Transfer Matrix Method, in *Proceedings International Conference Vibrations in Rotating Machinery*, Institute Mechanical Engineers, p. 599.
5. Hammons, T.J. (1977) Effect of Three Phase System Faults and Faulty Synchronization on the Mechanical Stressing of Large Turbogenerator Drives, *Revue General de l'Electricite*, vol. 86, p. 558.
6. Linke, W. (1912) Über Schaltvorgänge bei elektrischen Maschinen und Apparaten, *Archiv für Elektrotechnik*, vol. 1, no. 2, p. 69.
7. Rao, J.S. (1996) *Rotor Dynamics*, New Age International.
8. Rao, J.S., Rao, D.K. and Sarma, K.B. (1980) *The Transient Response of Turbo-Alternator Rotor Systems under Short Circuit Conditions, Vibrations of Rotating Machinery*, Institute of Mechanical Engineers, Cambridge, England, p. 271.
9. Saling, K.H. and Schwinder, Th. (1966) Variations in Design and Principles of Dimensioning of Rotors for Hydro Electric Generators, Bending Critical Speeds, Structural Resources and Torsional Vibration Stresses; Evaluation Criteria for the Vibration of Bearing and Rotor Unbalance, Report 138, CIGRE.
10. Schwibinger, P., Nordmann, R., Wohlrab, P., Steigleder, K. and Wutsdorff, P. (1987) Torsional Vibrations in Turbogenerators due to Network Disturbances, in *Rotating Machinery Dynamics*, ASME DE-Vol. 2, p. 501.
11. Stodola, A. (1917) *Schweiz Bauztg.*, vol. 69, pp. 93, 229.

Chapter 16
Finite Element Methods for Rotor Dynamics

In Chapter 13, we discussed the development of finite element methods for stationary systems. Rotor dynamics lagged behind stationary structures developments and in development of finite element methods for rotor dynamics, it was no exception.

16.1 Nelson's Beam Element

The finite element method for rotors was first developed by Ruhl and Booker [29]. Nelson and McVaugh [16] extended this to include gyroscopic effects. The effects of axial torque were included by Zorzi and Nelson [37] and Nelson [15] gave rotor dynamics elements with the Timoshenko beam theory. For details on the finite element method, see also [1,4,10,11,19].

Nelson's generalized Timoshenko beam element including torsional motion is derived here briefly. Figure 16.1 shows a 2-noded shaft element with 10 degrees of freedom. The kinetic energy including the torsional motion α is

$$\begin{aligned} T^s &= \frac{1}{2}\int_0^L \{\rho A(\dot{v}^2+\dot{w}^2)+I_D(\dot{\theta}^2+\dot{\phi}^2)\}d\hat{x} \\ &\quad -\frac{1}{2}\int_0^L \{I_P(\omega+\dot{\alpha})(\theta\dot{\phi}-\phi\dot{\theta})-I_P(\omega+\dot{\alpha}^2)\}d\hat{x} \end{aligned} \tag{16.1}$$

where I_D and I_P are the diametral and polar mass moments of inertia per unit length of the shaft element respectively. The potential energy including shear deformation and torsional deflection is given by

$$\begin{aligned} U^s &= \frac{1}{2}\int_0^L EI(\theta'^2+\phi'^2)d\hat{x}+\frac{1}{2}\int_0^L K'GA[(v'-\phi)^2+(w'+\theta)^2]d\hat{x} \\ &\quad +\frac{1}{2}\int_0^L GI_P\alpha'^2 d\hat{x} \end{aligned} \tag{16.2}$$

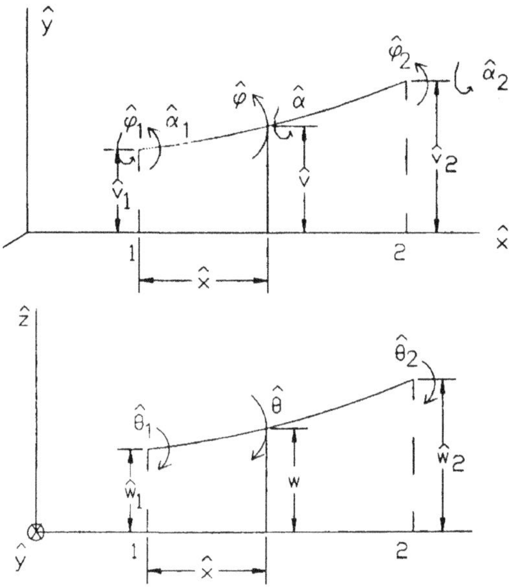

Fig. 16.1 Nelson's generalized Timoshenko beam rotor element

where K' is the shear correction factor given by

$$K' = \frac{6(1+\nu)(1+m^2)^2}{(7+6\nu)(1+m^2)^2 + (20+12\nu)m^2} \tag{16.3}$$

with ν as Poisson's ratio and m as the ratio of inner diameter to outer diameter of the shaft element.

The shape functions are assumed as

$$\begin{Bmatrix} v(\hat{x}) \\ w(\hat{x}) \end{Bmatrix} = [\Psi(\hat{x})]\{\hat{d}\}$$

$$\begin{Bmatrix} \theta(\hat{x}) \\ \phi(\hat{x}) \end{Bmatrix} = [\Phi(\hat{x})]\{\hat{d}\}$$

$$\alpha(\hat{x}) = [\Theta(\hat{x})]\{\hat{d}\} \tag{16.4}$$

where

$$[\Psi(\hat{x})] = \begin{bmatrix} \psi_1 & 0 & 0 & \psi_2 & 0 & \psi_3 & 0 & 0 & \psi_4 & 0 \\ 0 & \psi_1 & -\psi_2 & 0 & 0 & 0 & \psi_3 & -\psi_4 & 0 & 0 \end{bmatrix}$$

$$[\Phi(\hat{x})] = \begin{bmatrix} 0 & -\phi_1 & \phi_2 & 0 & 0 & 0 & -\phi_3 & \phi_4 & 0 & 0 \\ \phi_1 & 0 & 0 & \phi_2 & 0 & \phi_3 & 0 & 0 & \phi_4 & 0 \end{bmatrix}$$

$$[\Theta(\hat{x})] = \begin{bmatrix} 0 & 0 & 0 & 0 & \theta_1 & 0 & 0 & 0 & 0 & \theta_2 \end{bmatrix} \tag{16.5}$$

and

$$\{\hat{d}\}^T = \left\{ \hat{v}_1 \; \hat{w}_1 \; \hat{\theta}_1 \; \hat{\phi}_1 \; \hat{\alpha}_1 \; \hat{v}_2 \; \hat{w}_2 \; \hat{\theta}_2 \; \hat{\phi}_2 \; \hat{\alpha}_2 \right\} \tag{16.6}$$

The individual terms in (16.5) are given by

$$\begin{aligned}
\psi_1 &= \frac{1}{1+\Phi}\left[2a^3 - 3a^2 + 1 + \Phi(1-a)\right] \\
\psi_2 &= \frac{L}{1+\Phi}\left[a^3 - 2a^2 + a + \frac{1}{2}\Phi a(1-a)\right] \\
\psi_3 &= \frac{1}{1+\Phi}[-2a^3 + 3a^2 + \Phi a] \\
\psi_4 &= \frac{La}{2}(a-1)
\end{aligned} \tag{16.7}$$

$$\begin{aligned}
\phi_1 &= \frac{6a(a-1)}{L(1+\Phi)} \\
\phi_2 &= \frac{1}{(1+\Phi)}[3a^2 - 4a + 1 + \Phi(1-a)] \\
\phi_3 &= \frac{6a(1-a)}{L(1+\Phi)} \\
\phi_4 &= \frac{1}{(1+\Phi)}[3a^2 - 2a + \Phi a]
\end{aligned} \tag{16.8}$$

$$\begin{aligned}
\theta_1 &= 1 - a \\
\theta_2 &= a
\end{aligned} \tag{16.9}$$

where

$$\begin{aligned}
a &= \frac{\hat{x}}{L} \\
\Phi &= \frac{12EI}{K'GAL^2}
\end{aligned} \tag{16.10}$$

Following a Lagrangian approach, we can obtain

$$[M^s]\{\ddot{\hat{d}}^s\} + \omega[G^s]\{\dot{\hat{d}}^s\} + [K^s]\{\hat{d}^s\} = \{F_s^s\} \tag{16.11}$$

where

$$\begin{aligned}
[M^s] &= [M^s_T] + [M^s{}_R] + [M^s_\theta] \\
[M^s_T] &= [M^s_T]_0 + \Phi[M^s_T]_1 + \Phi^2[M^s_T]_2 \\
[M^s_R] &= [M^s{}_R]_0 + \Phi[M^s_R]_1 + \Phi^2[M^s_R]_2 \\
[G^s] &= [G^s]_0 + \Phi[G^s]_1 + \Phi^2[G^s]_2 \\
[K^s] &= [K^s]_0 + \Phi[K^s]_1 + [K^s_\theta]
\end{aligned} \tag{16.12}$$

The matrices in the above equation are given by:

Mass Matrices due to Translation

$$[M^s_T]_0 = m_T \begin{bmatrix}
312 & & & & & & & & & \\
0 & 312 & & & & & & & & \\
0 & -44L & 8L^2 & & & & & & & \\
44L & 0 & 0 & 8L^2 & & & & & & \\
0 & 0 & 0 & 0 & 0 & & & & & \\
108 & 0 & 0 & 26L & 0 & 312 & & & & \\
0 & 108 & -26L & 0 & 0 & 0 & 312 & & & \\
0 & 26L & -6L^2 & 0 & 0 & 0 & 44L & 8L^2 & & \\
-26L & 0 & 0 & -6L^2 & 0 & -44L & 0 & 0 & 8L^2 & \\
0 & 0 & 0 & 0 & 0 & 0 & 0 & 0 & 0 & 0
\end{bmatrix}_{\text{sym}}$$

$$[M^s_T]_1 = m_T \begin{bmatrix}
588 & & & & & & & & & \\
0 & 588 & & & & & & & & \\
0 & -77L & 14L^2 & & & & & & & \\
77L & 0 & 0 & 14L^2 & & & & & & \\
0 & 0 & 0 & 0 & 0 & & & & & \\
252 & 0 & 0 & 63L & 0 & 588 & & & & \\
0 & 252 & -63L & 0 & 0 & 0 & 588 & & & \\
0 & 63L & -14L^2 & 0 & 0 & 0 & 77L & 14L^2 & & \\
-63L & 0 & 0 & -14L^2 & 0 & -77L & 0 & 0 & 14L^2 & \\
0 & 0 & 0 & 0 & 0 & 0 & 0 & 0 & 0 & 0
\end{bmatrix}_{\text{sym}}$$

$$[M^s_T]_2 = m_T \begin{bmatrix}
280 & & & & & & & & & \\
0 & 280 & & & & & & & & \\
0 & -35L & 7L^2 & & & & & & & \\
35L & 0 & 0 & 7L^2 & & & & & & \\
0 & 0 & 0 & 0 & 0 & & & & & \\
140 & 0 & 0 & 35L & 0 & 280 & & & & \\
0 & 140 & -35L & 0 & 0 & 0 & 280 & & & \\
0 & 35L & -7L^2 & 0 & 0 & 0 & 35L & 7L^2 & & \\
-35L & 0 & 0 & -7L^2 & 0 & -35L & 0 & 0 & 7L^2 & \\
0 & 0 & 0 & 0 & 0 & 0 & 0 & 0 & 0 & 0
\end{bmatrix}_{\text{sym}} \tag{16.13}$$

where

$$m_T = \frac{\rho\pi L({r_o}^2 - {r_i}^2)}{840(1+\Phi)^2}$$

Mass Matrices due to Rotation

$$[M_R^s]_0 = m_R \begin{bmatrix} 36 &&&&&&&&& \\ 0 & 36 &&&&&&&& \\ 0 & -3L & 4L^2 &&&&&&& \\ 3L & 0 & 0 & 4L^2 &&&&&& \\ 0 & 0 & 0 & 0 & 0 &&&&& \\ -36 & 0 & 0 & -3L & 0 & 36 &&&& \\ 0 & -36 & 3L & 0 & 0 & 0 & 36 &&& \\ 0 & -3L & -L^2 & 0 & 0 & 0 & 3L & 4L^2 && \\ 3L & 0 & 0 & -L^2 & 0 & -3L & 0 & 0 & 4L^2 & \\ 0 & 0 & 0 & 0 & 0 & 0 & 0 & 0 & 0 & 0 \end{bmatrix}_{\text{sym}}$$

$$[M_R^s]_1 = m_R \begin{bmatrix} 0 &&&&&&&&& \\ 0 & 0 &&&&&&&& \\ 0 & 15L & 5L^2 &&&&&&& \\ -15L & 0 & 0 & 5L^2 &&&&&& \\ 0 & 0 & 0 & 0 & 0 &&&&& \\ 0 & 0 & 0 & 15L & 0 & 0 &&&& \\ 0 & 0 & -15L & 0 & 0 & 0 & 0 &&& \\ 0 & 15L & -5L^2 & 0 & 0 & 0 & 15L & 5L^2 && \\ -15L & 0 & 0 & -5L^2 & 0 & 15L & 0 & 0 & 5L^2 & \\ 0 & 0 & 0 & 0 & 0 & 0 & 0 & 0 & 0 & 0 \end{bmatrix}_{\text{sym}}$$

$$[M_R^s]_2 = m_R \begin{bmatrix} 0 &&&&&&&&& \\ 0 & 0 &&&&&&&& \\ 0 & 0 & 10L^2 &&&&&&& \\ 0 & 0 & 0 & 10L^2 &&&&&& \\ 0 & 0 & 0 & 0 & 0 &&&&& \\ 0 & 0 & 0 & 0 & 0 & 0 &&&& \\ 0 & 0 & 0 & 0 & 0 & 0 & 0 &&& \\ 0 & 0 & 5L^2 & 0 & 0 & 0 & 0 & 10L^2 && \\ 0 & 0 & 0 & 5L^2 & 0 & 0 & 0 & 0 & 10L^2 & \\ 0 & 0 & 0 & 0 & 0 & 0 & 0 & 0 & 0 & 0 \end{bmatrix}_{\text{sym}} \tag{16.14}$$

where

$$m_R = \frac{\rho\pi L({r_o}^4 - {r_i}^4)}{120L(1+\Phi)^2}$$

Gyroscopic Matrices

$$[G^s]_0 = \frac{2m_R}{L}\begin{bmatrix} 0 & & & & & & & & & \\ -360 & 0 & & & & & & & & \\ 3L & 0 & 0 & & & & & & & \\ 0 & 3L & -4L^2 & 0 & & & & & & \\ 0 & 0 & 0 & 0 & 0 & & & & & \\ 0 & -36 & 3L & 0 & 0 & 0 & & & & \\ 36 & 0 & 0 & 3L & 0 & -36 & 0 & & & \\ 3L & 0 & 0 & -L^2 & 0 & -3L & 0 & 0 & & \\ 0 & 3L & L^2 & 0 & 0 & 0 & -3L & -4L^2 & 0 & \\ 0 & 0 & 0 & 0 & 0 & 0 & 0 & 0 & 0 & 0 \end{bmatrix}_{\text{sym}}$$

$$[G^s]_1 = \frac{2m_R}{L}\begin{bmatrix} 0 & & & & & & & & & \\ 0 & 0 & & & & & & & & \\ -15L & 0 & 0 & & & & & & & \\ 0 & -15L & -5L^2 & 0 & & & & & & \\ 0 & 0 & 0 & 0 & 0 & & & & & \\ 0 & 0 & -15L & 0 & 0 & 0 & & & & \\ 0 & 0 & 0 & -15L & 0 & 0 & 0 & & & \\ -15L & 0 & 0 & -5L^2 & 0 & 15L & 0 & 0 & & \\ 0 & -15L & 5L^2 & 0 & 0 & 0 & 15L & -5L^2 & 0 & \\ 0 & 0 & 0 & 0 & 0 & 0 & 0 & 0 & 0 & 0 \end{bmatrix}_{\text{sym}}$$

$$[G^s]_2 = \frac{2m_R}{L}\begin{bmatrix} 0 & & & & & & & & & \\ 0 & 0 & & & & & & & & \\ 0 & 0 & 0 & & & & & & & \\ 0 & 0 & -10L^2 & 0 & & & & & & \\ 0 & 0 & 0 & 0 & 0 & & & & & \\ 0 & 0 & 0 & 0 & 0 & 0 & & & & \\ 0 & 0 & 0 & 0 & 0 & 0 & 0 & & & \\ 0 & 0 & 0 & 5L^2 & 0 & 0 & 0 & 0 & & \\ 0 & 0 & -5L^2 & 0 & 0 & 0 & 0 & -10L^2 & 0 & \\ 0 & 0 & 0 & 0 & 0 & 0 & 0 & 0 & 0 & 0 \end{bmatrix}_{\text{sym}} \tag{16.15}$$

Stiffness Matrices

$$[K^s]_0 = k^s \begin{bmatrix} 12 & & & & & & & & & \\ 0 & 12 & & & & & & & & \\ 0 & -6L & 4L^2 & & & & & & & \\ 6L & 0 & 0 & 4L^2 & & & & & & \\ 0 & 0 & 0 & 0 & 0 & & & & & \\ -12 & 0 & 0 & -6L & 0 & 12 & & & & \\ 0 & -12 & 6L & 0 & 0 & 0 & 12 & & & \\ 0 & -6L & 2L^2 & 0 & 0 & 0 & 6L & 4L^2 & & \\ 6L & 0 & 0 & 2L^2 & 0 & -6L & 0 & 0 & 4L^2 & \\ 0 & 0 & 0 & 0 & 0 & 0 & 0 & 0 & 0 & 0 \end{bmatrix}_{\text{sym}}$$

$$[K^s]_1 = k^s \begin{bmatrix} 0 & & & & & & & & & \\ 0 & 0 & & & & & & & & \\ 0 & 0 & L^2 & & & & & & & \\ 0 & 0 & 0 & L^2 & & & & & & \\ 0 & 0 & 0 & 0 & 0 & & & & & \\ 0 & 0 & 0 & 0 & 0 & 0 & & & & \\ 0 & 0 & 0 & 0 & 0 & 0 & 0 & & & \\ 0 & 0 & -L^2 & 0 & 0 & 0 & 0 & L^2 & & \\ 0 & 0 & 0 & -L^2 & 0 & 0 & 0 & 0 & L^2 & \\ 0 & 0 & 0 & 0 & 0 & 0 & 0 & 0 & 0 & 0 \end{bmatrix}_{\text{sym}} \tag{16.16}$$

where

$$k^s = \frac{EI}{L^3(1+\Phi)}$$

Torsional Mass Matrix

$$[M_\theta^s]_1 = \frac{1}{6} I_P{}^s L \begin{bmatrix} 0 & & & & & & & & & \\ 0 & 0 & & & & & & & & \\ 0 & 0 & 0 & & & & & & & \\ 0 & 0 & 0 & 0 & & & & & & \\ 0 & 0 & 0 & 0 & 2 & & & & & \\ 0 & 0 & 0 & 0 & 0 & 0 & & & & \\ 0 & 0 & 0 & 0 & 0 & 0 & 0 & & & \\ 0 & 0 & 0 & 0 & 0 & 0 & 0 & 0 & & \\ 0 & 0 & 0 & 0 & 0 & 0 & 0 & 0 & 0 & \\ 0 & 0 & 0 & 0 & 1 & 0 & 0 & 0 & 0 & 2 \end{bmatrix}_{\text{sym}}$$

Torsional Stiffness Matrix

$$[k_\theta^s]_1 = \frac{1}{6} I_P{}^s L \begin{bmatrix} 0 & & & & & & & & & \\ 0 & 0 & & & & & & & & \\ 0 & 0 & 0 & & & & & & & \\ 0 & 0 & 0 & 0 & & & & & & \\ 0 & 0 & 0 & 0 & 1 & & & & & \\ 0 & 0 & 0 & 0 & 0 & 0 & & & & \\ 0 & 0 & 0 & 0 & 0 & 0 & 0 & & & \\ 0 & 0 & 0 & 0 & 0 & 0 & 0 & 0 & & \\ 0 & 0 & 0 & 0 & 0 & 0 & 0 & 0 & 0 & \\ 0 & 0 & 0 & 0 & -1 & 0 & 0 & 0 & 0 & 1 \end{bmatrix}_{\text{sym}} \tag{16.17}$$

The bearing element matrix is

$$\begin{bmatrix} C_{yy} & C_{yz} \\ C_{zy} & C_{zz} \end{bmatrix} \begin{Bmatrix} \dot{v}^b \\ \dot{w}^b \end{Bmatrix} + \begin{bmatrix} K_{yy} & K_{yz} \\ K_{zy} & K_{zz} \end{bmatrix} \begin{Bmatrix} v^b \\ w^b \end{Bmatrix} = \{F_s{}^b\} \tag{16.18}$$

For a flexible torsional coupling, the strain energy is

$$U_\theta = \frac{1}{2} k_\theta (\alpha_{i+1} - \alpha_i)^2 \tag{16.19}$$

and the coupling matrix is

$$\begin{bmatrix} k_\theta & -k_\theta \\ -k_\theta & k_\theta \end{bmatrix} \begin{Bmatrix} \alpha_i \\ \alpha_{i+1} \end{Bmatrix} = \{F_s{}^c\} \tag{16.20}$$

The system equations of motion are then

$$[M_s]\{\ddot{d}_s\} + [\omega_1[G_s] + [C_s]]\{\dot{d}_s\} + [K_s]\{d_s\} = \{F_s\} \tag{16.21}$$

In the cases where there is a gear pair

$$[G_s] = \begin{bmatrix} [G_1{}^d] & \\ & \frac{\omega_2}{\omega_1}[G_2{}^d] \end{bmatrix}$$

(see also (16.25) and in a dual rotor), the gyroscopic matrix for the second rotor is $\omega_2/\omega_1[G_s]$, etc.

Here, the right-hand side is set to zero, to determine the eigen-values and eigen-vectors. For an undamped system, the eigen-values can be directly obtained for synchronous whirl with either of the shafts of a dual rotor system or a general rotor system. For damped systems, it is necessary to draw the Campbell diagram to obtain the critical speeds.

The finite element method has a distinct advantage in assembling the elemental equations without any recourse to intershaft bearing conditions as in the case of transfer matrix method (see also [9]).

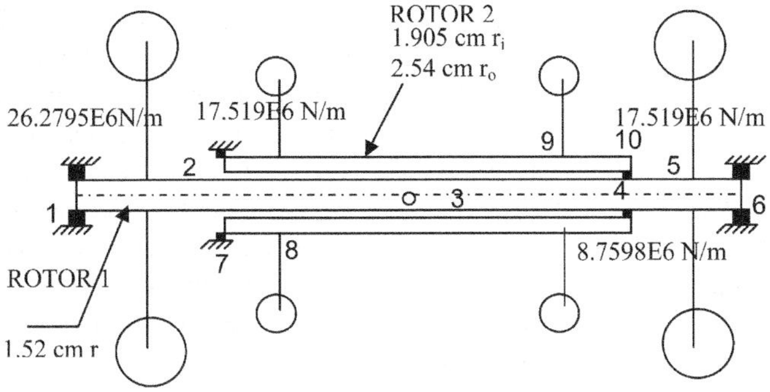

Distances: 1-2=7.62 cm; 2-3=17.78 cm; 3-4=15.24 cm;
4-5=5-6=7-8=9-10=5.08cm;8-9=15.24 cm
Masses: 2 = 4.904; 5 =10.02034; 8 = 0.01469; 9 = 2.227 kg
I_P: 2 = 0.02712; 5 = 0.02304; 8 = 0.01469; 9 =0.0972 kgm^2
$I_D = I_P/2$ $E = 206.9$ GPa; $\rho = 8.304$ kg/m^3 $\omega_2 = 1.5\ \omega_1$

Fig. 16.2 Dual rotor system of Rajan et al. [17]

Table 16.1 Critical speeds (rad/s) of dual rotor

Mode No.	
$\omega = \Omega_1\ 1^L$	863
$\omega = \Omega_1\ 2^L$	1,606
$\omega = \Omega_1\ 3^L$	2,283
$\omega = -\Omega_1\ 1^L$	660
$\omega = -\Omega_1\ 2^L$	1,423
$\omega = -\Omega_1\ 3^L$	2,125
1^T	112,175
2^T	568,019

Consider the dual rotor of Rajan et al. [17] given in Figure 16.2.

The critical speeds obtained are given in Table 16.1. For complete results, see Rao et al. [23].

16.2 Geared Rotors and Chaos

Lund [12] considered coupling in the lateral and torsional in a geared system. Iida et al. [6] studied a simple geared system including coupling in the torsional and flexural vibration. David and Park [2] made a gearmesh stiffness formulation in truncated Fourier series and studied the nonlinear disk response due to unbalance. Schwibinger et al. [30] carried out studies on three different model shafts without rotation, shafts rotating supported on ball bearings and shafts supported on journal bearings and showed good correlation between theory and experiment. Schwibinger and Nordmann [31] later extended this study to include stability analysis.

Kahraman et al. [8] developed a finite element model of a geared rotor system on flexible bearings, including rotary inertia, axial loading, and stiffness and damping of the gear mesh. We will illustrate the extension of finite element method in Section 16.1 for geared rotors following Rao et al. [23]; see also [24, 32, 33].

The gearmesh model is shown in Figure 16.3, here the teeth are replaced by equivalent stiffness and damping along the pressure line. For the present damping is ignored. The gearmesh force is

$$F_h = k_h({v^g}_2 - {v^g}_1)\sin\phi_p + k_h({w^g}_2 - {w^g}_1)\cos\phi_p - k_h(r_1{\alpha^g}_1 + r_2{\alpha^g}_2) \quad (16.22)$$

On gear 1, the components of the gear mesh force can be expressed as

$$\begin{aligned} F_{hv1} &= F_h\sin\phi_p = k_h\begin{bmatrix} -S^2 & -SC & 0 & 0 & -r_1S & S^2 & SC & 0 & 0 & -r_2S \end{bmatrix}\{q^g\} \\ F_{hw1} &= F_h\cos\phi_p = k_h\begin{bmatrix} -SC & -C^2 & 0 & 0 & -r_1C & SC & C^2 & 0 & 0 & -r_2C \end{bmatrix}\{q^g\} \\ T_{h1} &= r_1F_h = k_h\begin{bmatrix} -r_1S & -r_1C & 0 & 0 & -{r_1}^2 & r_1S & r_1C & 0 & 0 & -r_1r_2 \end{bmatrix}\{q^g\} \end{aligned} \quad (16.23)$$

Similarly, on gear 2

$$\begin{aligned} F_{hv2} &= -F_h\sin\phi_p = k_h\begin{bmatrix} S^2 & SC & 0 & 0 & r_1S & -S^2 & -SC & 0 & 0 & r_2S \end{bmatrix}\{q^g\} \\ F_{hw2} &= -F_h\cos\phi_p = k_h\begin{bmatrix} SC & C^2 & 0 & 0 & r_1C & -SC & -C^2 & 0 & 0 & r_2C \end{bmatrix}\{q^g\} \\ T_{h2} &= r_2F_h = k_h\begin{bmatrix} -r_2S & -r_2C & 0 & 0 & -r_1r_2 & r_2S & r_2C & 0 & 0 & -{r_2}^2 \end{bmatrix}\{q^g\} \end{aligned} \quad (16.24)$$

For the gear pair, we can obtain

$$\begin{bmatrix} [{M^g}_1] & \\ & [{M^g}_2] \end{bmatrix}\{\ddot{q}^g\} + \Omega_1\begin{bmatrix} [{G^g}_1] & \\ & \frac{\Omega_2}{\Omega_1}[{G^g}_1] \end{bmatrix}\{\dot{q}^g\} - k_h\begin{bmatrix} [S_1] \\ [S_2] \end{bmatrix}\{q^g\} = \{{F^g}_s\} \quad (16.25)$$

where

$$[S_1] = \begin{bmatrix} -S^2 & -SC & 0 & 0 & -r_1S & S^2 & SC & 0 & 0 & -r_2S \\ -SC & -C^2 & 0 & 0 & -r_1C & SC & C^2 & 0 & 0 & -r_2C \\ 0 & 0 & 0 & 0 & 0 & 0 & 0 & 0 & 0 & 0 \\ 0 & 0 & 0 & 0 & 0 & 0 & 0 & 0 & 0 & 0 \\ -r_1S & -r_1C & 0 & 0 & -{r_1}^2 & r_1S & r_1C & 0 & 0 & -r_1r_2 \end{bmatrix}$$

$$[S_2] = \begin{bmatrix} S^2 & SC & 0 & 0 & r_1S & -S^2 & -SC & 0 & 0 & r_2S \\ SC & C^2 & 0 & 0 & r_1C & -SC & -C^2 & 0 & 0 & r_2C \\ 0 & 0 & 0 & 0 & 0 & 0 & 0 & 0 & 0 & 0 \\ 0 & 0 & 0 & 0 & 0 & 0 & 0 & 0 & 0 & 0 \\ -r_2S & -r_2C & 0 & 0 & -r_1r_2 & r_2S & r_2C & 0 & 0 & -{r_2}^2 \end{bmatrix}$$

and

$$\{q^g\}^T = \{\ v_1{}^g \quad w_1{}^g \quad \theta_1{}^g \quad \phi_1{}^g \quad \alpha_1{}^g \quad v_2{}^g \quad w_2{}^g \quad \theta_2{}^g \quad \phi_2{}^g \quad \alpha_2{}^g\ \}$$

In Section 14.3, we briefly discussed the linear stiffness and damping coefficients of oil film bearings. In reality these forces are nonlinear as shown by Mohan and Hahn [13]. Without a derivation, we will give here the nonlinear relations for the radial and tangential components of short squeeze film bearings commonly employed by aircraft engine designers to provide additional damping. These forces are

$$\begin{aligned} F_r &= \frac{\mu R L^3}{{c_r}^2}(\varepsilon\dot{\phi}A^{11} + \dot{\varepsilon}A^{02}) \\ F_\phi &= \frac{\mu R L^3}{{c_r}^2}(\varepsilon\dot{\phi}A^{20} + \dot{\varepsilon}A^{11}) \end{aligned} \tag{16.26}$$

where

$$\begin{aligned} A^{ij} &= \int_{\phi_g+\pi}^{\phi_g+2\pi} \frac{\sin^i\theta\cos^j\theta}{(1-\varepsilon\cos\theta)^3}d\theta \\ \phi_g &= \tan^{-1}\frac{\dot{\varepsilon}}{-\varepsilon\dot{\phi}} \end{aligned} \tag{16.27}$$

The horizontal and vertical components of these forces are

$$\begin{aligned} F_v &= \frac{F_r v - F_\phi w}{\sqrt{v^2+w^2}} \\ F_w &= \frac{F_r w + F_\phi v}{\sqrt{v^2+w^2}} \end{aligned} \tag{16.28}$$

Consider a simple gear pair mounted on squeeze film damper supports. The equations of motion are

$$[M]\{\ddot{q}\} + (c_h[S_h] + [C_c])\{\dot{q}\} + (k_h[S_h] + [K_c])\{q\} = \{Q\} \tag{16.29}$$

where $\{q\} = \{v_1\ w_1\ \alpha_1\ v_2\ w_2\ \alpha_2\}^T$. Here,

$$[S_h] = \begin{bmatrix} S^2 & SC & r_1S & -S^2 & -SC & r_2S \\ SC & C^2 & r_1C & -SC & -C^2 & r_2C \\ r_1S & r_1C & {r_1}^2 & -r_1S & -r_1C & r_1r_2 \\ -S^2 & -SC & -r_1S & S^2 & SC & -r_2S \\ -SC & -C^2 & -r_1C & SC & C^2 & r_2C \\ -r_2S & r_2C & r_1r_2 & -r_2S & r_2C & {r_2}^2 \end{bmatrix}$$

$$\{Q\} = \begin{Bmatrix} F_{v1} + M_1 e_1 \Omega_1{}^2 \cos(\Omega_1 t + \psi_1) \\ F_{w1} + M_1 e_1 \Omega_1{}^2 \sin(\Omega_1 t + \psi_1) - M_1 g \\ 0 \\ F_{v2} + M_2 e_2 \Omega_2{}^2 \cos(\Omega_2 t + \psi_2) \\ F_{w2} + M_2 e_2 \Omega_2{}^2 \sin(\Omega_2 t + \psi_2) - M_2 g \\ 0 \end{Bmatrix}$$

The non-zero elements of $[M]$, $[C_c]$ and $[K_c]$ are

$$\begin{aligned} M_{11} &= M_{22} = M_1 \\ M_{44} &= M_{55} = M_2 \\ M_{33} &= I_{p1} \\ M_{66} &= I_{p2} \\ C_{c11} &= C_{c22} = C_1 \\ C_{c44} &= C_{c55} = C_2 \\ K_{c11} &= K_{c22} = K_1 \\ K_{c44} &= K_{c55} = K_2 \end{aligned}$$

In (16.29) the first two were divided by $M_1 c_r \Omega_1{}^2$, the third one by $I_{p1}\Omega_1{}^2$, the fourth and fifth by $M_2 c_r \Omega_1{}^2$ and the sixth by $I_{p2}\Omega_1{}^2$ to obtain the non-dimensional form as

$$[I]\{\bar{q}''\} + \left(\frac{2\zeta_h}{\Omega_r}[\bar{S}_h] + \frac{2\zeta}{\Omega_r}[\bar{C}_c]\right)\{\bar{q}'\} + \left(\frac{\zeta_K}{\Omega_r{}^2}[\bar{S}_h] + \frac{1}{\Omega_r{}^2}[\bar{K}_c]\right)\{\bar{q}\} = \{\bar{Q}\} \tag{16.30}$$

where $\{\bar{q}\} = \{\,\bar{v}_1\ \bar{w}_1\ \bar{\alpha}_1\ \bar{v}_2\ \bar{w}_2\ \bar{\alpha}_2\,\}^T$.

$$[\bar{S}_h] = \begin{bmatrix} S^2 & SC & \frac{S}{\xi_c} & -S^2 & -SC & \frac{\eta_r S}{\xi_c} \\ SC & C^2 & \frac{C}{\xi_c} & -SC & -C^2 & \frac{\eta_r C}{\xi_c} \\ \xi_I \xi_c S & \xi_I \xi_c C & \xi_I & -\xi_I \xi_c S & -\xi_I \xi_c C & \xi_I \eta_r \\ \frac{-S^2}{\eta_M} & \frac{-SC}{\eta_M} & \frac{-S}{\eta_M \xi_c} & \frac{S^2}{\eta_M} & \frac{SC}{\eta_M} & \frac{-S}{\eta_M \xi_c} \\ \frac{-SC}{\eta_M} & \frac{-C^2}{\eta_M} & \frac{-C}{\eta_M \xi_c} & \frac{SC}{\eta_M} & \frac{C^2}{\eta_M} & \frac{\eta_r C}{\eta_M \xi_c} \\ \frac{\xi_I \xi_c \eta_r S}{\eta_I} & \frac{\xi_I \xi_c \eta_r C}{\eta_I} & \frac{\xi_I \eta_r}{\eta_I} & \frac{-\xi_I \xi_c \eta_r S}{\eta_I} & \frac{-\xi_I \xi_c \eta_r C}{\eta_I} & \frac{\xi_I \eta_r{}^2}{\eta_I} \end{bmatrix}$$

$$\{\bar{Q}\} = \begin{Bmatrix} \frac{S_D}{\Omega_r} f_{v1} + U_1 \cos(\tau + \psi_1) \\ \frac{S_D}{\Omega_r} f_{w1} + U_1 \sin(\tau + \psi_1) - \frac{W_g}{\Omega_r{}^2} \\ 0 \\ \frac{S_D}{\eta_M \Omega_r} f_{v2} + \frac{U_2}{\eta_r{}^2} \cos(-\frac{\tau}{\eta_r} + \psi_2) \\ \frac{S_D}{\eta_M \Omega_r} f_{w2} + \frac{U_2}{\eta_r{}^2} \sin(-\frac{\tau}{\eta_r} + \psi_2) - \frac{W_g}{\Omega_r{}^2} \\ 0 \end{Bmatrix}$$

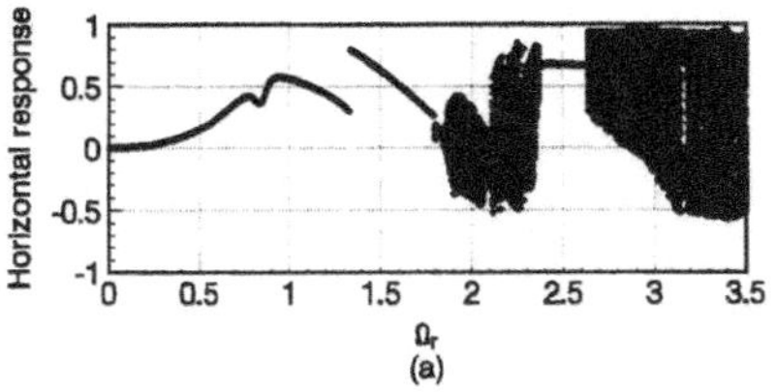

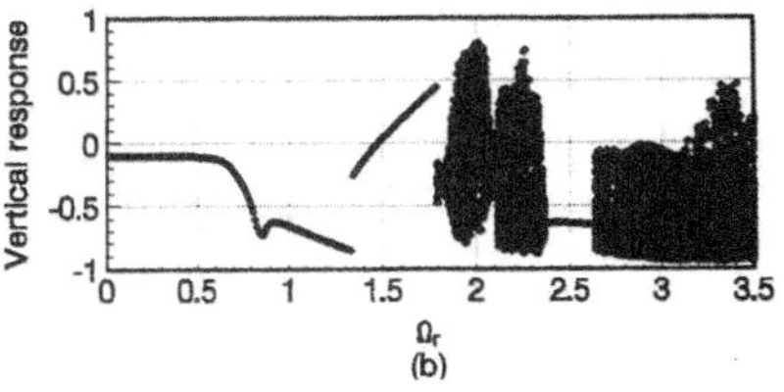

Fig. 16.3 Bifurcation maps of driving gear response for $U_1 = U_2 = 0.45$

$$f_v = \frac{f_r \bar{v} - f_\phi \bar{w}}{\sqrt{\bar{v}^2 + \bar{w}^2}} f_w = \frac{f_r \bar{w} + f_\phi \bar{v}}{\sqrt{\bar{v}^2 + \bar{w}^2}}$$

$$f_r = -(\varepsilon\phi' A^{11} + \varepsilon' A^{02}) f_\phi = -(\varepsilon\phi' A^{20} + \varepsilon' A^{11})$$

$$\bar{C}_{c11} = \bar{C}_{c22} = 1$$

$$\bar{C}_{c44} = \bar{C}_{c55} = \frac{\eta_c}{\eta_M} \text{rest} \bar{C} = 0$$

$$\bar{K}_{c11} = \bar{K}_{c22} = 1$$

$$\bar{K}_{c44} = \bar{K}_{c55} = \frac{\eta_K}{\eta_M} \text{rest} \bar{K} = 0$$

Newmark's method is used to determine the response by Shiau et al. [33]. For general study on Chaos, refer to [14]. The gear pair has the following data:

$$\xi_I = 2, \quad \xi_c = 10^{-5}, \quad \eta_M = \eta_C = \eta_K = \eta_r = \eta_I = \xi_K = 1$$

$$\phi_p = 22.5^0, \quad \zeta_h = \zeta = 0.0005, \quad S_D = 0.03$$

$$W_g = 0.1, \psi_1 = \psi_2 = 0$$

The critical speeds are found to be 0 (rigid body torsional), 0.79 (coupled) 1.0, 1.0, 1.0 (lateral) and 2.52 (coupled).

Effect of Imbalance: In all cases, initial conditions are kept the same, viz., zero for both response and velocities. After reaching a steady state regime in the time domain, 100 consecutive one-per-rotation speed marks which appear in on the response are recorded. For periodic response all these marks appear at a constant position for any number of rotations. For a steady synchronous response, there will be only one mark; if the response is half synchronous, there will be two marks etc., If the response is chaotic, multiple marks result from these consecutive 100 positions [7]. These positions for both the vertical and horizontal response of the system are given as bifurcation maps.

With larger imbalance $U_1 = U_2 = 0.45$, the influence on the chaotic range becomes significant as shown in Figures 16.3 and 16.4.

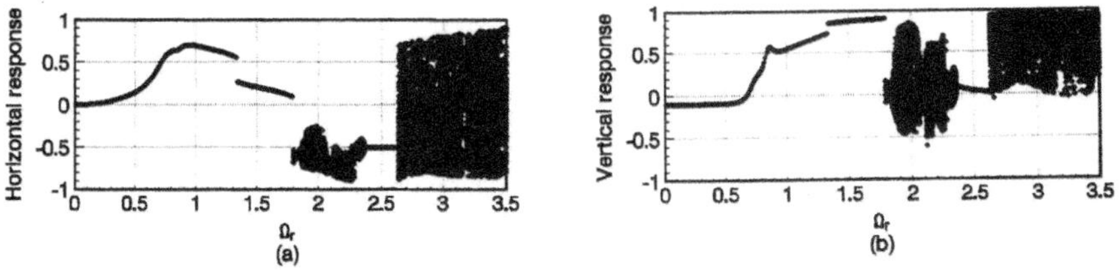

Fig. 16.4 Bifurcation maps of driven gear response for $U_1 = U_2 = 0.45$

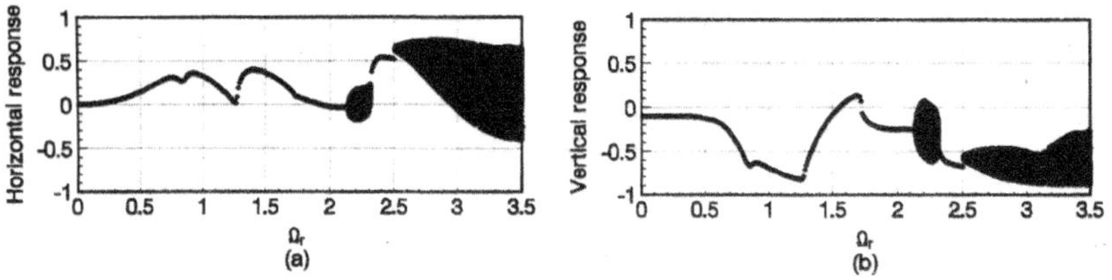

Fig. 16.5 Bifurcation maps of driving gear response for $S_D = 0.1$

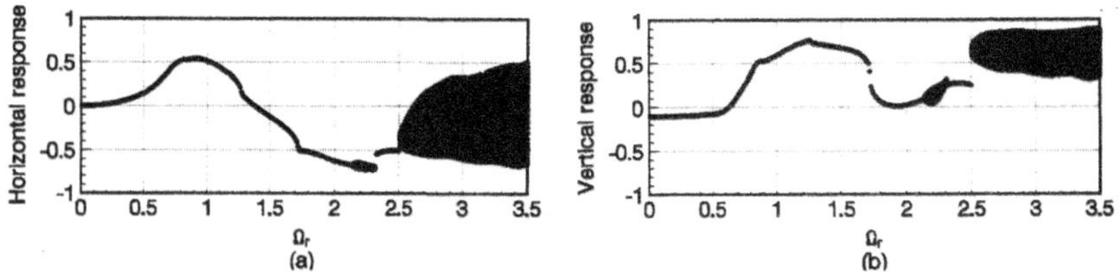

Fig. 16.6 Bifurcation maps of driven gear response for $S_D = 0.1$

The chaotic region is divided into two wider parts on both sides of the second coupled mode region. The response is periodic for $\Omega_r < 1.75$. Beyond this region, the response becomes chaotic with intermittent zones of periodic solutions. It is also clear from these figures that the response is periodic around all the critical speeds 0.79, 1.0 and 2.52.

For $S_D = 0.1$, Figures 16.5 and 16.6 give bifurcation maps which show that the response is periodic for $\Omega_r < 2.13$ and that the chaotic regime has intermittent zones of periodic response. Comparing these figures with Figures 16.3 and 16.4 we find that the major range of chaos for $S_D = 0.03$ is 1.8 to 2.38 and above 2.64 and for $S_D = 0.1$, it is 2.13 to 2.3 and above 2.5.

This is the greatest advantage of a squeeze film damper in increasing the range of operation in light-weight high-speed applications, particularly aircraft engines.

As another example of geared rotor systems, let us consider a geared turbo-generator set [18], given in Figure 16.7.

Here $M_{10} = 7.45$, $M_2 = 525.7$, $M_3 = 116.04$, $M_7 = 5.0$ and $M_6 = 726.4$ kg; $I_{T10} = 0.0745$, $I_{T2} = 16.1$, $I_{T3} = 3.115$, $I_{T7} = 0.002$ and $I_{T6} = 56.95$ kgm^2; $I_{P10} = 0.149$, $I_{P2} = 32.2$, $I_{P3} = 6.23$, $I_{P7} = 0.004$ and $I_{P6} = 113.9$ kgm^2.

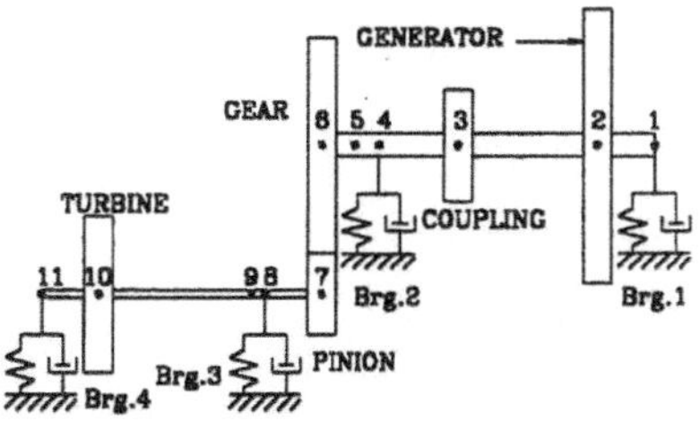

Fig. 16.7 Geared turbo-generator set

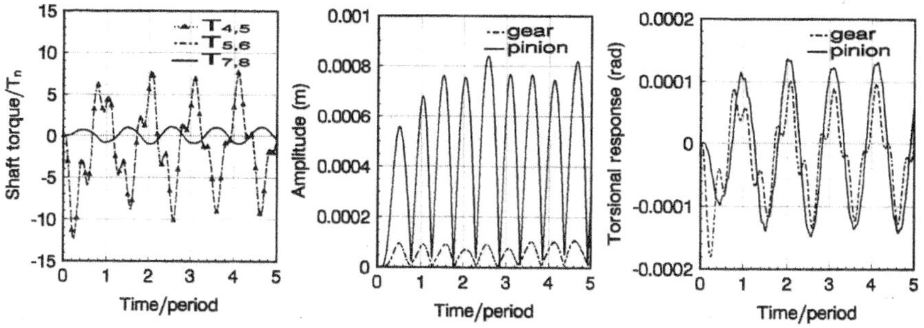

Fig. 16.8 Shaft torques, lateral and torsional response without damper

The pinion has 23 teeth with $r_7 = 0.03567$ m while the gear has 328 teeth with $r_8 = 0.5086$ m, $k_h = 10^8$ N/m and $\phi_p = 22.5°$. Shaft lengths between nodes are taken as $l_{12} = 0.1$, $l_{23} = 4.24$, $l_{34} = 1.16$, $l_{45} = l_{56} = 0.15$, $l_{78} = 0.3$, $l_{89} = 0.05$, $l_{9,10} = 4.95$, $l_{10,11} = 0.1$ m, radius between nodes 1 and 3 = 0.15 m, between nodes 3 and 6 = 0.11 m and for turbine shaft = 0.075 m. Young's modulus is 207 GPa, shear modulus is 79.5 GPa and density is 7800 kg/m^3.

Bearings 1 and 2 have $k_{yy} = 1.839 \times 10^8$, $kzz = 2.044 \times 10^8$ and bearings 3 and 4 $k_{yy} = 1.010 \times 10^7$ and $k_{zz} = 4.160 \times 10^7$ N/m, Damping in all bearings is 3000 Ns/m. Nominal torque $T_n = 318.31$ N/m. The short circuiting torque is $318.31 \times 12.353e^{-33.97t} \sin \omega t + 318.31 \times 0.5e^{-5.45t} \sin 2\omega t$.

The coupled natural frequencies for $\Omega_1 = 1500$ RPM are 73.1, 77.2, 108.5, 116.0 150.5, 236.8, 313.5 319.6, 353.5, 362.9, 371.9, 396.6 and 466.2 rad/s. The response without any squeeze film damper is shown in Figure 16.8.

Several arrangements of dampers are tried; here one arrangement is discussed to study the effectiveness of the dampers on short circuit currents. The damper chosen has a radial clearance $c_r = 152.4$ microns and viscosity of oil $\mu = 0.015016$ Ns/m^2. The damper was located at 9 near to the pinion where the maximum response is expected. Figure 16.9 shows the response and effectiveness of the damper when compared with Figure 16.8.

Finite element methods, whether for stationary or rotor dynamics applications were of fairly recent origin. Though beam elements became popular with Nelson's Timoshenko element in 1980, they remained one dimensional till recently until the

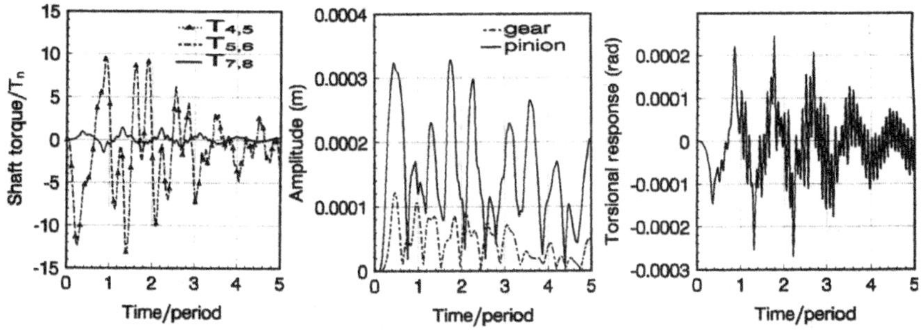

Fig. 16.9 Shaft torques, lateral and torsional response with I-9 damper

end of the 20th century. Meanwhile the structural models including rotating blades became fully solid and capable of adapting to any geometry of discontinuities; the rotor dynamics models remained rudimentary Euler beam models with an extension of the Timoshenko effect. Gyroscopic effects split the natural frequencies and brought in the effect of speed and Campbell diagrams into the design. However, the centrifugal loads and the influence of speed on natural frequencies remained unexplored.

Most of the research in the 20th century concentrated on special effects of rotors which are not present in stationary structures, such as unbalance, oil film supports, seals, looseness, asymmetry, variable inertia, misalignment and associated instabilities. Though the new culture is to model the structure, mesh (finite element model) and analyze in CFD, Heat Transfer, Thermo-Mechanical analysis, Elasticity, Elasto-Plasticity, Contacts, Nonlinearities, Composites, Stress Stiffening and Spin Softening effects etc. rotors however remained as simple beam models. Making a beam model of an aircraft engine with two spools or more, drum-like structures in place of disks, and so on, makes the designer's life difficult and time consuming and expensive for the equipment manufacturers. Therefore solid models have advantages of bringing the design process to model, mesh and go rather than reduce to beams, introduce gyroscopic effects, etc., and perform simple analysis with a smaller number of elements. Recent developments in the first decade are discussed in the next section.

16.3 Solid Rotors

The idea of solid or shell models of stationary structures became evident when the casings and foundations became a part of the rotating machine and test beds became flexible when coupled with rotors that interfered in the validating process. Stephenson and Rouch [34] used axisymmetric solid finite elements with matrix reduction in their analysis; Yu et al. [36] modeled shafts orbiting with 3-D solid finite elements. Neither of these included the effects of rotation, stiffening and softening

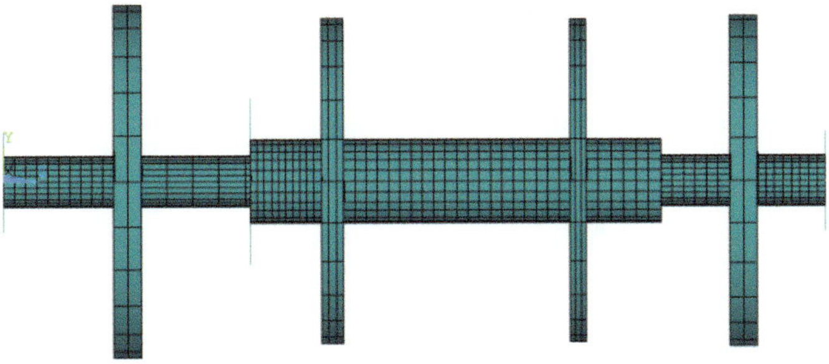

Fig. 16.10 A solid rotor model of Figure 16.2

effects that go with solid elements. Solid rotor dynamics analysis was first presented by Rao [20]. Rao et al. [28] provided the details subsequently.

The twin spool rotor system in Figure 16.2 is illustrated here for solid rotor modeling. An equivalent solid rotor model for the example of Figure 16.2 is made as shown in Figure 16.10. The four rotors with given masses and inertia properties are replaced by disks (1) 10.626 cm dia 1.7 cm wide, (2) 9.7344 cm dia 1.38 cm wide, (3) 9.682 cm dia 0.978 cm wide and (4) 9.9954 cm dia 1.66 cm wide consecutively to give identical properties as shown. It may be noted that Figure 16.2 can represent in a unique manner an equivalent beam model of the solid model of Figure 16.10, even though several other solid models can be derived for Figure 16.2 beam model. This in fact is the main limitation of beam model analysis as an equivalent derived beam model may represent the dynamics of different solid models. An actual physical model in solid form eliminates this approximation. Rao et al. [28] studied several different models, here some case studies are discussed. We begin with a *solid rotor model with lumped masses and inertias*.

The solid model is constructed using SOLID-45 elements with eight nodes each having three degrees of freedom of translation. The bearings are simulated by COMBIN-14 elements. Both the rotors are subjected to spin by using a "useracel" subroutine. By default, this model also has a spin-softening effect. A stress stiffening effect is not included as the masses are lumped. Mass21 elements are adopted for simulating the disks. The masses on the outer rotor are related by RBE3 (constraint equation). The total number of elements in this model is 5304 and the nodes are 6568. The model is shown in Figure 16.11. The Campbell diagram obtained for this model is given in Figure 16.12.

The results from a beam model with gyroscopic effects are also included in Figure 16.12. The model in Figure 16.11 is similar to the beam model except for the "spin softening" effects included in the system. The eigen-value problem with spin softening effect is defined by

$$|([K] - {\omega_s}^2[M]) - \omega^2[M]| = 0$$

$$|[\bar{K}] - \omega^2[M]| = 0 \tag{16.31}$$

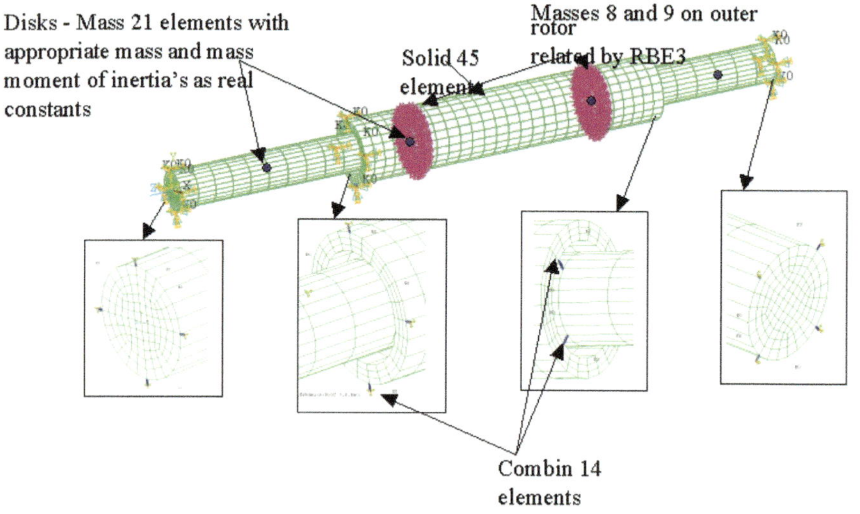

Fig. 16.11 Solid rotor model with disks modeled as Mass21 elements

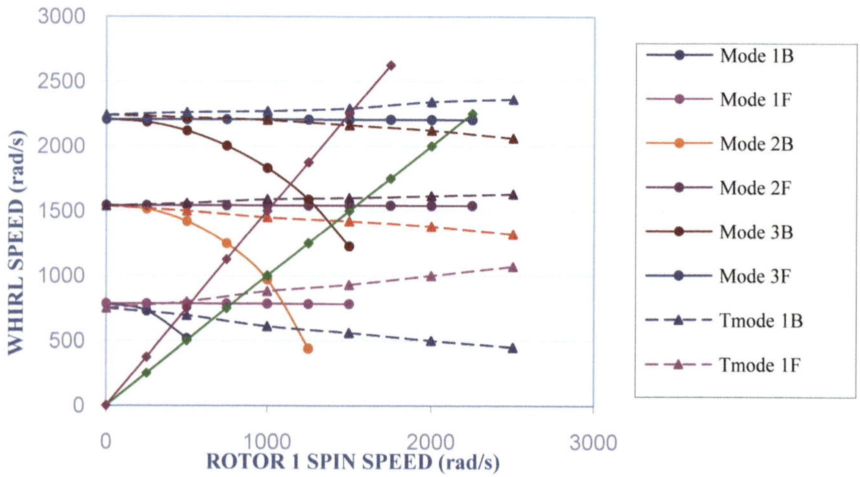

Fig. 16.12 Campbell diagram for rotor in Figure 16.11

where $[K]$ is the stiffness matrix, $[M]$ is the mass matrix, ω is the natural frequency and ω_s is the spin speed and the effective stiffness matrix is denoted by a bar above K.

With the conventional gyroscopic effects included in a beam model without spin softening effect, the backward whirl decreases in a linear manner as discussed before and also depicted in Figure 16.12. When spin softening effects are included, the backward whirl natural frequency decreases with spin speed and the effective

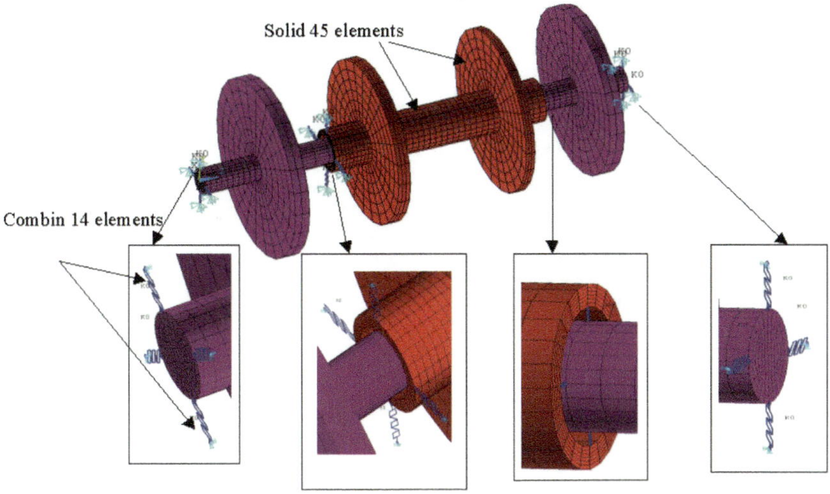

Fig. 16.13 Full solid rotor model

stiffness becomes zero when the spin speed becomes the natural frequency of the stationary shaft as in Figure 16.12. Therefore the backward whirl natural frequency drops at a faster rate when spin softening is accounted for and disappears at a spin speed equal to the natural frequency under stationary conditions. This is a new result from the spin softening effects due to spin speed included in the model. Since the stress stiffening effect is absent and there is no disk effect forward whirl critical speed is not affected in the solid model in the absence of gyroscopic effect from disks. Only the backward whirl is significantly affected according to equation (16.31).

In the next model, we consider a solid rotor model with disks Figure 16.13 and no stress stiffening. Here both the shafts and disks are modeled using Solid45 elements. The other modeling features are the same as the previous case. The elements are 12,680 in all with 15,367 nodes. The Campbell diagram for this case is shown in Figure 16.14.

Here there is no stress stiffening and the disk spin softening decreases even the forward whirl critical speed unlike in the case of Figure 16.12. Next a stress-stiffening effect is also included then Figure 16.15 shows the Campbell diagram.

Both forward and backward whirl frequencies increased over the previous case due to stress stiffening. The increase in the forward whirl frequencies is more predominant than that of the backward whirl frequencies under the influence of stress stiffening. The forward whirl frequencies do not vanish as in the previous case at higher spin speeds due to the stress stiffening effect. In reality we cannot expect the forward whirl modes to disappear. Thus the solid rotor models become more practical in applications.

Rao and Sreenivas [25] have extended the two rotor system to a three level with the casing included in the analysis. Such an analysis enabled practical engine rotor

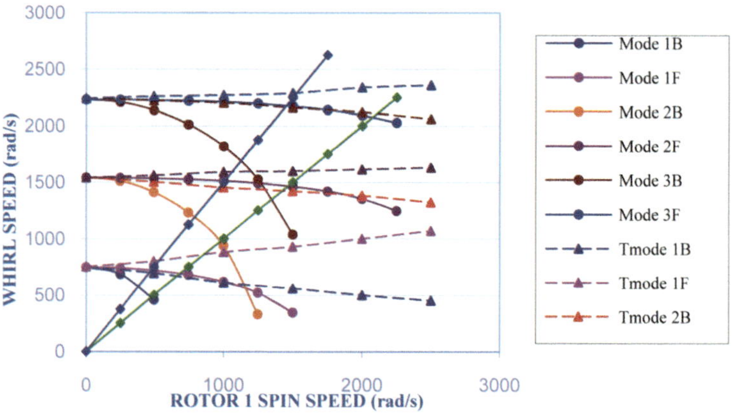

Fig. 16.14 Campbell diagram for rotor in Figure 16.13

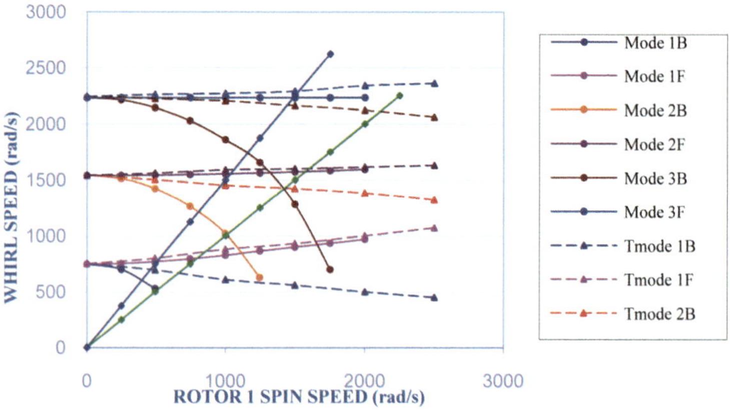

Fig. 16.15 Campbell diagram with stress stiffening Included

dynamics analysis. Surial and Kaushal [35] modeled an industrial gas turbine engine and validated it at Rolls Royce Canada. Due to the complexity of the engine structure the model has been divided into four substructures using super-elements. The first super-element represents the engine casing, the second the low pressure rotor (LP rotor), the third the intermediate pressure rotor (IP rotor), and the fourth the high pressure rotor (HP rotor). In Figure 16.15 two practical aerospace applications are illustrated.

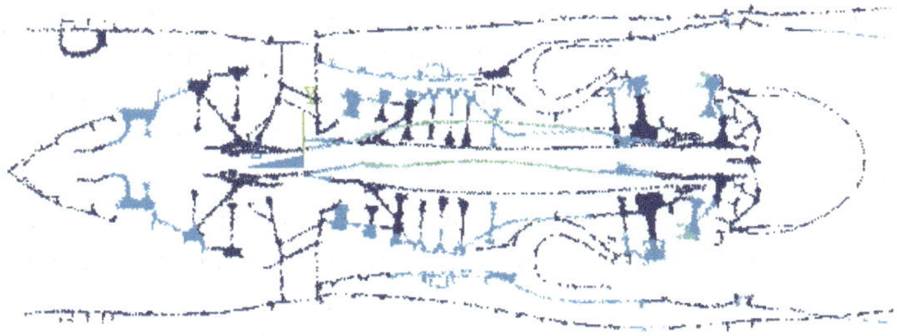

Fig. 16.16 Mode shape of a typical two spool aircraft engine

16.4 Two Spool Aircraft Engine [21]

Rotor dynamics plays an important role in engine design. In recent years, considerable advances have taken place from simple beam model analysis. The support characteristics are generally the main issue, from bearings, dampers and also seals all of which play significant roles in engine dynamics. The bearings in an engine are supported in flexible structures, the front frame of the fan, the middle frame, etc. The casing itself is 1.5 to 2 mm thick in advanced light engines and they add to considerable flexibility of the system. Therefore, it has become a necessity to be able to couple rotors and stators into a single model. Moreover, it is a complex process to determine the stiffness of the frames supporting the bearings for simulation of the rotor.

Making approximate beam models of the engine spools for analysis purposes has therefore become obsolete and one uses now solid elements directly in the model. Thus, one can use the same solid model and mesh adopted for structural and thermal analysis. A typical mode shape of an engine is shown in Figure 16.16.

The whirl amplitudes at critical speeds will be quite large and the blade tips can rub at these speeds. Taking into account, the transient thermal and steady load growths of the casing and the radial displacements of different stages of the rotor under centrifugal and gas loads and whirling amplitudes at critical speeds, overall clearances required can be fixed.

16.5 Cryogenic Pump Rotor Dynamic Analysis

High speed cryogenic pumps are employed in Geo Synchronous Satellite Launch Vehicles. They use liquid hydrogen LH_2 and oxygen LOX as propellants. The CAD model of such a pump is shown in Figure 16.17 [27]. The rotor and the casing are meshed with eight-noded brick elements (see Figure 16.18). Bearings and seals see Figure 16.19 are simulated by 12×12 matrix elements accounting for all translational and rotational degrees of freedom between two nodes. Total number of elements is 379,779 and the total number of nodes is 534,115.

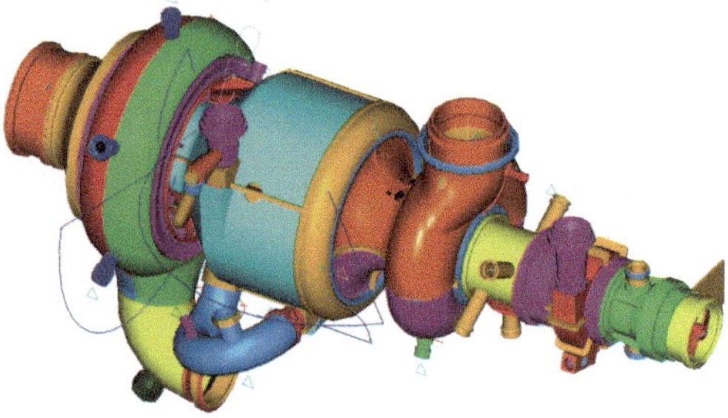

Fig. 16.17 CAD model of the turbo pump

Fig. 16.18 FE model of the turbo pump

The bearings (deep groove ball) are taken linear for first cut analysis and stiffnesses taken as 6, 5.6, 5.6, 5.04, 5.04 and 3.27 $\times$ 10^4 N/mm for bearing numbers 1 to 6 respectively based on static bearing reactions. (The stiffness is subsequently considered nonlinear depending on the reaction forces.) There are eight seals and their properties are determined using bulk flow models. The stiffness and damping values are curve fitted; typically the properties of first seal are given by

$$K_{xxlb1} = 4.738452E - 12 \times \text{RPM}^3 + 2.261960E - 06 \times \text{RPM}^2 + 2.659990E - 03 \times \text{RPM} + 2.958842E + 02\ \text{N/mm}$$

The boundary conditions adopted for the rotor alone analysis are given in Figure 16.19.

The other end of the bearing and the seal free nodes are constrained in all degrees of freedom. All the inlets and the outlets of the turbo-pump are constrained in all degrees of freedom as shown in Figure 16.20 for combined rotor-casing analysis.

Stress stiffening and spin softening effects are included. The results are post processed and the different forward and backward modes are identified at different speeds. Eigen-values are extracted using the Block Lanczos method and the Campbell diagrams obtained thus are given in Figures 16.21 and 16.22 for the rotor alone

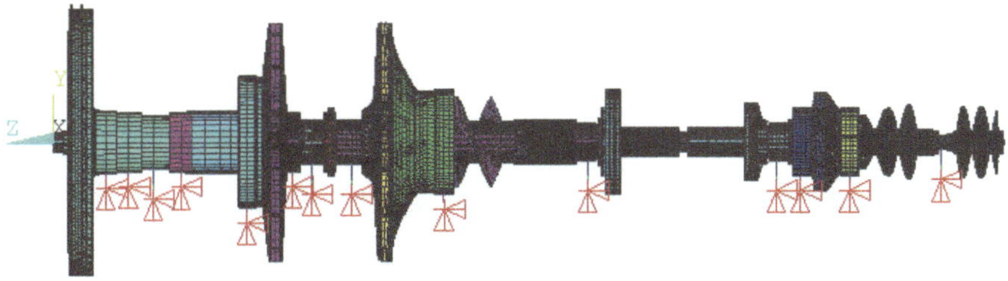

Fig. 16.19 Boundary conditions for rotor; bearing locations 1 to 6 are as marked, the remaining 8 are seal locations

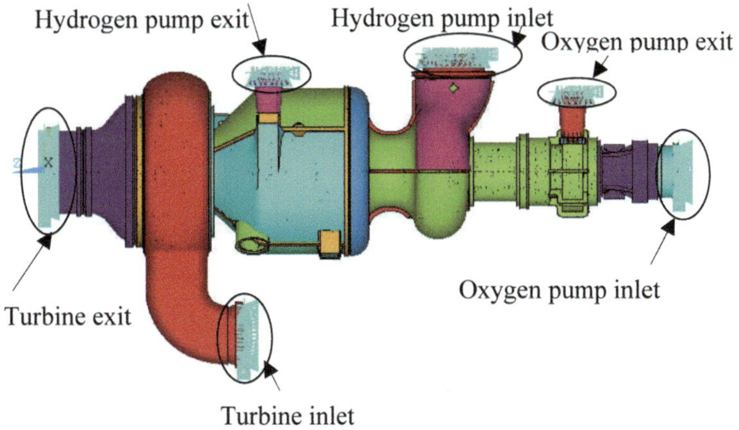

Fig. 16.20 Boundary conditions for rotor casing

and rotor plus casing respectively. Here, only bending modes are considered while plotting the Campbell diagrams, torsional and axial modes are excluded.

It is observed that the first forward critical speed of the rotor is at 28000 RPM and the second forward critical speed is at 55000 RPM. The first and second backward critical speeds are observed at 20000 and 44000 RPM respectively. The critical speeds dropped significantly to 23500 and 39000 RPM when the casing effect is included.

The unbalance distribution in gm-mm is taken as given in Figure 16.23. Macros are written to apply the unbalance load in accordance to the following relation:

$$F_x = u\omega^2 \cos\beta \cos\omega t + iu\omega^2 \sin\beta \sin\omega t$$
$$F_y = u\omega^2 \sin\beta \cos\omega t - iu\omega^2 \cos\beta \sin\omega t \tag{16.32}$$

where u is the unbalance and β is the phase angle.

Bearing reaction forces obtained for the rotor alone case in the y direction are given in Figure 16.24.

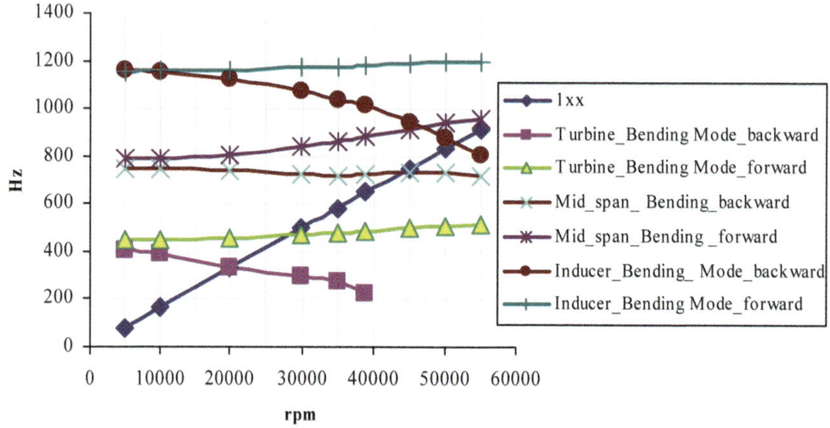

Fig. 16.21 Rotor alone case Campbell diagram

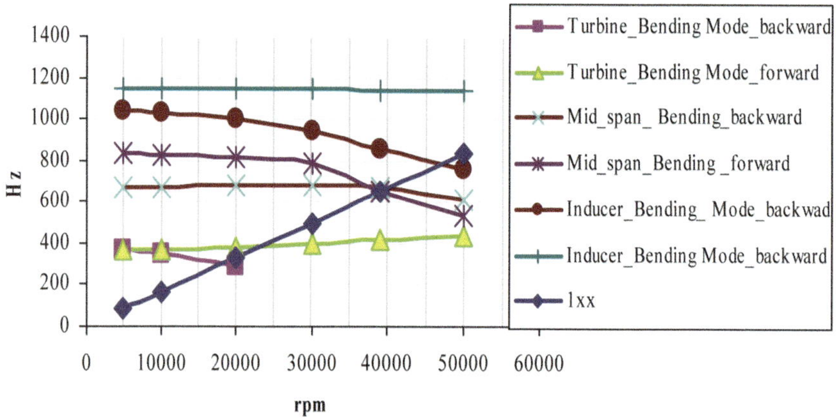

Fig. 16.22 Rotor-casing Campbell diagram

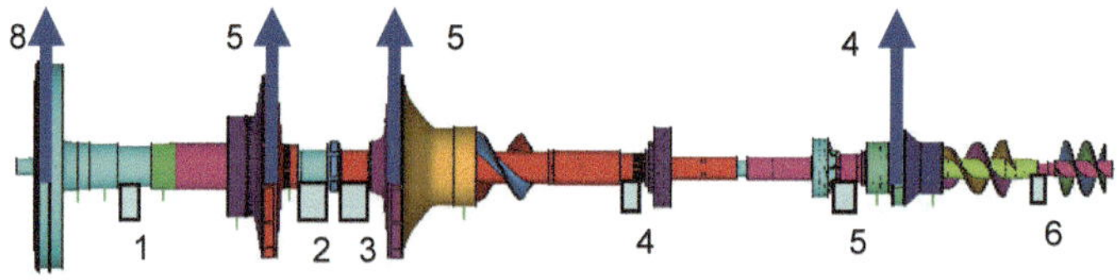

Fig. 16.23 Unbalance distribution

It is observed that peak responses occur at critical speeds as predicted in the rotor alone case Campbell diagram Figure 16.21. Bearing 1 suffers maximum reaction force 2150 N at the first forward critical speed. The response in this region is not correct because the bearing stiffness is significantly affected by the response. In

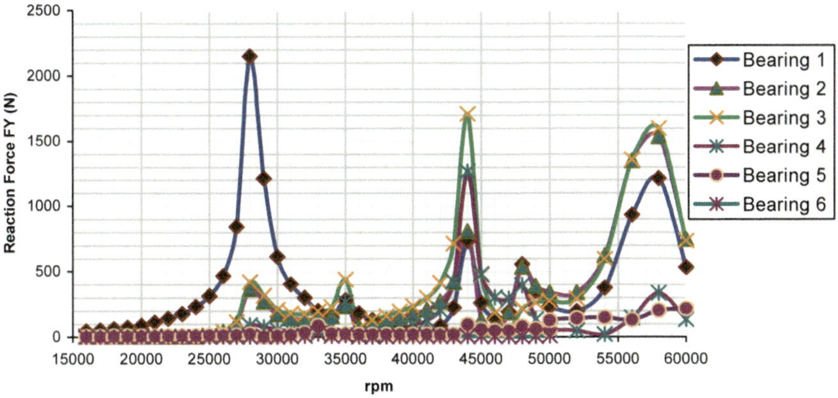

Fig. 16.24 Bearing reactions forces in y direction

reality the bearing stiffness is nonlinear and given by

$$k = 1.289{n_z}^{0.666}d^{0.333}F^{0.333}\ \mathrm{N/\mu m} \tag{16.33}$$

where n_z is the number of balls in the ball bearing = 9, d is the diameter of the ball = 11.12 mm and F is the reaction force at the desired frequency = 2150 N.

Substituting these values, we get $k = 16.363 \times 10^7$ N/m. This is 2.727 times more than the stiffness used before. This calls for consideration of bearing nonlinearity in the analysis. Nonlinear spring systems were first considered by Duffing [3]. An iterative process is to be adopted to determine the response in the critical speed region to match the starting value of stiffness and the bearing reaction force with the results in the final step. The casing stiffness, internal pressure etc. amongst other parameters will influence the end result, for the present it is assumed that the average value of starting stiffness is 6×10^7 N/m and the evaluated stiffness 16.36×10^7 N/m, viz., 12×10^7 N/m will be applicable to assess the influence of nonlinearity of the bearing. With this, the Campbell diagram obtained is shown in Figure 16.25.

The first forward critical speed is now at 34,000 RPM and the second forward critical speed is above 55,000 RPM. Thus, the first forward critical speed increased from 28,000 RPM by 21.4%.

Rao et al. [27] have also considered the influence of internal pressure, stability analysis and the influence of dead bands using the solid model for rotor and casing.

Rao and Sreenivas [26] conducted several transient studies of Jeffcott type solid rotor dynamics analysis to study the instability and gravity effects. These transient studies were extended by Rao [22] to consider high angular acceleration of the pump in Figure 16.17 to reach full speed (50000 RPM) in three seconds. The mass of the rotor is about 8 kg. The rotor is assumed to be balanced to a quality G2.0. The permissible eccentricity of this system (at 50,000 RPM) is 0.000398 mm. Hence the residual unbalance in the system, during coasting up is about 0.0032 kg-mm. This unbalance was applied to the turbine rotor of the pump disk 8 as shown in

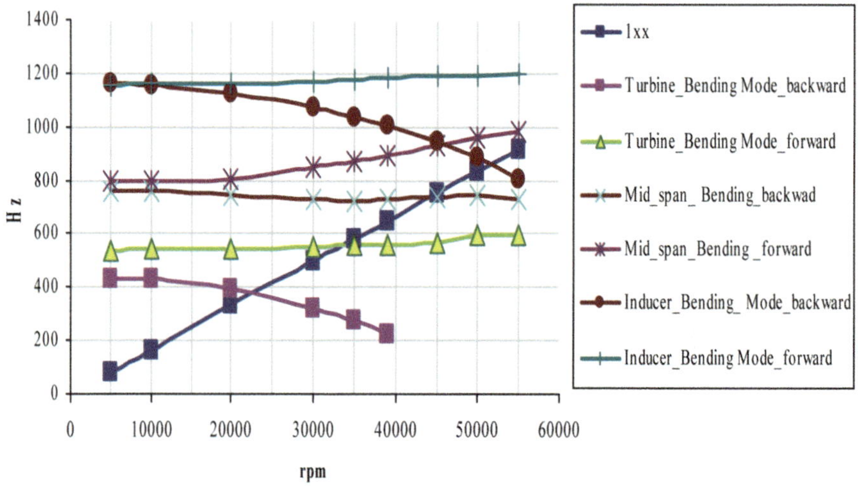

Fig. 16.25 Rotor alone Campbell diagram with revised stiffness for bearing 1

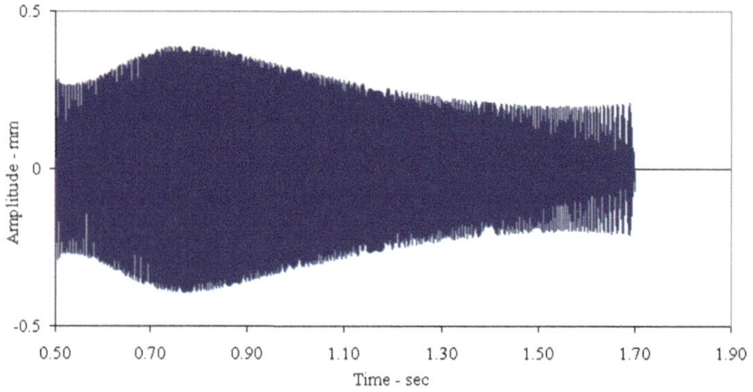

Fig. 16.26 Coast up response of a high speed Turbopump with acceleration 1800 rad/sec^2

Figure 16.23 and the pump is allowed to coast up from about 10,000 to about 30,000 RPM. Figure 16.26 shows the time domain response of the rotor, while coasting up through its first critical speed of 228 Hz. It may be seen that the peak response occurs at around 0.795 seconds. The mode shape at peak response speed obtained is shown in Figure 16.27.

Solid rotor models for rotor dynamics analysis have simplified the analysis by removing approximation in deriving beam models. The solid models have enabled inclusion of significant and important stress stiffening and spin softening affects which otherwise have been neglected. These speed effects are important in high speed rotating machinery that is becoming increasingly common. Another significant advantage is the gyroscopic effects automatically being included in the solid models and there is no need to do an artificial split and look for backward whirl ef-

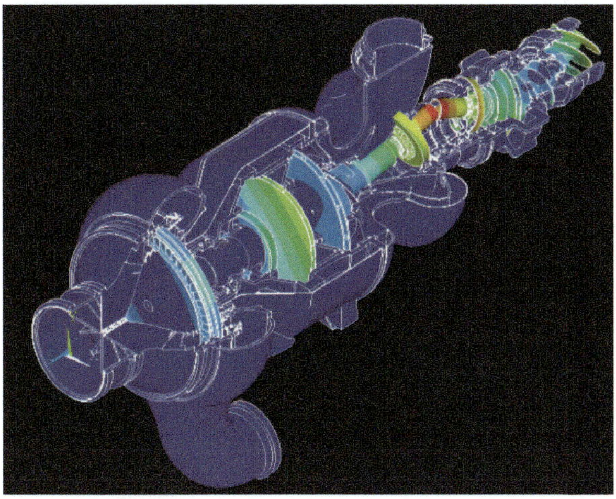

Fig. 16.27 Mode shape of the cryogenic pump

fects. Solid models indeed predict the backward whirl modes more precisely. They may play an important role in coupled rotors experiencing both forward and backward whirls in higher modes.

References

1. Childs, D.W. (1993) *Turbomachinery Rotordynamics: Phenomena, Modeling, and Analysis*, Wiley Inter Science.
2. David, J.W. and Park, N.G. (1986) The Vibration Problem in Gear Coupled Rotor Systems, *Bulletin JSME*, vol. 29, no. 252, p. 297.
3. Duffing, G. (1918) *Erzwungene Schwingungen bei veranderlicher Eigenfrequenz*, F. Vieweg u. Sohn, Braunschweig.
4. Ehrich, F.F. (1992) *Handbook of Rotor Dynamics*, McGraw-Hill.
5. Ehrich, F.F. (1995) Nonlinear Phenomena in Dynamic Response of Rotors in Anisotropic Mounting Systems, *ASME Special 50th Anniversary Design Issue*, vol. 117, p. 154.
6. Iida, H., Tamura, A., Kikuchi, K. and Agata, H. (1980) Coupled Torsional Flexural Vibration of a Shaft in a Geared System of Rotors, *Bulletin JSME*, vol. 23, no. 186, p. 2111.
7. Goldman, P. and Muszynska, A. (1994) Chaotic Behavior of Rotor/Stator Systems with Rubs, *ASME Journal of Engineering for Gas Turbines and Power*, vol. 116, p. 692.
8. Kahraman, A., Nevjat Ozguven, H., Houser, D.R. and Zakrajsek, J.J. (1982) Dynamic Analysis of Geared Rotors by Finite Elements, *J. Mech Des., ASME*, vol. 114, p. 507.
9. Kiciński, J. (2006) *Rotor Dynamics, Institute of Fluid Flow Machinery*, Polish Academy of Sciences, Gdansk.
10. Lalanne, M. and Ferraris, G. (1990) *Rotordynamics Prediction in Engineering*, John Wiley and Sons, New York.
11. Lalanne, M. and Queau, J.P. (1979) Calcul par elements finis du comportment dynamique des chaines cinematiques de reducteur, Societe Nationale des Industries Aerospatiale.
12. Lund, J.W. (1978) Critical Speed, Stability and Response of a Geared Train of Rotors, *Journal of Mechanical Design, ASME*, vol. 100, p. 535.

13. Mohan, S. and Hahn, E.J. (1974) Design of squeeze film damper supports for rigid rotors, *Journal of Engineering for Industry, Trans. ASME*, p. 976.
14. Moon, F.C. (1987) *Chaotic Vibrations*, John Wiley.
15. Nelson, H.D. (1980) Finite Rotating Shaft Element Using Timoshenko Beam Theory, *Journal of Mechanical Design, ASME*, vol. 102, p. 793.
16. Nelson, H.D. and McVaugh, J.M. (1976) The Dynamics of Rotor Bearing Systems Using Finite Elements, *Journal of Engineering for Industry, ASME*, vol. 98, p. 593.
17. Rajan, M., Nelson, H.D. and Chen, W.J. (1986) Parameters Sensitivity in the Dynamics of Rotor-Bearing Systems, *Journal Vib. Acoustic. Stress Rel. Des., ASME*, vol. 108, p. 197.
18. Rao, J.S. (1992) Life Estimation of Gear Transmission Unit in a Turbine Generator Set due to Short Circuits, *Mechanism and Machine Theory*, vol. 27, no. 3, p. 283.
19. Rao, J.S. (1996) *Rotor Dynamics*, New Age International.
20. Rao, J.S. (2002) Rotor Dynamics Comes of Age, Keynote address, in *Proceedings Sixth IFToMM International Conference Rotor Dynamics*, Sydney, September 30–October 3, vol. I, p. 15.
21. Rao, J.S. (2003) Recent Developments in Structural Design Aspects of Aircraft Engines, in *Proceedings National Conference on Association of Machines and Mechanisms*, IIT, New Delhi, 18–19 December 2003, Professor B.M. Belgaumkar Memorial and Inaugural Lecture.
22. Rao, J.S. (2006) Transient Dynamics of Solid Rotors under high angular accelerations, *Advances in Vibration Engineering, Journal of Vibration Institute of India*, vol. 5, no. 1, p. 25.
23. Rao, J.S., Shiau, T.N. and Chang, J.R. (1995) Coupled Bending-Torsion Vibration of Geared Rotors, in *Proceedings 1995 Design Engineering Technical Conferences*, ASME DE-Vol. 84-2, p. 977.
24. Rao, J.S., Shiau, T.N. and Chang, J.R. (1998) Theoretical Analysis of Lateral Response due to Torsional Excitation of Geared Rotors, *Mechanism and Machine Theory*, vol. 33, no. 6, p. 761.
25. Rao, J.S. and Sreenivas, R. (2003) Dynamics of a Three Level Rotor System Using Solid Elements, ASME GT 2003-38783.
26. Rao, J.S. and Sreenivas, R. (2004) Dynamics of Asymmetric Rotors Using Solid Models, *Advances in Vibration Engineering, Journal of Vibration Institute of India*, vol. 3, no. 3, p. 272.
27. Rao, J.S., Sreenivas, R. and George, P, (2004) Dynamics of High Speed Cryo Pump Rotors, in *Proceedings 8th International I Mech E Conference on Vibrations in Rotating Machinery*, C623/103/2004, p. 467.
28. Rao, J.S., Sreenivas, R. and Veeresh, C.V., (2002) Solid Rotor Dynamics, in *Proceedings Fourteenth US National Congress of Theoretical and Applied Mechanics*, Blacksburgh, VA, 23–28 June; (2003) *Advances in Vibration Engineering, Journal of Vibration Institute of India*, vol. 2, no. 4, p. 305.
29. Ruhl, R.L. and Booker, J.F. (1972) A Finite Element Model for Distributed Parameter Turborotor Systems, *Journal of Engineering for Industry, Trans. ASME*, vol. 94, p. 126.
30. Schwibinger, P., Neumer, T., Zurbes, A. and Nordmann, R. (1988) The Influence of Torsional Lateral Coupling in Geared Rotor Systems on Its Eigen Values, Modes and Unbalance Vibrations, in *Proceedings I Mech E Conference Vibrations in Rotating Machinery*, Edinburgh, C295/88, p. 279.
31. Schwibinger, P. and Nordmann, R. (1988) The Influence of Torsional-Lateral Coupling on the Stability Behavior of Geared Rotor Systems, *Journal of Engineering for Gas Turbines for Power, ASME*, vol. 110, p. 563.
32. Shiau, T.N., Rao, J.S. and Chang, J.R. (1996) Dynamic Coupling in Simple Geared Rotor Bearing System, in *Vibrations in Rotating Machinery*, Instn. of Mech. Engrs., C500/127/96, p. 599.
33. Shiau, T.N., Rao, J.S., Chang, J.R. and Siu-Tong, C. (1999) Dynamic Behavior of Geared Rotors, *Journal of Engineering for Gas Turbines and Power, Trans. ASME*, vol. 121, no. 3, p. 494.
34. Stephenson, R.W. and Rouch, K.E. (1993) Modeling rotating shafts using axi-symmetric solid finite element with matrix reduction, *ASME Journal of Vibration & Acoustics*, vol. 115, p. 484.

35. Surial, A. and Kaushal, A. (2005) Dynamic Analysis of a Variable Speed Industrial Gas Turbine Engine and Drivetrain – Analysis and Testing, *Advances in Vibration Engineering*, vol. 4, no. 3, p. 279.
36. Yu, J., Craggs, A. and Mioduchowski, A. (1999) Modeling of shaft orbiting with 3-D solid finite elements, *International Journal of Rotating Machinery*, vol. 5, p. 53.
37. Zorzi, E.S. and Nelson, H.D. (1980) The Dynamics of Rotor Bearing Systems with Axial Torque: A Finite Element Approach, *ASME Journal of Mechanical Design*, vol. 102, p. 158.

Chapter 17
Bladed Disks

Bladed-disk vibrations were well studied because of the critical fatigue problems. They are the most stressed systems in machines. Campbell [5], Stodola [52] and Sezawa [48] are amongst the first few who studied the bladed-disks. Kroon [25] applied difference calculus to the case of lashed blades to determine the blade stresses. Though no vibrations were considered, this paper is the first attempt to point out that the whole blade group should be considered. Smith [51] made a two-dimensional free vibrational analysis in the tangential direction using a dynamic stiffness matrix method on a six and twenty bladed group. His contribution was most significant since the group frequencies and mode shapes were determined for the first time as shown in Figure 17.1.

Myklestad's adaption of the Holzer method was used by Jarret and Warner [20] to determine the natural frequencies of rotating tapered twisted beams, the influence of lashing wires and shroud were included as constraints. Prohl [34] presented a method of calculating natural frequencies, mode shapes and bending stresses for 3D free and forced vibrations in the tangential and axial directions. In a companion paper by Weaver and Prohl [55], Prohl's method was used to calculate the natural frequencies and mode shapes of a simple blade group. A method of analysis for a laced group of rotating blades was given by Deak and Baird [8]). Rao [35] gave these methods in his book on *Turbomachine Blade Vibration*.

Armstrong et al. [1] provided a solution for bladed-disks using the method of receptances described by Bishop and Johnson [3]. A detailed study of vibrations of bladed-discs was given by Ewins [13]. Ewins and Rao [14] presented the effect of damping on bladed-disc vibrations. The influence of coupling between bending and torsion and rotation are included in the analysis by Thomas and Subuncu [54]. Irretier [18] further generalized by taking coupled bending-bending-torsion motions. Later Irretier [19] gave a spectral analysis of mistuned bladed-disc assemblies by component mode synthesis. Friction related mistuning issues were studied by Muszynska and Jones [29].

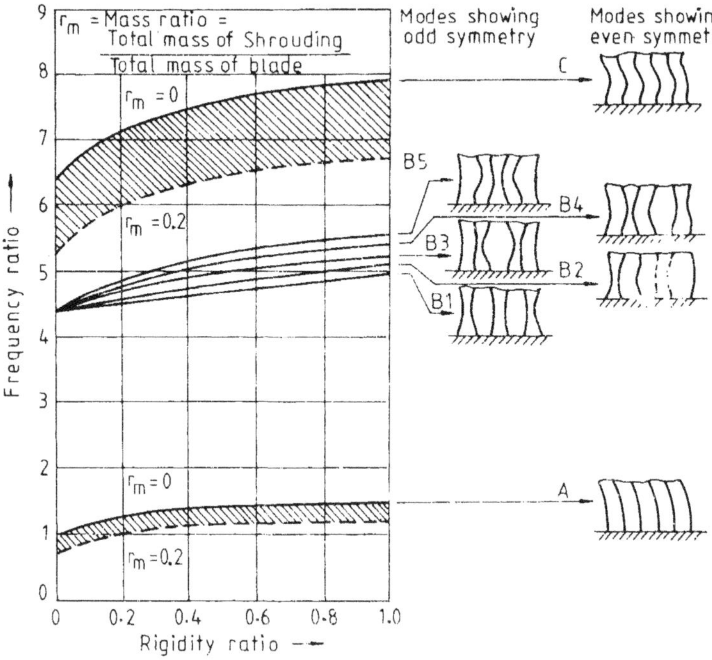

Fig. 17.1 Smith's results for a six bladed packet [51]

17.1 Armstrong's Analysis for Tuned Systems

The receptance coupling method is essentially the application of dynamic equilibrium and compatibility between two connected components. Such a method when properly developed becomes useful in studying the dynamic characteristics of structures involving different components that are governed by displacement fields of different nature, such as plates and beams.

If a structure responds with $x(t)$ under a force $F(t)$ the relation $x(t) = \alpha F(t)$ defines the receptance α. For a disk, we can determine the following:

- Direct Receptance between force and displacement ${}_N\alpha_w$.
- Cross Receptance between force and slope ${}_N\alpha_\phi$.
- Cross Receptance between moment and displacement ${}_C\alpha_w$.
- Direct Receptance between moment and slope ${}_C\alpha_\phi$.

These values can be determined for a given disk in closed form, series form or Point receptances. Similarly for the blade:

- Direct Receptance between force and displacement ${}_N\beta_w$.
- Cross Receptance between force and slope ${}_N\beta_\phi$.
- Cross Receptance between moment and displacement ${}_C\beta_w$.
- Direct Receptance between moment and slope ${}_C\beta_\phi$.

The receptances between the blade root and the blade tip, and receptances at the blade tip, can be determined in a convenient form. These are not dealt with here.

Consider first a disk represented by B with one blade represented by C. Only two degrees of freedom are considered. Let N and C represent the force and couple with response x and θ. The displacement field is written using the corresponding receptances

$$X_B = {}_N\alpha_w N_B + {}_C\alpha_w C_B$$
$$\theta_B = {}_N\alpha_\phi N_B + {}_C\alpha_\phi C_B \tag{17.1}$$

and

$$X_C = {}_N\beta_w N_C + {}_C\beta_w C_C$$
$$\theta_C = {}_N\beta_\phi N_C + {}_C\beta_\phi C_C \tag{17.2}$$

Let N_1, C_1 be the forces and couples between the blade and disk. Then applying conditions of equilibrium and compatibility

$$N_1 = N_B + N_C$$
$$C_1 = C_B + C_C$$
$$X = X_B = X_C$$
$$\theta = \theta_B = \theta_C \tag{17.3}$$

If there no applied forces between the blade and disk, then

$$N_B = -N_C$$
$$C_B = -C_C \tag{17.4}$$

We can then obtain the frequency equation

$$\begin{bmatrix} {}_N\alpha_w + {}_N\beta_w & {}_C\alpha_w + {}_C\beta_w \\ {}_N\alpha_\phi + {}_N\beta_\phi & {}_C\alpha_\phi + {}_C\beta_\phi \end{bmatrix} \begin{Bmatrix} N_B \\ C_B \end{Bmatrix} = 0 \tag{17.5}$$

If there are M identical blades (tuned system), we assume that the assembly is vibrating in a mode with n nodal diameters such that the reaction forces and couples between the blades and disk are each distributed around the rim of the disk given by

$$f_j = -2\varepsilon A_0 \cos\frac{2\pi nj}{M} \tag{17.6}$$

where A_0 is the amplitude of force distribution and 2ε is the angular width of the blade. We can similarly write an expression for the couple C. For the disk, the distributed forces acting on the rim are

$$F(\theta) = 0 \qquad \left\{\frac{2\pi m}{M} + \varepsilon\right\} < \theta < \left\{\frac{2\pi(m+1)}{M} - \varepsilon\right\}$$

$$= A_0 \cos\frac{2\pi mn}{M} \qquad \left\{\frac{2\pi m}{M} - \varepsilon\right\} < \theta < \left\{\frac{2\pi m}{M} + \varepsilon\right\}$$

$$m = 1, 2, \ldots, M \tag{17.7}$$

For the disk, we consider the force as a set of distributed forces acting on its rim, and for $n < m$, we can show that

$$F_D(\theta) = \varepsilon M A_0 \cos n\theta \tag{17.8}$$

For the location of jth blade, this force is

$$F_D(\theta) = \varepsilon M A_0 \cos\frac{2\pi nj}{M} \tag{17.9}$$

Combining equations (17.6) and (17.9), it can be seen that the disk force is $M/2$ times that of the blade forces. That is the disk receptances should be modified as

$$_N\alpha_w\}_{\text{mod}} = \frac{1}{2} M_N\alpha_w \tag{17.10}$$

Therefore the frequency determinant from (17.5) is

$$\Delta = \begin{vmatrix} \frac{1}{2}M_N\alpha_w + _N\beta_w & \frac{1}{2}M_C\alpha_w + _C\beta_w \\ \frac{1}{2}\text{M}_N\alpha_\phi + _N\beta_\phi & \frac{1}{2}M_C\alpha_\phi + _C\beta_\phi \end{vmatrix} \tag{17.11}$$

17.2 Ewins' Analysis

Ewins [13] extended the above analysis to six degrees of freedom, three each translational and rotational at each blade-disk junction. Let vectors $\{F\}$ and $\{f\}$ represent the dynamic reaction forces and couples which act upon the disk and the blade respectively at each junction, and vectors $\{X\}$ and $\{x\}$ represent the corresponding displacements and rotations. They are

$$\{F\} = \begin{Bmatrix} N_a \\ N_t \\ N_c \\ C_a \\ C_t \\ C_r \end{Bmatrix}; \quad \{X\} = \begin{Bmatrix} w_a \\ w_t \\ w_c \\ \phi_a \\ \phi_t \\ \phi_r \end{Bmatrix}; \quad \{f\} = \begin{Bmatrix} N_a \\ N_t \\ N_c \\ C_a \\ C_t \\ C_r \end{Bmatrix}; \quad \{x\} = \begin{Bmatrix} w_a \\ w_t \\ w_c \\ \phi_a \\ \phi_t \\ \phi_r \end{Bmatrix} \tag{17.12}$$

Suffixes a, t and r represent disk axial direction, tangential direction and radial direction measured from arbitrarily fixed $\theta = 0$, junction j is located at angle θ_j.

The displacements of the rim of the disk at point j, X_j will be the sum of the responses of each of the M system of forces, F_k and may be written as

$$\{X\}_j = \sum_{k=1}^{M} [\alpha]_{jk} \{F\}_k \tag{17.13}$$

where $[\alpha]_{jk}$ is the point receptance matrix relating to response at j due to the force at k. Consider now the blade attached at point j, we have

$$\{x\}_j = [\beta]_j \{f\}_j \tag{17.14}$$

Assuming the blades to be rigidly attached to the disk, we have

$$\begin{aligned} \{X\}_j &= \{x\}_j, \quad j = 1, 2, \ldots, M \\ \{F\}_j &= -\{f\}_j, \quad j = 1, 2, \ldots, M \end{aligned} \tag{17.15}$$

From the above equations we get

$$\sum_{k=1}^{M} [\alpha]_{jk} \{F\}_k + [\beta]_j \{F\}_j = 0; \quad j = 1, 2, \ldots, M \tag{17.16}$$

The above leads to the frequency determinant

$$[\Delta]_{6M \times 6M} \{F\}_{6M \times 1} = 0 \tag{17.17}$$

When the system is mistuned by allowing small-dimensional variations between one blade and the next, some of the modes split into distinct but similar pairs. In general a random mistuning of blades will cause all the so-called double modes to split in this way. It is most convenient to select the 1F blade cantilever frequency to describe the state of mistuning. Individual blade frequencies f_1, f_2, are given by

$$f_i = f(\theta)\}_{\theta = \theta_i}, \quad \theta_i = \frac{2\pi i}{M} \tag{17.18}$$

where $f(\theta)$ is of the form

$$f(\theta) = f_{\text{mean}} \left\{ 1 + a_0 + \sum_{j=1}^{M} a_j \cos(n_j \theta + \phi_j) \right\} \tag{17.19}$$

Consider the following bladed-disk assembly:

- Blade length = 80.26 mm;
- Width 1 = 3.05 mm;
- Width 2 = 10.78 mm;
- Disk thickness = 5.59 mm;
- Disk radius = 127 mm;

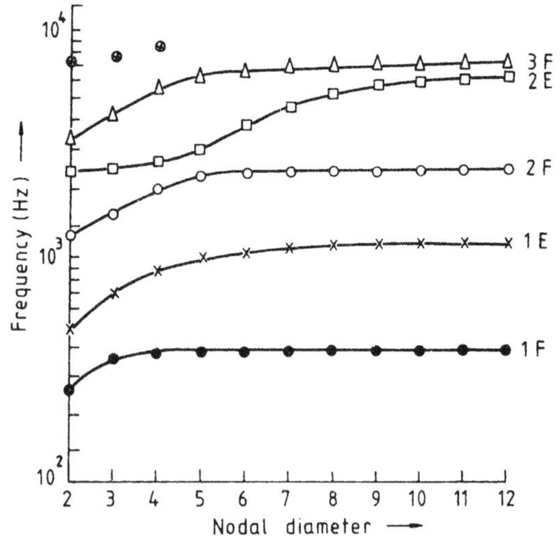

Fig. 17.2 Frequency vs. nodal diametral plot for a 24 bladed disk assembly

Table 17.1 Mistuning pattern

Blade Number	Mistune Ratio	Blade Number	Mistune Ratio
1	1.00625	13	0.98625
2	0.99250	14	0.97500
3	0.99750	15	0.99875
4	0.99500	16	1.00375
5	1.00625	17	1.03125
6	0.98125	18	1.04500
7	1.01375	19	1.00375
8	1.00250	20	1.00250
9	0.98875	21	0.99250
10	0.99000	22	0.97500
11	1.00125	23	0.99875
12	0.98375	24	0.99625

- Number of blades = 24;
- Stagger angle = 45°;
- Young's modulus = 207 GPa;
- Poisson's ratio = 0.287;

The tuned system natural frequencies are given in Figure 17.2.

The mistuning pattern is assumed as given in Table 17.1. The natural frequencies obtained for the same are given in Table 17.2.

Cottney and Ewins [6] considered the shroud in bladed-disk assemblies; Ewins [12] and Ewins and Rao [14] included blade damping on the forced vibrations of bladed-disks. Rao et al. [41] programmed the bladed-disk assembly vibration char-

Table 17.2 1F natural frequencies of mistuned bladed-disk

No. of Nodal Diameters	Frequencies in Hz		
	Tuned System	Mistuned System	
2	262.848	262.193	263.663
3	354.368	352.990	353.940
4	373.228	370.785	371.584
5	379.790	376.722	378.082
6	382.927	379.347	383.785
7	384.683	383.188	390.057
8	385.751	382.234	384.319
9	386.422	384.432	388.710
10	386.836	379.924	386.777
11	387.061	386.136	386.390
12	387.133	400.024	–

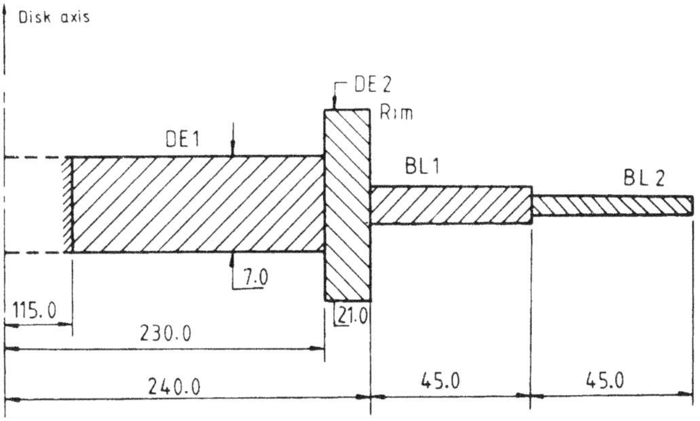

Fig. 17.3 Turbine stage geometry

acteristics and verified the natural frequencies by experiments on an Orpheus aircraft engine. The geometry of a gas turbine bladed disk assembly of the engine is shown in Figure 17.3. The stage of mass 24.2 kg has 125 blades and density is taken as 7833 kg/m^3 Young's modulus is 207 GPa and Poisson's ratio is 0.287. The area, I_{xx}, I_{yy} and the stagger angle for the blade elements 1 and 2 respectively are 61.08 mm^2, 147.909, 2123.9 mm^4, 15°; 42.315 mm^2, 124.13, 1793.14 mm^4, 25°. Figure 17.4 shows the analytical and experimental results of the stage.

The finite element method is ideal for bladed-disk type problems; Kirkhope and Wilson [24] were amongst the earlier ones to use this method. Elchuri et al. [11] used commercially developed codes for cyclically symmetric structures and Swaminatham et al. [53] considered a bladed disk vibration analysis. Reference may also be made to Rzadkowski [46].

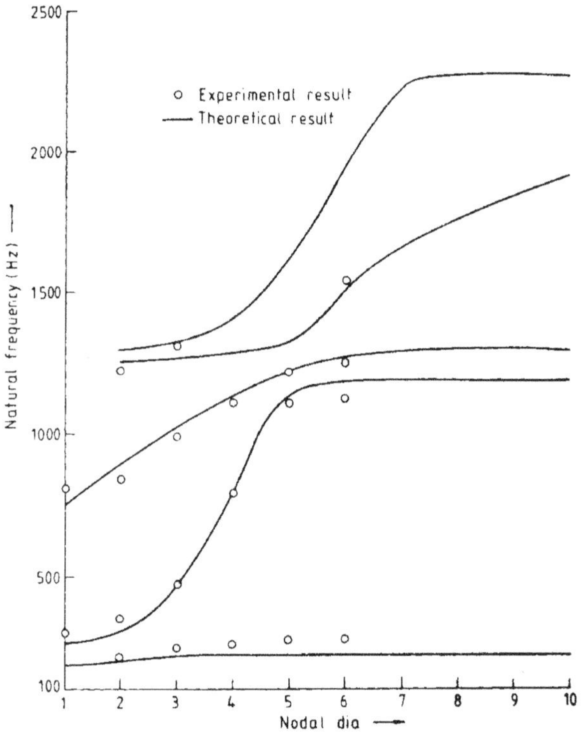

Fig. 17.4 Theoretical and experimental results of the orpheous engine turbine stage

17.3 Mistuning Arrangement

When the blades are mistuned, it is usually found that excessively large vibrations occur when compared to perfectly tuned systems. Whitehead [56] first addressed this problem by developing algebraic expressions for the maximum response based on the number of blades in a bladed-disk. Recently, in [57] he considered example cases in which his proposed maximum would occur. The theoretical limit for maximum magnification has been shown to be $(1/2)(1+\sqrt{N_B})$ where N_B is the number of blades.

Several attempts have been made to determine the maximum response in a bladed disk design. Dye and Henry [10] used Monte-Carlo simulations with a simple numerical model based on the measured system parameters to statistically estimate the maximum response in a bladed-disk. Several methods [49], were proposed to determine the statistical distribution of mistuned forced response amplitudes from which the maximum response can be obtained.

Kenyon and Griffin [21] identified two primary physical mechanisms associated with increases in forced response amplitudes due to mistuning in a single family of bladed-disk modes, frequency splitting and mode distortion. Frequency splitting de-

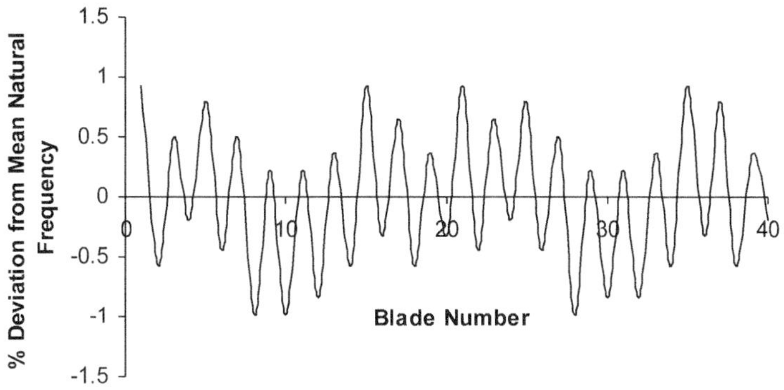

Fig. 17.5 Mistuning pattern

scribes the separation of the repeated natural frequencies that occur in tuned bladed disks into two distinct natural frequencies. Mode distortion refers to harmonic content in the mode shape of a mistuned system in addition to the fundamental wave form of the normally tuned mode. Kenyon et al. [23] recently developed a theory for predicting the maximum forced response in a bladed disk from mode distortion in a single structural mode. The maximum response derived there was robust, i.e., if a bladed-disk is intentionally mistuned to exhibit that maximum response, then small, unintentional random mistuning will not significantly alter the response amplitude of the bladed disk.

As mistuning is inevitable, several authors have worked out the maximum level response increase in forced vibration of bladed disk assemblies, the lowest reported value is 21% in [10] and the highest reported value is 110% in [47]. Kenyon and Griffin [22] demonstrated experimentally the maximum mistuned bladed disk forced response. The theoretical result for maximum amplitude magnification predicted using Kenyon et al.'s [23] formulation is 1.918 and the experimental result was close to this prediction.

The general practice of measuring mistuning of bladed-disks is to determine each blade's natural frequency; the difference of each blade's natural frequency from the mean value is taken as a measure of mistuning. When it comes to integral bladed-disks or blisks, this approach fails. Feiner and Griffin [15] have recently proposed a method of mistuning identification of Bladed Disks using a fundamental mistuning model.

Rao [36] adopted a modal distortion technique [23] to minimize the mistuning effects and control resonant response. The pattern chosen here to minimize the response in 3ND mode is given in Figure 17.5.

The mistuning pattern is designed in the chosen case, to have the frequency values of consecutive N_B number of blades, say 40, arranged in a manner of highs and lows consecutively. This will form a sine wave consisting of $(1/2)N_B$ cycles around the disk in one rotation. The blades are further organized such that even harmonics, 2 and 4 are present in one rotation around the disk as shown in Figure 17.5.

It may be noted that the range of frequencies in Figure 17.5 is quite small; the maximum difference in the individual blade is only 2% of the mean unlike the 48.5% obtained by Kenyon and Griffin [21] for their worst mistuning case. Such a large deviation can never occur in practice, particularly for aircraft engine blades which use very high precision manufacturing processes.

In the CAD model of the bladed-disk, 2,49,760 8-noded solid elements, are used with 3,36,280 nodes. There are 40 sectors. Modal analysis of the pre-stressed structure at the operating speed is performed. For the first bending of the blade, 20 pairs of roots are obtained for a perfectly tuned system, one pair each for different orders of nodal diameters. 3ND mode (around 760 Hz) is found to be close to 3 per rev excitation near the operating speed.

The mistuning is applied by changing, say the density, from blade to blade to obtain the frequency pattern as given in Figure 17.5. Full system modal analysis gives 40 split roots because of mistuning. The first four (0 to 3) nodal diameter mode shapes of the tuned and mistuned cases are shown in Figure 17.6.

It can be seen from Figures 17.6a to d, all the modes are distorted. In the case of 0-ND mode, the basic umbrella mode exists with mistuning (see Figure 17.6a). The 1-ND mode is a little more distorted than the 0-ND mode; however, the nodal diameter is clearly visible.

The 2-ND mode is fairly well distorted; it is a bit difficult to see clearly two nodal lines, see Figure 17.6c. For the 3-ND mode, however, the shape is completely destroyed, if at all only two nodal lines can be somehow seen there, see Figure 17.6d.

It may be noted here that in all cases, the natural frequencies of mistuned bladed disk are in the same range of the tuned system; only the mode shapes are distorted. Therefore, resonance takes place when an excitation is applied at the resonance frequency, only the effectiveness will gradually decrease as the mode shape gets distorted and becomes a minimum when the mode shape is fully distorted as in 3-ND mode.

We will consider the case of 3 per rev excitation from flow path acting on the blades. It may be noted here that the excitation from nozzles is a multiple of running speed times the number of nozzles, therefore 3 per rev excitation on the blades is unlikely in a practical case.

For this purpose, harmonic excitation is made on the blades at the desired frequency and phase angle between successive blades. The excitation applied in an axial direction is taken in the form of a harmonic with unit force and frequency corresponding to the natural frequency. A phase angle 0.47124 radians is used from blade to blade consecutively so that the 40th blade has 6π radians phase from the first one. 1% damping is used. The maximum amplitude in the mistuned case is found to be 44% more than the tuned case. The lowest value reported in the literature is 21% by Dye and Henry [10]. The worst magnification according to Whitehead [56] is $(1/2)(1+\sqrt{40}) = 3.66$.

The response in the axial direction is found to be 19% over the tuned case. In the hoop direction the magnification is 14% over the tuned case. Net bending deflection is magnified by around 17.5%, which is lower than the 21% value reported by Dye and Henry [10].

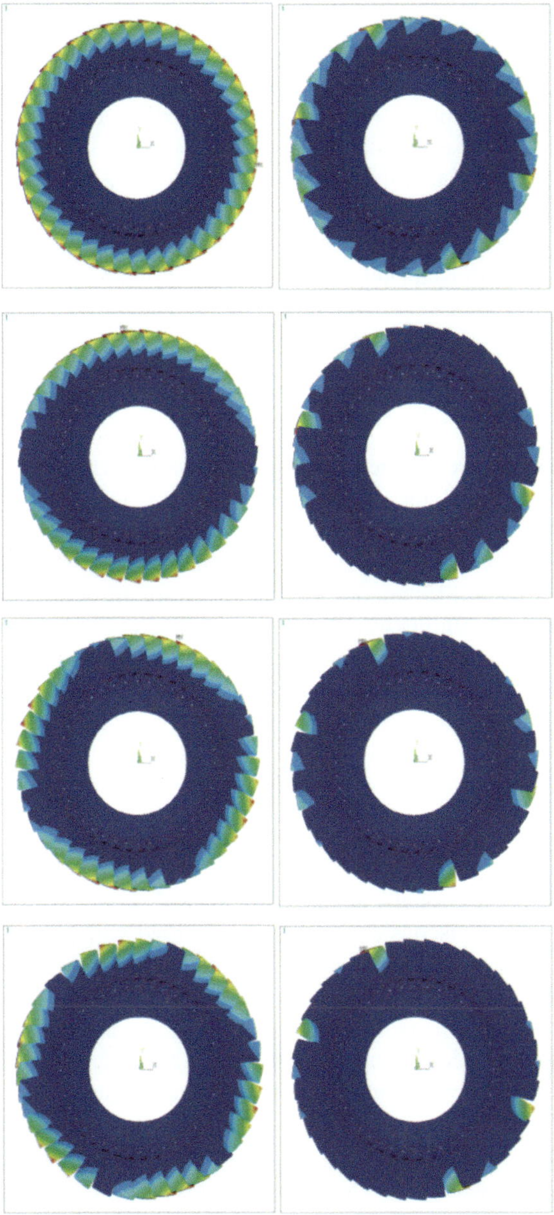

Fig. 17.6 (a) Zero ND tuned and mistuned mode shapes. (b) One ND tuned and mistuned mode shapes. (c) Two ND tuned and mistuned mode shapes. (d) Three ND tuned and mistuned mode shapes

In practice 3× excitation from flow path in axial flow aircraft engine compressors is less likely in comparison to the 3× excitation arising out of rotor misalignment and other causes. For this purpose, transient excitation is made on the disk at the desired frequency and phase angle between the disk sectors.

Maximum von-Mises stress obtained for a unit load in the tuned case is found to be 10.21 N/mm^2. Maximum von-Mises stress obtained for a unit load in the mistuned case is 0.84 N/mm^2. Thus, the stress value in the mistuned case is 12 times less than the stress value in the tuned case. Thus proper designed mistuning giving modal distortion can help in decreasing the stresses.

17.4 Damping

Turbine blades have very little damping and therefore when they go through a resonance at critical speed, the stresses can easily get magnified by 100 times or even higher of steady stress. These very high resonant stresses are responsible for fatigue damage. Therefore, damping has been identified long ago as a key parameter in blade design. Rowett [45] conducted tests on elastic hysteresis in steel. Effects of friction and loose mounting were studied by Hansen et al. [17].

Lazan [26], see also Nashif et al. [30], measured hysteresis in simple tension tests and defined the loss of energy per cycle by

$$D = J\left(\frac{\sigma}{\sigma_e}\right)^n \tag{17.20}$$

where J and n are material properties and σ_e is endurance limit.

Sinha and Griffin [50] studied analytically the effects of static friction on the forced response of frictionally damped turbine blades.

Usually a simple viscous damping model is used with equivalent damping determined from a test. Such a linear model is inadequate since material damping is highly dependent on the state of stress in the blade. A nonlinear damping model was quantified through experiments by Rao et al. [37]; the equivalent viscous damping is expressed as a function of strain amplitude at a reference point in a given mode of vibration at a given speed of rotation. The experiment consisted of spinning the rotor in vacuum and providing excitation through distributed electromagnets in place of steam or gas nozzles, Figure 17.7 gives the test apparatus and Figure 17.8 measured equivalent viscous damping as a function of reference strain amplitude at a given RPM and in a given mode of vibration.

Centrifugal load is simulated by means of thermal expansion to avoid rotation and simplify the test rigs. Rieger and Beck [44] performed such tests for EPRI.

In Section 13.4, the finite element method given by Rao et al. [42] was discussed to study the friction damping between blade root and disk by using contact elements. They have also designed and built a test rig simulating the centrifugal load by means

Fig. 17.7 Measurement of damping in rotating blades

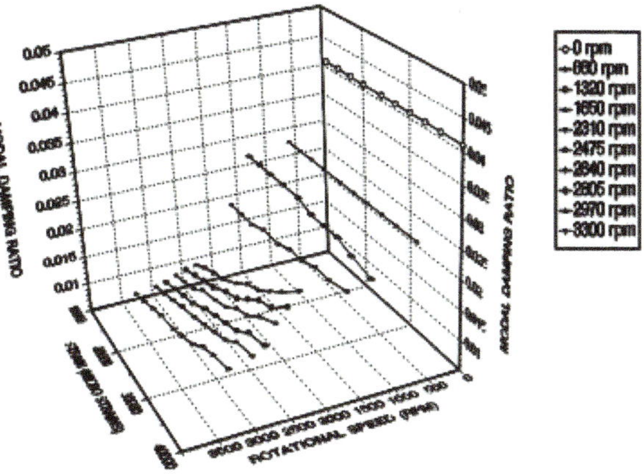

Fig. 17.8 Equivalent viscous damping as a function of reference strain amplitude measured at a given RPM for a given mode

of cryogenic liquid cooling on a blade pair mounted in the frame contracted by thermal cooling. This avoided expensive rotation simulation, see Figure 17.9.

Rao and Saldanha [40] developed an analytical procedure using Lazan's hysteresis law [26]. Briefly the steps followed are:

The total damping energy D_0 (Nm) is given by

$$D_0 = \int_0^v D dv \tag{17.21}$$

where v is the volume. The loss factor η is

$$\eta = \frac{D_0}{2\pi W_0} \tag{17.22}$$

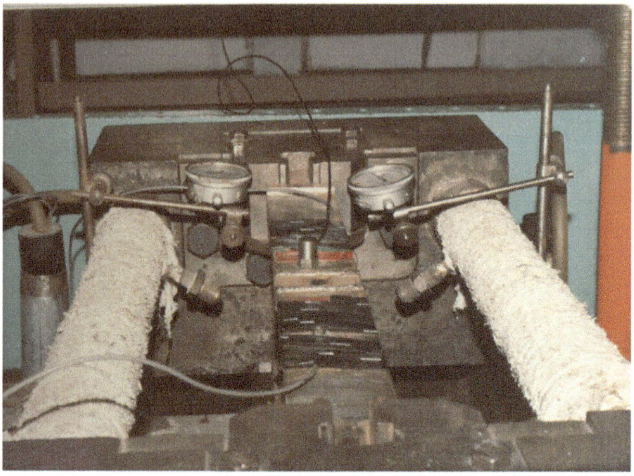

Fig. 17.9 Blade pair in tension simulated by thermal cooling

where W_0 is the total strain energy (Nm). Then, equivalent viscous damping C (N-s/m) is

$$C = \frac{\eta K}{\omega} \tag{17.23}$$

where ω the natural frequency (rad/s) and K is the modal stiffness (N/m).

For increased strain amplitudes, the orthonormal reference strain amplitudes, stress and strain energy are multiplied by a factor F to obtain the equivalent viscous damping C_e at various strain amplitudes as given below:

$$\begin{aligned}
\varepsilon' &= \varepsilon F \\
W_0' &= W_0 \times F^2 \\
\eta' &= \frac{D_0'}{2\pi W_0'} \\
C_e' &= \frac{\eta' K}{\omega_n} F^2 \\
\xi &= \frac{C_e}{2\sqrt{Km}} F^2
\end{aligned} \tag{17.24}$$

A plot of equivalent viscous damping ratio as function of reference strain amplitude in the chosen mode of vibration defines the nonlinear damping model [38]. The friction damping characteristic is obtained by determining the transient response due to an impulse excitation at a suitable point on the blade to simulate the desired mode of vibration and to assess the decay curve. Typical transient response blade and a nonlinear friction damping model of the blade are shown in Figure 17.10.

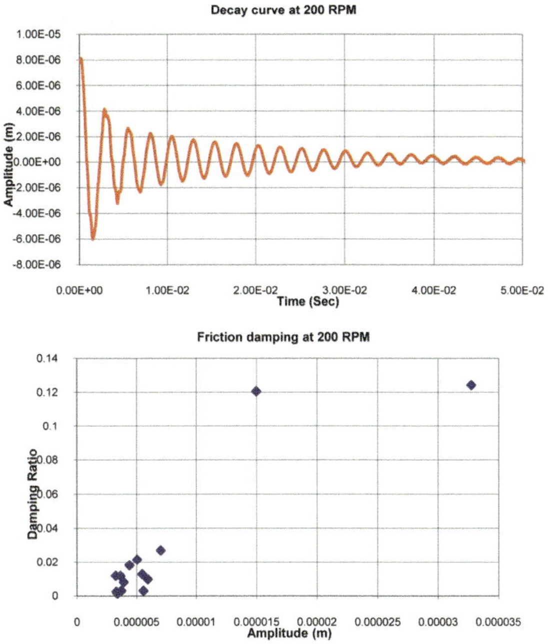

Fig. 17.10 Impulse response of bladed-disk and damping ratio

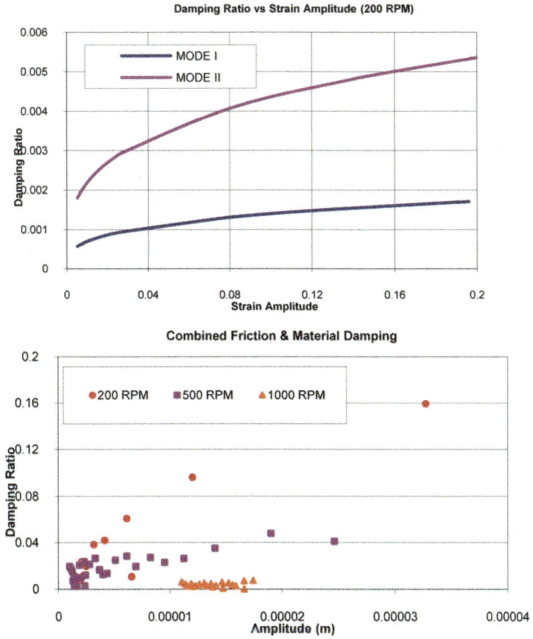

Fig. 17.11 Material and combined material and friction damping

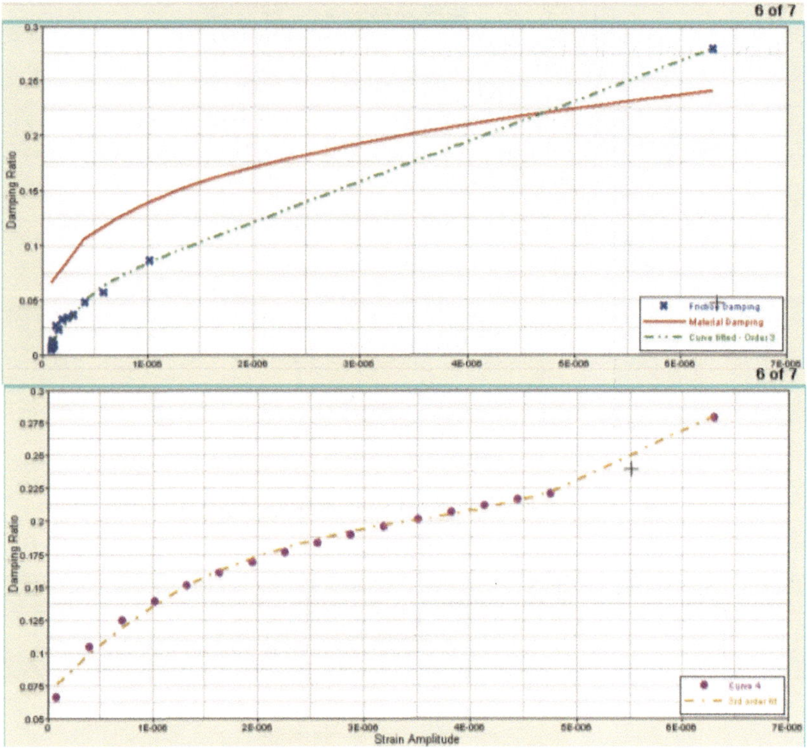

Fig. 17.12 Material and friction dampings and the resultant

Rao et al. [39] determined the material damping and combined material friction damping of a bladed disk as a function of reference strain amplitude in a given mode of vibration at an operating speed as shown in Figure 17.11.

An alternate method is to compare the material damping and friction damping separately and retain which-ever contributes maximum value as shown in Figure 17.12.

17.5 Micro-Slip Damping (Fretting Fatigue)

Referring to Figure 17.8 we find that strain amplitudes (response) decrease considerably as the speeds go up when the blade and disk get locked in the root. Because of the tightness at high speeds, slip amplitudes get reduced and the friction is governed by contacts at asperity level rather than global Coulomb's laws.

When two surfaces in contact slide against each other, an elastic-plastic deformation called micro-slip may occur before macro-slip takes place that can be considered governed by Coulomb's laws. Courtney-Pratt and Eisner [7] are amongst the

Fig. 17.13 Fretting fatigue due to micro-slip in a compressor dovetail

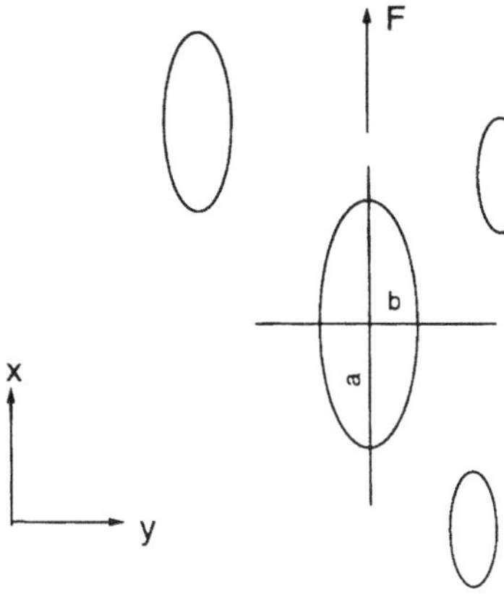

Fig. 17.14 Schematic of contact

earlier people who worked on micro-slip theory. The frictional-load displacement relationship during micro-slip is similar to the stress strain relationship for brittle materials (hysteresis). This microslip is important from the point of fretting fatigue that is commonly observed in turbomachine blades. Figure 17.13 shows fretting-fatigue in a turbine blade root. Burdekin et al. [4] proposed a theoretical model for micro-slip based on reasonable physical properties, assuming that the contacting asperities are substituted by prismatic rods of equal stiffness. Hagman [16] proposed a theoretical model with contacting asperities replaced by spherical bodies of constant radius. Olofsson [31] and Olofsson and Hagman [32] expanded this model to include oscillating displacements and elliptical shaped asperities.

Following Olofsson and Hagman [32], consider a flat smooth surface in contact with a rough flat surface shown in Figure 17.14. The frictional load is parallel to the x axis.

The following assumptions are made:

1. Shape of asperities is ellipsoidal.

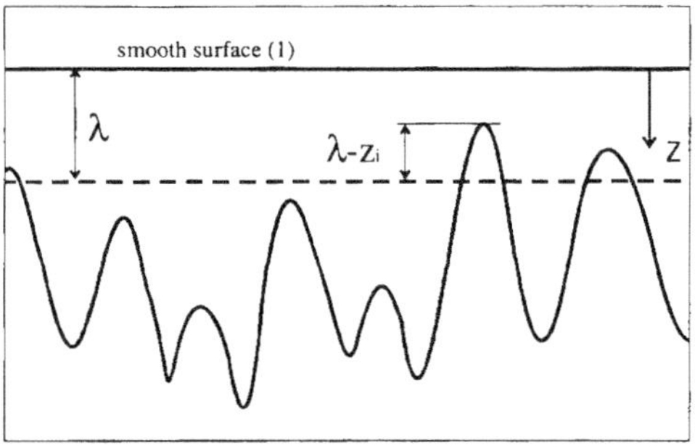

Fig. 17.15 Flat surface in contact with rough surface

2. Height distribution of asperities is uniform.
3. Surface contact is elastic and the behavior of an individual asperity follows Hertz theory for elliptical contacts.
4. All asperities have their semi-axes a and b in the same global x and y directions, respectively.
5. Contacting asperities have the same constant ovality ratio $k = a/b, a < b$ and $k = b/a, b < a$

The surface is brought into contact with a normal approach λ, see Figure 17.14. The normal load, P_i, for an asperity at depth z_i and the major semi-axis c for that asperity is expressed as

$$P_i = \frac{2\pi E'}{k\left(\frac{9}{2\varepsilon R}\right)^{1/2}}\left(\frac{\lambda - z_i}{\kappa}\right)^{3/2} \tag{17.25}$$

$$c = \left(\frac{3\varepsilon P_i R}{k^2 \pi E'}\right)^{1/3} \tag{17.26}$$

In the above:

- E' is the composite modulus of elasticity given by

$$\frac{1}{E'} = \frac{1 - \nu_1{}^2}{E_1} + \frac{1 - \nu_2{}^2}{E_2}$$

- R is the curvature sum of elliptical contact given by

$$\frac{1}{R} = \frac{1}{r_{2x}} + \frac{1}{r_{2y}}$$

with r as radius of curvature and ε are complete elliptic integrals of the first and second kind with argument $e = \sqrt{1 - k^2}$.

The number of asperities in contact, N, is assumed to increase linearly with the approach of the two surfaces. Thus

$$N = Cz \tag{17.27}$$

where C is a surface parameter which relates the number of contacts per unit area and z the approach of the surfaces. The normal load for the approach λ can be expressed as

$$P = CA \int_0^{\lambda} P_i dz = \frac{4CA\kappa\pi E'}{5k\left(\frac{9}{2\varepsilon R}\right)^{1/2}} \left(\frac{\lambda}{\kappa}\right)^{5/2} \tag{17.28}$$

where A is the apparent area of contact.

Constant Normal Load and Increasing Frictional Load: The displacement of elastic ellipsoidal bodies in contact was studied by Deresiewicz [9]. The force-displacement relationship for an individual asperity, i, can be expressed as

$$F_i = \mu P_i \left(1 - \left(1 - \frac{16cG'}{3\mu P_i} \frac{\delta_d}{\phi}\right)^{5/2}\right) \tag{17.29}$$

where G' is the composite shear modulus given by

$$\frac{1}{G'} = \frac{2-\nu}{G_1} + \frac{2-\nu_2}{G_2}$$

and

$$\phi = \left[\frac{4}{\pi(2-\nu_2)}\right]\left[\left(1 - \frac{\nu_2}{e^2}\right)\kappa + \frac{\nu_2\varepsilon}{e^2}\right], \quad a < b$$

$$\phi = \left[\frac{4}{\pi(2-\nu_2)}\right]\left[\left(1 - \nu_2 + \frac{\nu_2}{e^2}\right)\kappa - \frac{\nu_2\varepsilon}{e^2}\right], \quad b < a \tag{17.30}$$

F_i will deflect upto $F_i = \mu P_i$. Equation (17.29) gives the limit deflection δ_{Li}

$$\delta_{Li} = \frac{3\mu P_i}{16cG'}\phi = \frac{\pi\mu E'\phi}{8G'}\left(\frac{\lambda - z_i}{\kappa}\right) \tag{17.31}$$

Equation (17.31) gives the limit height of the asperities, z_{Li}. Asperities higher than z_{Li} will stick and asperities lower than z_{Li} will slip.

$$z_{Li} = \lambda - \frac{8G'\delta}{\pi\mu E'\phi} \tag{17.32}$$

The total frictional load becomes

$$F = F_{\text{spring}} + F_{\text{slip}} = \int_0^{z_{Li}} F_i CA dz + \int_{z_{Li}}^{\lambda} \mu P_i CA dz \tag{17.33}$$

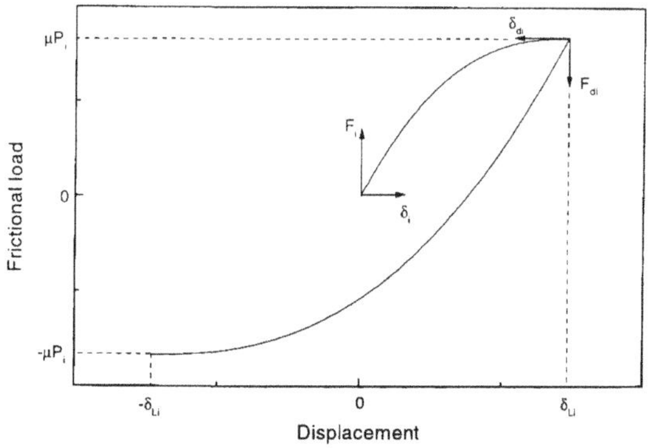

Fig. 17.16 Frictional load vs. displacement for an individual ellipsoidal body

where F_{spring} is the frictional load from the active asperities which have not reached their limiting tangential deflection and F_{slip} is the contribution from asperities which have reached their limiting tangential deflection.

The total frictional load is obtained from using equations (17.25), (17.29) and (17.32) in equation (17.33)

$$F = \frac{4}{5}\mu C A\kappa \frac{\pi E'}{k\left(\frac{9}{2\varepsilon R}\right)^{1/2}}\left[\left(\frac{\lambda}{\kappa}\right)^{5/2} - \left(\frac{\lambda}{\kappa} - \frac{8G'\delta}{\mu\pi E'\phi}\right)^{5/2}\right]$$

$$= \mu P\left[1 - \left(1 - \frac{8G'\delta\kappa}{\mu\pi E'\phi\lambda}\right)^{5/2}\right] \tag{17.34}$$

Equation (17.34) is valid until

$$\delta_{\max} = \frac{\lambda\mu\pi E'\phi}{8\kappa G'} \tag{17.35}$$

Constant Normal Load and Oscillating Frictional Load: Suppose that after reaching a value F^*, the frictional load F is reduced; the force displacement relationship under unloading for an individual asperity i can be expressed as (see Figure 17.16)

$$F_{di} = 2\mu P_i\left[1 - \left(1 - \frac{16cG'\delta_{di}}{2 \times 3\mu P_i\phi}\right)^{3/2}\right] \tag{17.36}$$

The corresponding limit deflection for unloading is twice that for loading. The maximum height of the asperities z_{dli} for which they will slip is

$$z_{dLi} = \lambda - \frac{4G'\delta}{\pi\mu E'\phi} \tag{17.37}$$

The sense of slip must be reversed [28] but its absolute magnitude is not altered during unloading. Then the slip part of the tangential load during unloading is twice that for loading. The equation for the frictional load during unloading is

$$F_d = \int_0^{z_{dLi}} F_{di} C A dz + \int_{z_{dLi}}^{\lambda} 2\mu P_i C A dz$$

$$= 2\mu P\left[1 - \left(1 - \frac{4G'\delta_d \kappa}{\mu\pi E'\phi\lambda}\right)^{5/2}\right] \tag{17.38}$$

where δ_d is the reduction in the initially loaded displacement, δ^*, and F_d is the reduction in the initially applied load, F^*.

The equation for frictional load transformed to the original co-ordinate system is

$$F_r = F^* - 2\mu P\left[1 - \left(1 - \frac{4G'\kappa\,(\delta^* - \delta)}{\mu\pi E'\phi\lambda}\right)^{\frac{5}{2}}\right] \tag{17.39}$$

Suppose now that the frictional load is oscillating between F^* and $-F^*$. The situation at $F = -F^*$ is identical with that at $F = F^*$, except for the reversal of sign. Hence the frictional load then becomes

$$F_s = -F_r(-\delta) = -F^* + 2\mu P\left[1 - \left(1 - \frac{4G'\kappa\,(\delta^* + \delta)}{\mu\pi E'\phi\lambda}\right)^{\frac{5}{2}}\right] \tag{17.40}$$

Energy Dissipation: The area enclosed by the curves F_s and F_r gives the energy dissipation during micro-slip per cycle. Integration gives the energy dissipation, W as

$$W = \int_{-\delta*}^{\delta*} (F_s - F_r)\, d\delta$$

$$= \frac{4}{7}\left[14\mu P\delta^* + \frac{P\mu^2\pi E'\phi}{2G'\left(\frac{\lambda}{\kappa}\right)^{5/2}}\left(\frac{\mu\pi E'\left(\frac{\lambda}{\kappa}\right)\phi - 8G'\delta^*}{\mu\pi E'\phi}\right)^{7/2} - \frac{P\mu^2\pi E'\phi}{2G'}\frac{\lambda}{\kappa} - 7F^*\delta^*\right] \tag{17.41}$$

Now consider when $\phi = 1$, $a = b$ (asperities modeled as spheres, then $\kappa = \varepsilon = \pi/2$, equations (17.34), (17.38) and (17.41) become

$$F = \mu P\left[1 - \left(1 - \frac{4G'\delta}{\mu E'\lambda}\right)^{5/2}\right] \tag{17.34a}$$

$$F_d = 2\mu P\left[1 - \left(1 - \frac{2G'\delta_d}{\mu E'\lambda}\right)^{5/2}\right] \tag{17.38a}$$

Table 17.3 Test data

Input Data	Case 1	Case 2
r_{2x} (μm)	1	100
r_{2y} (μm)	100	1
ν	0.3	0.3
μ	0.1	0.1
G' (N/m^2)	2.38×10^{10}	2.38×10^{10}
E' (N/m^2)	1.15×10^{11}	1.15×10^{11}
C (contacts/m^2/m)	10×10^{10}	10×10^{10}
A (mm^2)	100	100
P (N)	5000	5000

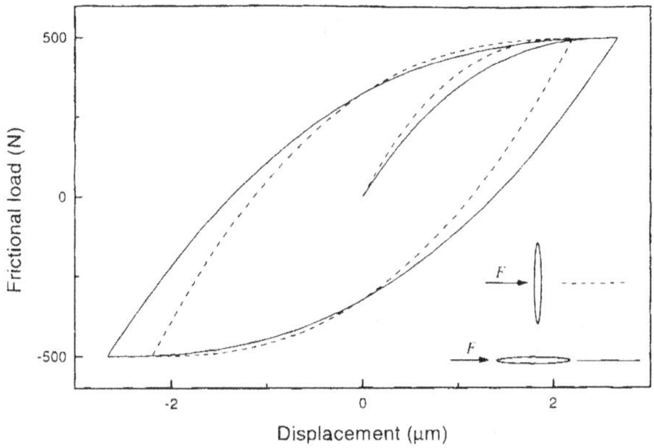

Fig. 17.17 Frictional load vs. displacement: full line — Case 1; dashed line - - - Case 2

$$W = \frac{4}{7}\left[14\mu P\delta^* + \frac{P}{\mu^{3/2}E'^{5/2}\lambda^{5/2}G'}(\mu E'\lambda - 4G'\delta^*)^{7/2} - \frac{\mu^2 E'\lambda}{G'} - 7F^*\delta^*\right] \tag{17.41a}$$

These are the same as the ones by Hagman [16] and Olofsson [31] for the case where the asperities are replaced with spheres. Olofsson gave two test cases; the data for these cases is given in Table 17.3.

Figure 17.17 gives the results from equations (17.34), (17.39) and (17.40).

Marquina et al Model: For the blade-disk junction case, Marquina et al. [27] adopted the above Olofsson and Hagman's micro-slip model into their approach. First the shear (tangential) stiffness is obtained from equation (17.34a) written as

$$F_t = \mu F_N\left[1 - \left(1 - \frac{4G'\delta}{\mu E'\lambda}\right)^n\right] \tag{17.42}$$

$$K_t = \frac{dF_t}{d\delta} = \frac{4F_N nG'}{E'\lambda}\left(1 - \frac{4G'\delta}{\mu E'\lambda}\right)^{n-1} \tag{17.43}$$

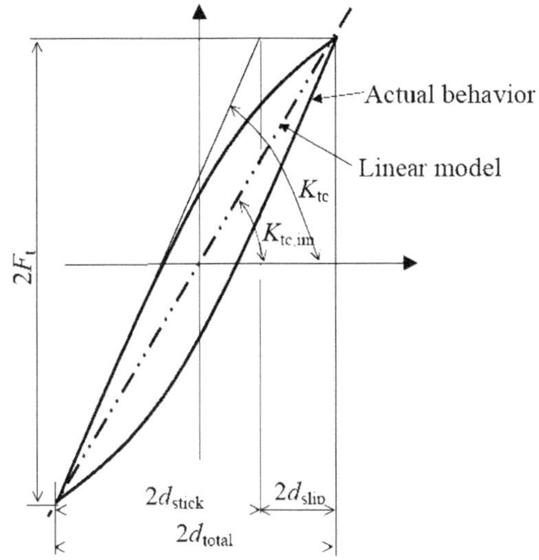

Fig. 17.18 Micro-slip damping model

Limiting δ to $\delta_{max} = (\lambda\mu E')/4G'$ the above equation (17.43) was simplified by Marquina et al. [27] and Olofsson's relation is then written as

$$F_t = \mu F_N \left[1 - \left(1 - \frac{K_t}{nF_N}\frac{\delta}{\mu}\right)^{5/2}\right] \tag{17.44}$$

Asai et al: Here the formulation is slightly different made in such a way as to verify Olofsson's formulation for blades. Their microslip damping model of two surfaces under contact with F_t and F_n as tangential and normal forces is given in Figure 17.18 [2]; the tangential contact stiffness is

$$K_{tc} = \frac{F_t}{d_{\text{stick}}} \tag{17.45}$$

For the linear model without hysteresis, the material property is Imaginary Tangential Contact Stiffness $K_{tc,im}$ given by

$$K_{tc,im} = \frac{F_t}{d_{\text{total}}} \tag{17.46}$$

In Figure 17.18, the total displacement is stick and slip as shown and given by

$$\begin{aligned} d_{\text{total}} &= d_{\text{stick}} + d_{\text{slip}} \\ &= \frac{F_t}{K_{tc}} + d_{\text{slip}} \end{aligned} \tag{17.47}$$

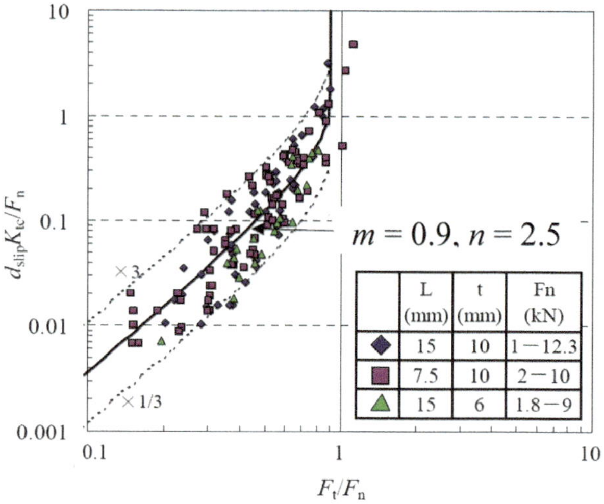

Fig. 17.19 Asai et al. [2] experimental result for micro-slip

If δ_{slip} is the displacement due to the normal force F_n (slip per unit normal force) and tangential stiffness k_{tc}, we define a parameter

$$\frac{d_{\text{slip}} K_{tc}}{F_n} = \frac{d_{\text{slip}}}{\delta_{\text{slip}}} \tag{17.48}$$

Under constant normal load, as the displacements are increasing, in Oloffson's model for oscillating displacements, the asperities are replaced by spheres with the same radius. It is assumed that the height distribution of the asperities is uniform and the behavior of an individual asperity follows Hertz theory. The resulting contact model is

$$\frac{F_t}{F_n} = m\left[1 - \left\{1 - \left(\frac{K_{tc} d_{\text{total}}}{nmF_n}\right)\right\}^n\right] \tag{17.49}$$

where n and m are constants. Using (17.47) the above becomes

$$\frac{F_t}{F_n} = m\left[1 - \left\{1 - \frac{1}{nm}\left(\frac{F_t}{F_n} - \frac{d_{\text{slip}} K_{tc}}{F_n}\right)\right\}^n\right] \tag{17.50}$$

Asai et al. [2] verified the above experimentally for the parameter $(d_{\text{slip}} K_{tc})/F_n$ as shown in Figure 17.19 for three different test specimens. Microslip occurs for large values of $F_n (F_t < F_n)$ as shown and d_{slip} values are in the range of 0.1 to 5 microns. m and n are obtained from the mean curve of experimental results.

Asai's experiments have shown that the Hagman and Olofsson elasto-plastic theory of contact provides a workable model for blades given by (17.34a). The problem however is highly nonlinear and not simple.

$$F = \mu P \left[1 - \left(1 - \frac{4G'\delta}{\mu E'\lambda} \right)^{5/2} \right] \tag{17.34a}$$

1. First of all the coefficient of friction μ at asperity level is not known and as given in Olofsson's relation it is dependent on tangential displacement δ.
2. Secondly the steady state condition for the penetration λ is not known. Marquina et al. [27] circumvented this problem by considering δ to be $\delta_{\max}$ in obtaining the relation in equation (17.44).

Otherwise the penetration is left to be determined. The penetration can often be achieved directly from the finite element code. If the penalty method is used to simulate contact stiffness, the penetration values can be unreliable. Instead a more reliable variable in finite element simulations, the contact pressure, can be used to calculate the penetration. An empirical relationship between the penetration and the contact pressure, P, can be adequately described by the following equation [43]:

$$\lambda = cP^m \tag{17.51}$$

where $c = 0.0014$ for ground/ground steel surfaces and $m = 0.5$ for most metallic materials and for normal contact pressures encountered in joints.

References

1. Armstrong, E.K., Christie, P.L. and Hague, W.M. (1965) Natural Frequencies of Bladed-Discs, *Proc. I Mech E*, vol. 180, part 31, p. 110.
2. Asai, K., Sakurai, S., Kudo, T. and Ozawa, N. (2009) Evaluation of Friction Damping in Dovetail Root Joints Based on Dissipation Energy on Contact Surfaces, ASME Turbo Expo, GT2009-59508.
3. Bishop, R.E.D. and Johnson, D.C. (1960) *Mechanics of Vibration*, Cambridge University Press.
4. Burdekin, N., Cowley, A. and Back, N. (1978) An Elastic Mechanism for the Micro-Sliding Characteristics between Contacting Machined Surfaces, *Journal Mech. Engng. Sci.*, vol. 20, p. 121.
5. Campbell, W. (1924) Prediction of Steam Turbine Disk Wheels from Axial Vibration, *ASME*, vol. 46, no. 1920.
6. Cottney, D.J. and Ewins, D.J. (1974) Towards the Efficient Vibration Analysis of Shrouded Bladed Disk Assemblies, *J. Engng Indus, Trans. ASME*, p. 1054.
7. Courtney-Pratt, J.S. and Eisner, E. (1957) The Effect of a Tangential Force on the Contact of Metallic Bodies, *Proc. Royal Soc., Series A*, p. 529.
8. Deak, A.L. and Baird, R.D. (1963) A Procedure for Calculating the Frequencies of Steam Turbine Exhaust Blades, *Trans ASME*, vol. 85, p. 324.
9. Deresiewicz, H. (1957) Oblique Contact of Non-Spherical Elastic Bodies, *Journal of Applied Mechanics* vol. 24, p. 623.
10. Dye, R.C.F. and Henry, T.A. (1969) Vibration Amplitudes of Compressor Blades Resulting from Scatter in Blade Natural Frequencies, *Journal of Engineering for Power*, vol. 91, p. 182.
11. Elchuri, V., Smith, G.C.C. and Gallo, A.M. (1984) NASTRAN Forced Vibration Analysis of Rotating Cyclic Structures, *J. Vib. Acoust. Stress Rel. Des., ASME*, vol. 1046, p. 224.
12. Ewins, D.J. (1969) The Effects of Detuning upon Forced Vibrations of Bladed-Disks, *Journal of Sound and Vibration*, vol. 9 p. 66.

13. Ewins, D.J. (1973) Vibration Characteristics of Bladed-Disc Assemblies, *J. Mech. Engng. Sci.*, vol. 15, No. 3, p. 165.
14. Ewins, D.J. and Rao, Y.V.K.S. (1975) Effect of Blade Damping on the Forced Vibration Response of Bladed Discs, SRC Report No. 1, Imperial College, London.
15. Feiner, D.M. and Griffin, J.H. (2004) Mistuning Identification of Bladed Disks Using a Fundamental Mistuning Model, Part I: Theory and Part II: Application, *Journal of Turbomachinery*, vol. 126, p. 150.
16. Hagman, L. (1993) Micro-slip and Surface Deformation, Licentiate Thesis, Royal Institute of Technology, Stockholm, Sweden, TRITA-MAE, 1993:5
17. Hansen, M.P. et al. (1953) A Method of Evaluating Loose-Blade Mounting as a Means of Suppressing Turbine and Compressor Blade Vibration, *Proc. SESA*, vol. 10, p. 103.
18. Irretier, H. (1979) Coupled Vibration of Blades in Bending-Bending-Torsion and Discs in Out-of-Plane and In-Plane Motion, ASME 79-DET-90.
19. Irretier, H. (1983) Spectral Analysis of Mistuned Bladed-Disc Assemblies by Component Mode Synthesis, in *Vibrations of Blades and Bladed-Disc Assemblies, ASME*, p. 115.
20. Jarret, G.W. and Warner, P.C. (1953) Vibration of Rotating Tapered Twisted Beams, *Journal of Applied Mechanics, ASME*, vol. 20, p. 381.
21. Kenyon, J.A. and Griffin, J.H. (2003a) Forced Response of Turbine Engine Bladed Disks and Sensitivity in Harmonic Mistuning, *Journal of Engineering for Power*, vol. 125, p. 352.
22. Kenyon, J.A. and Griffin, J.H. (2003b) Experimental Demonstration of Maximum Mistuned Bladed Disk Forced Response, *Journal of Turbomachinery*, vol. 125, p. 673.
23. Kenyon, J.A., Griffin, J.H. and Feiner, D.M, (2003) Maximum Bladed Disk Forced Response From Distortion of a Structural Mode, *Journal of Turbomachinery*, vol. 125, p. 673.
24. Kirkhope, J. and Wilson, G.J. (1971) Analysis of Coupled Blade-Disk Vibration in Axial Flow Turbines and Fans, in *Proceedings 12th AIAA/ASME Conf. Struct. Dyn.*, Anaheim, CA.
25. Kroon, R.P. (1964) Influence of Lashing and Centrifugal Force on Turbine-Blade Stresses, *Trans ASME*, vol. 56, p. 109.
26. Lazan, B.J. (1968) *Damping of Materials and Members in Structural Mechanics*, Pergamon Press.
27. Marquina, F.J., Coro, A., Gutierrez, A., Alonso, R., Ewins, D.J. and Girini, G. (2008) Friction Damping Modeling in High Stress Contact Areas Using Microslip Friction Model, in *Proceedings of ASME Turbo Expo 2008*, GT2008-50359.
28. Mindlin, R.D. and Deresiewicz, H. (1953) Elastic Spheres in Contact under Varying Oblique Forces, *Journal of Applied Mechanics*, vol. 20, p. 327.
29. Muszynska, A. and Jones, D.I.G. (1983) On Tuned Bladed-Disc Dynamics – Some Aspects of Friction Related Mistuning, *Journal of Sound and Vibration*, vol. 86, p. 107.
30. Nashif, A.D., Jones, D.I.G. and Henderson, J.P. (1985) *Vibration Damping*, John Wiley & Sons.
31. Olofsson, U. (1995) Cyclic Micro-Slip under Unlubricated Conditions, *Tribology International*, vol. 28, p. 207.
32. Olofsson, U. and Hagman, L. (1997) A model for microslip between flat surfaces based on deformation of ellipsoidal elastic bodies, *Tribology International*, vol. 30, no. 8, p. 599.
33. Petrov, E.P. and Ewins, D.J. (2003) Analysis of the Worst Mistuning Parameters in Bladed-Disk Assemblies, *Journal of Turbomachinery*, vol. 125, p. 623.
34. Prohl, M.A. (1958) A Method for Calculating Vibration Frequency and Stress of a Banded Group of Turbine Buckets, *ASME*, vol. 80, p. 169.
35. Rao, J.S. (1991) *Turbomachine Blade Vibration*, John Wiley & Sons.
36. Rao, J.S. (2006) Mistuning Of Bladed Disk Assemblies to Mitigate Resonance, *Advances in Vibration Engineering, Journal of Vibration Institute of India*, vol. 5, no. 1, p. 17.
37. Rao, J.S., Gupta, K. and Vyas, N.S. (1986) Blade Damping Measurement in a Spin Rig with Nozzle Passing Excitation Simulated by Electromagnets, *Shock & Vib Bulletin*, vol. 56, part 2, p. 109.
38. Rao, J.S., Narayan, R. and Ranjith, M.C. (2008) Lifing of Turbomachinery Blades – A Process Driven Approach, in *Proceedings ASME Turbo Expo*, Berlin, Germany, GT2008-50231.

39. Rao, J.S., Narayan, R., Ranjith, M.C. and Rejin, R. (2008) Blade Lifing with Material and Friction Damping, in *Proceedings The Future of Gas Turbine Technology, 4th International Conference*, Brussels, Belgium.
40. Rao, J.S. and Saldanha, A. (2003) Turbomachine Blade Damping, *Journal of Sound and Vibration*, vol. 262, no. 3, p. 731.
41. Rao, J.S., Shah, C.B., Ganesh, Ch.L. and Rao, Y.V.K.S. (1986) Vibration Characteristics of Aircraft Engine Blade-Disk Assembly, *Defence Science Journal*, vol. 36, p. 9.
42. Rao, J.S., Usmani, M.A.W. and Ramakrishnan, C.V. (1990) Interface Damping in Blade Attach Damping in Blade Attachment Region, in *Proceedings 3rd International Conference Rotor Dynamics*, Lyon, p. 185.
43. Reshetov, D.N. and Levina, Z.M. (1965) Machine Design for Contact Stiffness, *Machines and Tooling*, vol. 36, p. 15.
44. Rieger, N.F. and Beck, C.M. (1980) Damping Tests on Steam Turbine Blades, EPRI Project RP-1185.
45. Rowett, F.E. (1914) Elastic Hysteresis in Steel, in *Proceedings Roy. Soc.*, vol. 89.
46. Rzadkowski, R. (1998) *Dynamics of Rotor, Steam Turbine Rotor Blading, Part Two, Bladed Discs*, Maszyny Przepływowe Tom 22, Wrocław Ossolineum.
47. Sanliturk, K.Y., Imregun, M. and Ewins, D.J. (1992) Statistical Analysis of Random Mistuning of Bladed Assemblies, *Journal of Mechanical Engineering*, C432/110, p. 51.
48. Sezawa, K. (1933) Vibration of Turbine Blades in Packets, *Phil. Mag.*, p. 164.
49. Sinha, A. (1997) Computation of the Maximum Amplitude of a Mistuned Bladed Disk Assembly via Infinity Norm, in *Proceedings ASME International Mechanical Engineering Congress and Exposition*, vol. AD-55, p. 427.
50. Sinha, A. and Griffin, J. (1984) Effects of Static Friction on the Forced Response of Frictionally Damped Turbine Blades, *J. Engng. Power and Gas Turbines, ASME*, vol. 106, p. 65.
51. Smith, D.M. (1948) Vibrations of Turbine Blades in Packets, in *Proceedings 7th International Congress Applied Mechanics*, London, p. 178.
52. Stodola, A. (1910) *Dampf- und Gasturbinen*, Springer, Berlin. Translation (1927) *Steam and Gas Turbines*, McGraw-Hill.
53. Swaminatham, M. et al. (1986) On Modal Generation and Nodal Analysis of Flexible Bladed-Disc Assemblies, in *Proceedings ASME Vibrations Conference*, Bladed Disk-Assemblies, Cambridge, MA, vol. 6, p. 49.
54. Thomas, J. and Subuncu, M. (1979) Vibration Characteristics of Asymmetric Cross-section Bladed Discs under Rotation, ASME 79-DET-94.
55. Weaver, F.L. and Prohl, M.A. (1958) High Frequency Vibration of Steam Turbine Buckets, *Trans. ASME*, vol. 80, p. 181.
56. Whitehead, D.S. (1966) Effect of Mistuning on the Vibration of Turbomachine Blades Induced by Wakes, *Journal of Mechanical Engineering Science*, vol. 8, no. 1, p. 15.
57. Whitehead, D.S. (1998) The maximum Factor by Which Forced Vibration of Blades Can Increase due to Mistuning, *Journal of Engineering for Gas Turbines and Power*, vol. 120, p. 115.

Chapter 18
Lifing

Turbine bladed disks continue to fail because of fatigue caused by resonant stresses; one of the major failures in the 1990s was reported in a nuclear machine in Narora, India. Bearing failures caused machine trip and rubbing caused blade fatigue (see Figure 18.1).

Albert Wilhelm (1838) is recognized as the first person to record observations of metal fatigue. While working in the Mining and Forestry Office in Clausthal, Germany, in 1829, he observed, studied and reported the failure of iron mine-hoist chains arising from repeated small loadings, the first recorded account of metal fatigue, see [31].

Jean-Victor Poncelet, a French military engineer and later professor at the École d'Application in Metz published his monograph on *Introduction à la mécanique industrielle* in 1829. In his lectures at the military school at Metz around 1837-1839, he for the first time referred to metals under stress as being "tired". He introduced the notion of fatigue of metals characterized by a drop in durability of steel products under repeated variable loads [33].

Rankine [16] was one of the first engineers to recognize that fatigue failures of railway axles were caused by the initiation and growth of brittle cracks. In the early 1840s he examined many broken axles, particularly after the Versailles train crash in 1842 when a locomotive axle suddenly fractured. He showed that the axles had failed by progressive growth of a brittle crack from a shoulder or other stress concentration source on the shaft, such as a keyway. He was supported by a similar direct analysis of failed axles by Joseph Glynn (1844), where the axles failed by slow growth of a brittle crack in a process now known as metal fatigue.

British Rail experienced a series of fatigue failures of the railway axles; therefore, The Railway Inspectorate was formed in 1840 to investigate the accidents. The report was made in 1848. Their first investigation concerned the derailment of a train caused by the fall of a large casting from a wagon on a passenger train. The Howden rail crash on 7 August 1840 killed four passengers. After submission of the report in 1848, Eaton Hodgkinson in 1849 was granted a small sum of money to report to the UK Parliament on his work in ascertaining by direct experiment, the effects of continued changes of load upon iron structures and to what extent they

Fig. 18.1 Last stage steam turbine blade fatigue crack and propagation

could be loaded without danger to their ultimate security. This is the first attempt to understand fatigue phenomenon in a scientific manner. Braithwaite [4] reported on common service fatigue failures and coined the term *fatigue*.

Systematic fatigue testing was undertaken in 1860 by William Fairbairn and August Wöhler. Fairbairn [6] built a large-scale testing apparatus for the studies, partly funded by the Board of Trade. He studied the effects of repeated loading of wrought and cast iron girders, showing that fracture could occur by crack growth from incipient defects, a problem now known as fatigue.

Wöhler [34] summarized his work in several papers on railroad axles; see also Wöhler [35]. He concluded that cyclic stress range is more important than peak stress and introduces the concept of endurance limit. His work on fatigue marks the first systematic investigation of *S*-*N* Curves, also known as Wöhler curves, to characterize the fatigue behavior of materials. He showed clearly that fatigue occurs by crack growth from surface defects until the product can no longer support the applied load. The history of a fracture can be understood from a study of the fracture surface. He developed an apparatus for repeated loading of railway axles, mainly because many accidents were caused by sudden fatigue fracture. The presentation of his work at the Paris Exposition in 1867 brought world wide attention.

Towards the end of the 19th century, it was realized that a reciprocating steam engine causes many vibrations and fatigue failures and people began to look at pure rotating machines in the hope that a better design could eliminate vibration, noise and fatigue. Within a hundred years of its existence, the reciprocating steam engine had been challenged.

During the second century BC, Hero demonstrated the principle of a reaction turbine, but could not realize any useful work [15]. Despite the scientific revolution followed by the industrial revolution, James Watt tried to build a steam turbine and came to the conclusion that it could not be built given the state of contemporary technology.

Nearly 100 years after Watt built his steam engine, De Laval of Stockholm succeeded in building the first steam turbine (impulse turbine), see Chapter 8 where we discussed the rapid developments in rotating machinery.

The rapid developments of rotating machinery, steam or gas turbines in power generation, centrifugal compressors and pumps in oil and gas industry and aerospace applications and finally aircraft engines, particularly defense applications pushed the technology beyond recognition during the later half of the 20th century.

Whittle faced several fatigue failure problems in the development of his W.2 engine, in his own words, "Frequency of turbine blade failures was becoming the latest technological barrier to overcome", see [32]. The QE2 9th Stage Starboard HP Turbine Rotor experienced blade fatigue failures on 24 December 1968, in its maiden voyage [7]. Bladed disks are the most flexible elements in high speed rotating machinery. Due to rotation, the blade root gets tightened in the disk slot and transmits the centrifugal load. The mating contact surfaces could be just two for high pressure turbine blades a few centimeters long and may increase to six or more for low pressure 1 m long turbine blades. While the average stress in the mating areas is fully elastic and well below yield, the peak stress at singularities in the groove shape can reach yield values and into local plastic region. Last stage LP turbine blades are the most severely stressed blades in the system. Usually these are the limiting cases of blade design allowing the peak stresses to reach yield or just above yield conditions.

Failures can occur with crack initiation at the stress raiser location and propagation, for which two cases can be cited. The last stage blades in an Electricite de France B2 TG Set failed in Porcheville on August 22, 1977 during over-speed testing [8]. On 31 March 1993, Narora machine LP last stage blades suffered catastrophic failures [18]. The left (No. 1) engine of Boeing 777-300 A6-EMM failed because of blade fatigue on 31 January 2001 at Melbourne Airport. American Airlines Boeing 767 doing a high power GE CF6 engine run at Los Angeles airport had a #1 engine HPT failure on 2 June 2006. HPT let go and punctured the left wing, #2 engine, peppered the fuselage and set fire to the aircraft. The turbine disk exited the engine, sliced through the aircraft belly and lodged in the outboard side of the #2 engine. Most recently, the number two engine on the Qantas Airbus A380 experienced an uncontained failure after departing from Singapore for Sydney on November 4, 2010. Thus turbine blade failures continue to occur despite best design practices.

Turbine bladed-disks continue to fail because of fatigue caused by resonant stresses and thus considerable attention is given to life estimation and optimization to increase life. Life estimation process of turbine blades is captured in Figure 18.2.

Mean Stress Field: Blades are subjected to steady stress fields due to gas loads, thermal loads, and centrifugal loads under normal conditions of operation. The gas loads are determined from CFD analysis of the gas path, which have a steady part and an unsteady part at nozzle passing frequencies. Because of compression in the compressor flow path or the hot gas path in turbine blades, they are subjected to thermal loading during the transient period of start-up and shut-down; they will form a mean load at a given steady operation or an overall cyclic load for each start-up and shut-down operation. The blades are also subjected to mean loads due to centrifugal loading which could be substantial in a low pressure compressor and turbine blades that will push the structure into globally elastic and locally plastic conditions. Cyclic symmetry can be utilized in assessing these stress fields. The steady stress field

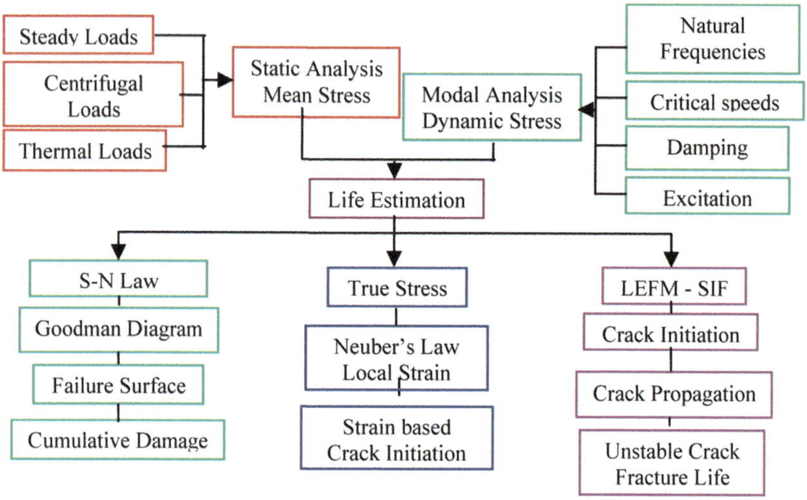

Fig. 18.2 Outline for life estimation

determination is well established and can be directly imported by the user at the start of life estimation to a recently developed tool, *TurboManager* [26].

Of particular interest is the result at the stress raiser location. Usually, the centrifugal loads lead to local plastic conditions in the root; an elasto-plastic analysis can give a nearly true picture in the stress raiser location.

Modal Analysis and Dynamic Stresses: Determination of natural frequencies and mode shapes as a function of speed is well known. The results are imported into *TurboManager* which plots the Campbell diagram as in Figure 18.3. The Campbell diagram identifies critical speeds at the resonances (natural frequencies crossing with excitation frequencies) through which the blade passes.

The resonant stress at critical speed depends on two factors, viz.: (1) excitation strength: The excitation strength is determined from a transient CFD analysis (not discussed here); (2) damping: this has been discussed in Section 17.4.

Determination of Stress Distribution around Resonance: Now, the question is how to determine the stress rise and fall at critical speed and the resonant stress magnitude. The best approach is to assume the unsteady force field to be steady and obtain the equivalent static stress field first. This equivalent static stress distribution is then multiplied by the quality factor $1/2\xi$ to obtain correct resonant stress; the only catch is we do not have the damping value. The general practice so far is to assume this damping value or determine the average damping value from an experiment. Here we have a nonlinear damping model as described in Section 17.4.

A nonlinear approach is used to determine the correct value of ξ by an iteration process to determine the resonant stress at all the critical speeds [22, 29]. This procedure is implemented in *TurboManager* so that the designer can determine the dynamic stress accurately for a given mode experiencing resonance at a critical speed.

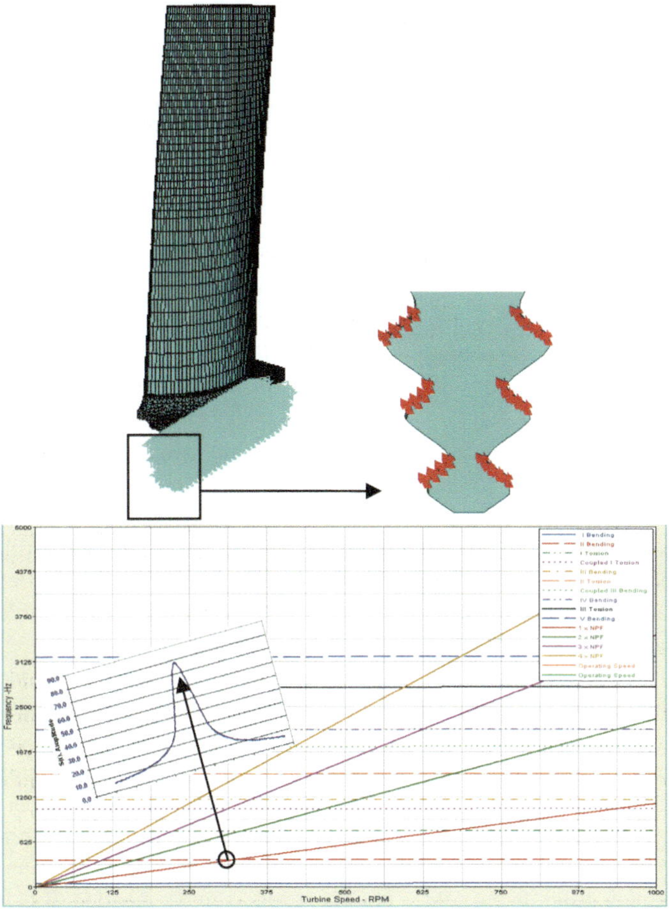

Fig. 18.3 FE model of a blade and its Campbell diagram showing its I critical speed

The stress response in this resonance region is then obtained by the dynamic magnifier relation $H(\omega)$. It may be noted here that the phase angle of the excitation force may vary from point to point on the blade surface and this phase should be accounted for in the stress estimation [24]. In those cases it may be easy to perform a forced vibration analysis with a frequency very close to zero, e.g., 0.01 rad/s. The result will be same as the equivalent static stress field.

18.1 High Cycle Fatigue (HCF) Life Estimation

In Figure 18.2 the life estimation is divided into three separate modules. All these three modules need mean and dynamic stress fields and the location where peak values occur. Historically it was HCF that was first developed as the failures of rail road axles initially occurred after sufficiently long hours of operation lasting months or years.

S-N Curves: Fatigue testing is done under controlled laboratory conditions; the actual condition of a mechanical component is far from the ideal conditions, e.g., surface finish, specimen size effect, stress concentration, temperature conditions. We also have to account for the reliability as the fatigue tests are statistical in nature as well as any special conditions such as corrosion. These factors are discussed in [2, 19]. *TurboManager* updates the material *S-N* curve to the component curve.

Effect of Mean Stress – Goodman Diagram: The mean stress has significant influence on fatigue and *S-N* curves. In the interim since Goodman suggested in 1899 a linear relationship using endurance limit and ultimate tensile strength of the material, several relations have been proposed [19].

Cumulative Damage: All structures are subjected to variable stress fields, both mean and alternating stresses. Typically a turbomachine blade crosses several critical speeds on the Campbell diagram with the alternating stress raising quickly and falling in a very sharp manner. It is important to determine the damage fraction while crossing each of the resonances from the start up of the engine till reaching the full operating speed or while performing a maneuver.

There are several linear and nonlinear cumulative damage rules; Rao et al. [22] presented calculations of damage using these rules and compared the results. The stress response before and after resonance and above endurance limit at each critical speed is divided into several steps and a linear or nonlinear cumulative damage is adopted to determine the damage fraction while crossing each critical speed. Knowing the acceleration with which the blade passes the resonance, each step through resonance is considered with the stress amplitude and the elapsed number of cycles in that step time period. Cumulative damage is then estimated to reach resonance and return to stress levels below the fatigue limit. Since the application of stress levels may play a significant role, a nonlinear rule is recommended. It is found that decreasing stress from resonant value to fatigue limit consumes more cycles of life; further while the machine is shut down, these damages could be different because acceleration rates will be different. It may be noted that the acceleration through the resonance may affect the resonant amplitude as well as speed at which resonance takes place and this may be incorporated in determining the resonant stress [26, 29]. The total damage fractions for each start up and shut down operation is next obtained to give the life in terms of start ups.

Rao and Peraiah [23] used high cycle stress based fatigue life analysis of a gear box in a turbo-generator set and Rao and Rzadkowski [27] considered tuned and mistuned turbine blades for lifing using linear and nonlinear cumulative damage theories.

18.2 Low Cycle Fatigue (Strain Based Life Estimation)

Blades are usually mounted on disks in turbomachines. Both the blades and disk contain mating grooves or notches which primarily act as stress raisers. The stress in this region of discontinuity is usually beyond elastic conditions. The strain based life estimation process is briefly given here [19].

Ludwik [12] defined the stress–strain relation beyond yield $\sigma = K\varepsilon^n$, where K is the strength coefficient and n is the strength exponent. For globally elastic and locally plastic structures, Neuber gave his hypothesis in 1961 which is used for relating the nominal and local cyclic stresses and strains. For fatigue loading, using fatigue stress concentration factor K_f, local stress and strain ranges are related to nominal stress range.

$$\Delta\varepsilon\Delta\sigma = \frac{1}{E}\left(K_f \Delta S\right)^2 \tag{18.1}$$

Based on Basquin's [3] relation of life to endurance and Manson's [13] and Coffin's [5] works, we have a life relation ($2N_i$ is number of stress reversals for failure) for crack initiation given by

$$\frac{1}{2}\Delta\varepsilon = \frac{1}{E}\sigma_f{}'(2N_i)^b + \varepsilon_f{}'(2N_i)^c \tag{18.2}$$

where ε'_f is the fatigue ductility coefficient and σ'_f is the fatigue strength coefficient.

Rao [21] applied this strain based life method to determine crack initiation life of a Francis turbine runner blade.

18.3 Linear Elastic Fracture Mechanics

The first analysis of fracture behavior of components was developed by Griffith [10]. He considered ideally brittle materials and assumed that incipient fracture occurs when the elastic energy supplied during an incremental increase in crack length is equal to or greater than the elastic energy at the crack tip. Usually, fracture mechanics helps in finding how a fatigue failure occurred through a study of crack surface and understanding how a crack has initiated and propagated until final rupture conditions. To prevent such fatigue failures, it is advisable to account for this at the design stage itself along with stress-based or strain-based methods.

Such a design process consists of the following steps; see Barsom and Rolfe [2] and Rao [19]:

1. Determine crack initiation threshold stress range and prevent it by ensuring no such stress occurs during the operation of the machine. This stress range has to be a very high value that can arise due to sudden loads, e.g., rubbing. If no initiation takes place, the question of crack propagation does not arise. However, a material defect or manufacturing blemish at the stress raiser location can decrease the threshold value considerably and moreover there is always a likelihood

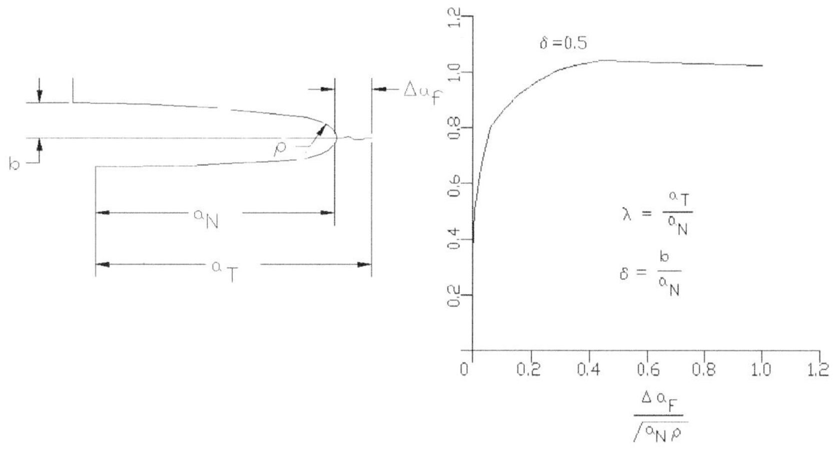

Fig. 18.4 Semi-elliptical notch model

of escaping the detection of such a defect when thousands of blades are manufactured. Therefore, it is customary to assume a defect of size less than the least count of crack detection machine to exist at the stress raiser location. In such an eventuality, a crack that has initiated or escaped detection can grow leading to a failure.

2. Determine the crack propagation threshold stress range and prevent crack propagation by ensuring the operating stress range is lower than the threshold. As a turbomachine has several stages, one stage or other will be operating closer to resonant conditions. Unfortunately the alternating stress range under operating conditions nearer to a blade's critical speed is sufficient to propagate a crack if it exists. Therefore this step becomes important at the design stage.
3. Determine the propagation of the crack with each cycle of loading and the rate at which crack front propagates (striation spacing) until unstable crack conditions are reached according to Griffith's theory. At operating conditions the propagation life can be considerable and the design should provide for as much crack propagation life as possible.

To determine the stress intensity factor from nominal stress values, the notch geometry is modeled, say by a semi-elliptical (b and a_N) notch as shown in Figure 18.4. Δa_f is starting crack length, assumed generally as the least count of the crack detection machine and a_T is the location of the crack tip, ρ is the radius at the notch. Conversion between stress intensity factor range ΔK and nominal stress range $\Delta\sigma$ is given by

$$\Delta K = 1.112\Delta\sigma\sqrt{\pi a_T}k\left(\frac{a}{b}\right)f_2(\lambda, \delta) \tag{18.3}$$

where $f_2(\lambda, \delta)$ and $k(a/b)$ are given in Figure 18.4 and Table 18.1.

Extensive tests conducted have shown that there is a threshold value of stress range for crack initiation, e.g., for HY-130 steels under a stress ratio $R = 0.1$ [2,19],

Table 18.1 k as a function of a/b

a/b	k
0.1	1.0
0.2	1.0
0.3	1.1
0.4	1.2
0.5	1.3
0.6	1.5
0.7	1.6
0.8	1.9
0.9	2.2
1.0	2.5

this value is defined by

$$\left(\Delta K_I/\sqrt{\rho}\right)_{th} = 586 \text{ MPa} \tag{18.4}$$

Using (18.3), the stress intensity factor range can be converted to get the stress range.

In a similar manner, there is a threshold value for an initiated crack to propagate, e.g., for martensitic, bainitic, ferrite-pearlite and austenitic steels given by the relation

$$\begin{aligned} \Delta K_{th} &= 7\,(1 - 0.85R)\,\text{MPa}\sqrt{\text{m}} \quad \text{for} \quad R > 0.1 \\ &= 6\text{MPa}\sqrt{\text{m}} \quad \text{for} \quad R < 0.1 \end{aligned} \tag{18.5}$$

An initiated crack propagates when the stress range exceeds the threshold value given above following Paris law, see [14] e.g., for ferrite-pearlite steels

$$\frac{da}{dN} = C(\Delta K)^m = 6.891 \times 10^{-6}(\Delta K)^3 \text{ microns/cycle} \tag{18.6}$$

until the stress intensity factor range becomes equal to fracture toughness of the material when it becomes unstable according to Griffith's law.

TurboManager provides capabilities as discussed above for stress-based life estimation, strain-based life estimation as well as linear fracture-mechanics-based life calculations. Typical stress distribution around a resonance and the fatigue failure surface obtained is shown in Figure 18.5. It also accounts for life estimation by considering the acceleration with which the blade is taken through the critical speed. Transient Response through critical speeds is shown in Figure 18.6 to determine life.

TurboManager also performs fracture mechanics lifing and the crack propagation for the blade in Figure 18.4 as demonstrated by Rao [20]. The striation spacing as a function of crack length and number of cycles and crack length as a function of number of cycles elapsed are displayed to the user (see Figure 18.7).

Influence of Plasticity on Crack Propagation: The root design for a low-pressure stage becomes crucial as the stresses in the notch zone are very high compared to the

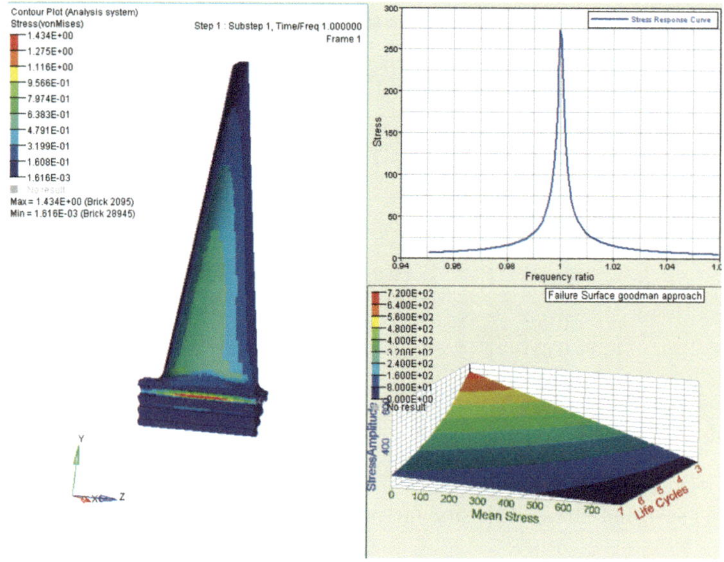

Fig. 18.5 Stress distribution around resonance and failure surface

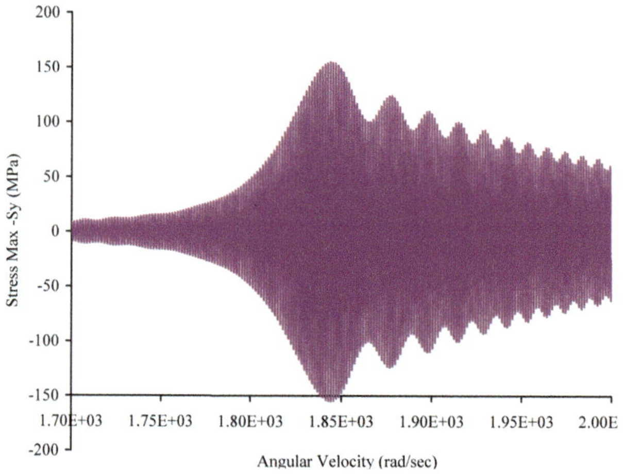

Fig. 18.6 Transient response through critical speed with accelerating blade

vane portion of the blade; Robertson and Walton [30] have included the blade root as well. While investigating a real life failure, the elastic stress field in this region of a last stage steam turbine blade was determined by Rao et al. [28]. Rao [18] used the centrifugal, steady state and dynamic steam bending stress information to determine the life by using linear elastic fracture mechanics principles. Conventionally the narrow field of plastic zone is neglected and the fracture mechanics analysis is conducted using a linear elastic approach. Rao et al. [25] included the plastic zone

Table 18.2 Elasto-plastic crack propagation calculations

a_{avg} m	ΔN cycles	$\Sigma \Delta N$ cycles	$\frac{\Delta a}{\Delta N}$ microns
0.00461	22515.2	22515.2	0.00444
0.00471	21978.1	44493.4	0.00455
0.00481	21464.8	65958.1	0.00466
0.00491	20973.6	86931.7	0.00477
0.00501	20503.2	107435.0	0.00488
0.00511	20052.4	127487.3	0.00499
0.00521	19619.9	147107.3	0.00510
0.00531	19204.7	166312.0	0.00521
0.00541	18805.8	185117.8	0.00532
0.00551	18422.3	203540.2	0.00543
0.00561	18053.3	221593.5	0.00554
Blade fails here under the influence of plasticity			
0.00661	15011.1	384520.6	0.00666
0.00761	12811.0	521988.0	0.00781
0.00861	11149.7	640601.1	0.00897
0.00961	9853.2	744717.3	0.01015
0.01061	8814.8	837357.6	0.01134
0.01161	7965.3	920699.3	0.01255
0.01261	7258.3	996361.7	0.01378
0.01361	6661.2	1065581.0	0.01501
0.01461	6150.5	1129320.0	0.01626
0.01561	5709.0	1188346.0	0.01752
0.01661	5323.8	1243275.0	0.01878
0.01761	4985.0	1294615.0	0.02006
0.01861	4684.6	1342784.0	0.02135
0.01961	4416.8	1388132.0	0.02264
0.02061	4176.4	1430957.0	0.02394
0.02161	3959.6	1471511.0	0.02525
0.02181	3918.8	1479369.0	0.02552
0.02201	3878.8	1487146.0	0.02578
0.02221	3839.5	1494844.0	0.02605
0.02241	3801.0	1502466.0	0.02631
0.02251	3782.0	1506248.0	0.02644
0.02261	3763.2	1510011.0	0.02657
0.02271	3744.5	1513755.0	0.02671
0.02281	3726.1	1517481.0	0.02684
0.02291	3707.8	1521189.0	0.02697
Blade fails here according to LEFM			

near the strain raiser (notch) where the crack initiates and its influence on crack propagation life is studied. This result is shown in Table 18.2.

Lifing of turbine blades or other mechanical components till recently required highly skilled engineers but today is being converted to a code culture where trained manpower can carry out the work in a routine manner, thus speeding up the design process. Lifing is a multifaceted technology involving CFD, thermal, thermo-mechanical, dynamics and modal analysis, hysteresis damping, macro and micro

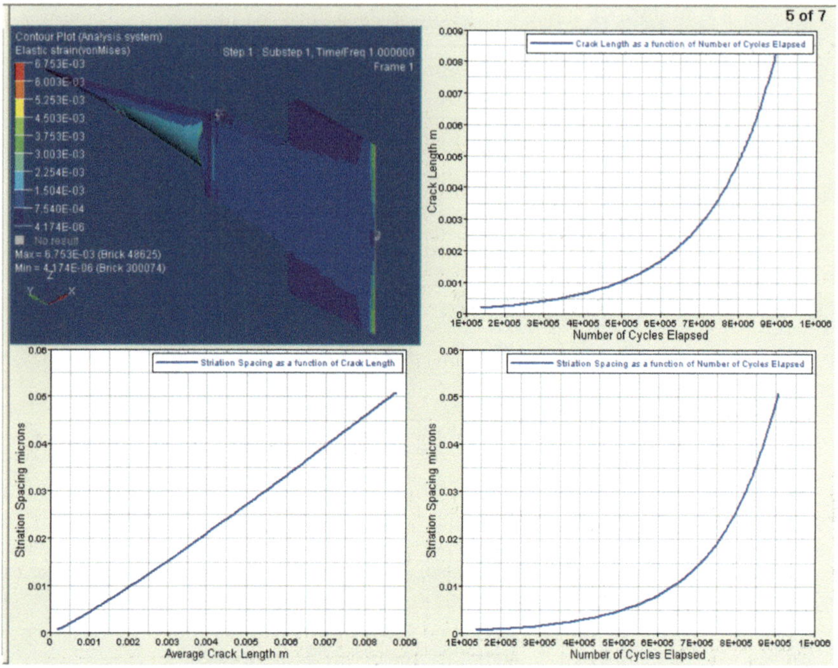

Fig. 18.7 Crack propagation study in TurboManager

displacement friction damping and then cumulative fatigue under elastic or globally-elastic but locally plastic structures with fracture mechanics capabilities. Most of these technologies are now automated to provide simulation without experimentation and thus decrease design cycle time.

The next obvious question a designer has – if a component like a turbine blade has the longest possible life and is it light enough; it obviously leads to optimization at a high scientific and engineering level. Today's technologies have already produced answers to these questions. We will as a final topic, discuss optimization that makes virtual simulation practical in the design stage before prototypes can be made and tested for final validation and constructing of a product.

References

1. Albert, W.A.J. (1838) Über Treibseile am Harz, *Archive für Mineralogie Geognosie Bergbau und Hüttenkunde*, vol. 10, p. 215.
2. Barsom, J.M. and Rolfe, S.T. (1987) *Fracture and Fatigue Control in Structures*, Prentice Hall, Englewood Cliffs.
3. Basquin, O.H. (1910) The Experimental Law of Endurance Tests, *Proc. ASTM*, vol. 10, p. 625.
4. Braithwaite, F. (1854) On the Fatigue and Consequent Fracture of Metals, in *Institution of Civil Engineers, Minutes of Proceedings*, p. 463.

5. Coffin, L.F. Jr. (1954) A Study of the Effects of Cyclic Thermal Stresses on a Ductile Material, *Trans ASME*, vol. 76, p. 931.
6. Fairbairn, W. (1864) Experiments to Determine the Effect of Impact, Vvibratory Action, and Long Continued Changes of Load on Wrought Iron Girders, *Philosophical Transactions of the Royal Society, London*, vol. 154, p. 311.
7. Fleeting, R. and Coats, R. (1970) Blade Failures in the HP Turbines of RMS Queen Elizabeth 2 and Their Rectification, *Trans. Instn. of Marine Engrs.*, vol. 82, p. 49.
8. Frank, W. (1982) *Schaden Speigel*, vol. 25, no. 1, 20.
9. Glynn, J. (1844) On the Causes of Fracture of the Axles of Railway Carriages, *Institution of Civil Engineers, Minutes of Proceedings*, p. 202.
10. Griffith, A.A. (1920) The Phenomenon of Rupture and Flaw in Solids, *Trans. Roy Soc.*, A-221.
11. Poncelet, Jean-Victor (1829) *Introduction à la mécanique industrielle*, École d'Application, Metz, France.
12. Ludwik, P. (1909) *Elemente der Technologischen Mechanik*, Springer Publication.
13. Manson, S.S. (1953) Behavior of Materials under Constant Thermal Stress, in *Proceedings Heat Transfer Symposium*, University of Michigan, Engng. Research Institute, p. 9.
14. Paris, P.C. and Erdogan, F. (1963) A Critical Analysis of Crack Propagation Laws, *Journal Basic Engng., Trans ASME*, vol. 85, p. 528.
15. Pederson, O. (1993) *Early Physics and Astronomy: A Historical Introduction*, 2nd. ed., Cambridge University Press.
16. Rankine, W.J.M. (1842) On the Causes of the Unexpected Breakage of the Journals of Railway Axles, and on the Means of Preventing such Accidents by Observing the Law of Ccontinuity in Their Construction, *Institution of Civil Engineers, Minutes of Proceedings*, p. 105.
17. Rao, J.S. (1995) Fracture Mechanics Analysis of a Steam Turbine Blade Failure, in *Proceedings 1995 Design Engng Technical Conferences*, ASME DE-Vol. 84-2, p. 117.
18. Rao, J.S. (1998) Application of Fracture Mechanics in the Failure Analysis of a Last Stage Steam Turbine Blade, *Mechanism and Machine Theory*, vol. 33, no. 5, p. 599.
19. Rao, J.S. (2001) *Turbine Blade Life Estimation*, Alpha Science.
20. Rao, J.S. (2009) Fracture Mechanics in TurboManager Quickens Blade Failure Investigations, *International Review of Aerospace Engineering (I.RE.AS.E)*, vol. 2, no. 6, p. 329.
21. Rao, J.S., Nimbekar, P.K., Misra, R. and Singh, A.K. (1998) Application of Local Stress-Strain Approach to Predict Fracture Initiation of a Francis Turbine Runner Blade, in *Proceedings ISROMAC-7*, Hawaii, vol. B, p. 674.
22. Rao, J.S., Pathak, A. and Chawla, A., (2001) Blade Life – A Comparison by Cumulative Damage Theories, *Journal of Engineering for Gas Turbines and Power*, vol. 123, no. 4, p. 886.
23. Rao, J.S. and Peraiah, K.Ch. (2001) Gearbox Failure of a Turbogenerator Set, in *Proceedings of ASME Turbo Expo 2001*, 2001-GT-0235.
24. Rao, J.S., Peraiah, K.Ch. and Uday, K.S. (2009) Estimation of Dynamic Stresses in Last Stage Steam Turbine Blades under Reverse Flow Conditions, *Advances in Vibration Engineering, Journal of Vibration Institute of India*, vol. 8, no. 1, p. 71.
25. Rao, J.S., Ramakrishnan, C.V., Gupta, K. and Singh, A. (2000) Elastic Plastic Fracture Mechanics of a LP Stage Steam Turbine Blade Root, in *Proceedings 5th ASME Annual Engineering Systems Design & Analysis*, Montreux.
26. Rao, J.S., Rejin, R., Suresh, S. and Narayan, R. (2009) A Procedure to Predict Influence of Acceleration and Damping of Blades Passing Through Critical Speeds on Fatigue Life, in *Proceedings of ASME Turbo Expo 2009: Power for Land, Sea and Air*, GT2009-59433.
27. Rao, J.S. and Rzadkowski, R. (2002) Life Estimation of Tuned and Mistuned Turbine Blades Using Linear and Nonlinear Cumulative Damage Theories, *Advances in Vibration Engineering, Journal of Vibration Institute of India*, vol. 1, no. 4, p. 322.
28. Rao, J.S., Singh, A.K. and Sharma, K.V.B., (1998) Blade and Rotor Dynamics in an Atomic Power Plant Accident, in *Proceedings Ninth World Congress Theory of Machine and Mechanisms*, Milan, Vol. II, p.1334.

29. Rao, J.S. and Vyas, N.S. (1996) Determination of Blade Stresses under Constant Speed and Transient Conditions with Nonlinear Damping, *Journal of Engng for Gas Turbines and Power, Trans ASME*, vol. 118, no. 2, p. 424.
30. Robertson, M.D. and Walton, D. (1990) Design Analysis of Steam Turbine Blade Roots under Centrifugal Loading, *Journal of Strain Analysis for Engineering Design*, vol. 25, no. 3, pp. 185–195.
31. Stephens, R.I. (2001) *Metal Fatigue in Engineering*, John Wiley & Sons, Inc.
32. St. Peter, J. (1999) *The History of Aircraft Gas Turbine Engine Development in the United States ... A Tradition of Excellence*, ASME Publication.
33. Tóth, L. and Yarema, S.Y. (2006) Formation of the Science of Fatigue of Metals. Part 1. 1825–1870, *Material Science*, vol. 42, no. 5, p. 673.
34. Wöhler, A. (1858–1870) Über die Festigkeitsversuche mit Eisen and Stahl, *Zeitschrift für Bauwesen*, vol. 8 (1858), p. 641; vol. 10 (1860), p. 583; vol. 13 (1863), p. 233; vol. 16 (1866), p. 67; vol. 20 (1870), p. 73.
35. Wöhler, A. (1867) Experiments on the Strength of Metals, *Engineering*, vol. 4, p. 160.

Chapter 19
Optimization

Optimization had its roots in the scientific revolution period and thus is one of the oldest sciences [8]. For industrial applications, however it remained dormant until recent times. The pioneering work of modern structural topology can be traced back to 1981 when Cheng and Olhoff, see Keng-Tuno [5], introduced the concept of microstructure to structural optimization in studying the optimum thickness design of a solid elastic plate for minimum compliance. A continuum approach to structural topology optimization was first introduced by Bendsøe and Kikuchi [1]. Optimization of finite element-based structures is acknowledged as a useful methodology for achieving important improvements in product design and is widely used in automotive and aerospace industries.

Prior to commercial code development, conventional optimization in structures was achieved by determining the strain energy density and identifying material areas where a removal can be made or a component is required to be strengthened and a DOE approach adopted. This is a tedious and time consuming process requiring skilled engineers. Many commercial Structural and Flow codes have now added optimization processes; however, topology optimization has received considerable attention in recent times.

Topology optimization uses the SIMP (Solid Isotropic Material with Penalization) method which is also called the *density method* [1–4]. In this method, the stiffness of the material is assumed to be linearly dependent on the density. The material density of each element is directly used as the design variable, and is normalized to have a value between 0 and 1, representing the state of void and solid, respectively.

The classical topology optimization setup involves the objective of minimizing the compliance with volume fraction as constraint. The compliance is the strain energy of the structure and can be considered a reciprocal measure for the stiffness of the structure. Volume fraction constraint specifies what fraction of the volume is to be removed. The remaining material is redistributed within the design space to obtain an optimal load path.

Topology optimization is now used as a combination of the Finite Element Method (FEM) with an optimization algorithm. As design parameters, every finite element gets a so-called relative density ρ, which may continuously vary between 0 and 1 and affects the elasticity tensor of a finite element as

$$\frac{E}{E_0} = \left(\frac{\rho}{\rho_0}\right)^p \tag{19.1}$$

where E_0 describes the nominal stiffness properties of the element. The task for the optimizer is to determine a density value for every element. The exponent p is a penalty parameter used to reach a result that is discrete as possible, by penalizing intermediate densities. If ρ tends to zero, the stiffness tends to zero too. This means, the element could be deleted because it is not important for the structure. If the density reaches a value of 1, the element is very important for the structure and may not be removed. This approach is called SIMP (Solid Isotropic Material with Penalization) [2].

To illustrate the capabilities of the topology optimization, a simple example problem is illustrated here: Figure 19.1 shows the meshed design-space of a "truss".

Here the entire space is considered design space. The task is to find a structure with a minimum volume such that the maximum displacement remains the same as the baseline. The baseline result is shown in Figure 19.2 with $E = E_0 = 210$ GPa: the maximum deflection at the node shown is 0.001826 mm. After three iterations the density distribution or elasticity distribution as displayed in Figure 19.3 is arrived at. The "red elements" indicate a density of 1. This means these elements are very important and have nominal stiffness properties. The "blue elements" have a density close to zero and therefore a very low stiffness and they can be removed. We can build the structure around the "red elements".

The optimization problem in a finite element and multi-body dynamics software for the design, analysis, and optimization of linear structures can be stated as

$$\begin{aligned}
&\text{Min } f(X) = f(X_1, X_2, \ldots, X_n) \text{ subject to} \\
&g_j(X) \leq 0, \quad j = 1, \ldots, m \\
&X_i^L \leq X_i \leq X_i^U, \quad i = 1, \ldots, n
\end{aligned}$$

where $f(X)$ is the objective function, $g(X)$ are the constraints, both of which are functions of the design variables. There are m constraints and n design variables. The type of design variables can be size, shape/topography, and topology.

Topology optimization is a mathematical technique that generates an optimized shape and material distribution for a set of loads and constraints within a given design space. The design space can be defined using shell or solid elements, or both. The classical topology optimization can be set up to solve the minimum compliance problem as well as the dual formulation with multiple constraints. Manufacturing constraints can be imposed using a minimum member size constraint, draw direction constraints, extrusion constraints, symmetry planes, pattern grouping, and pattern repetition, etc.

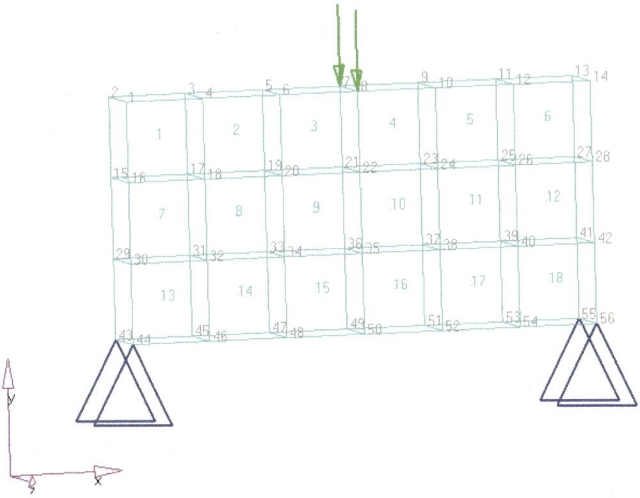

Fig. 19.1 Topology problem for a truss

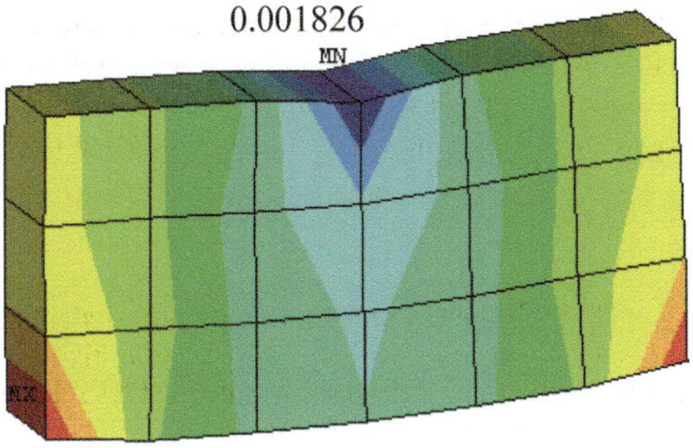

Fig. 19.2 Baseline result

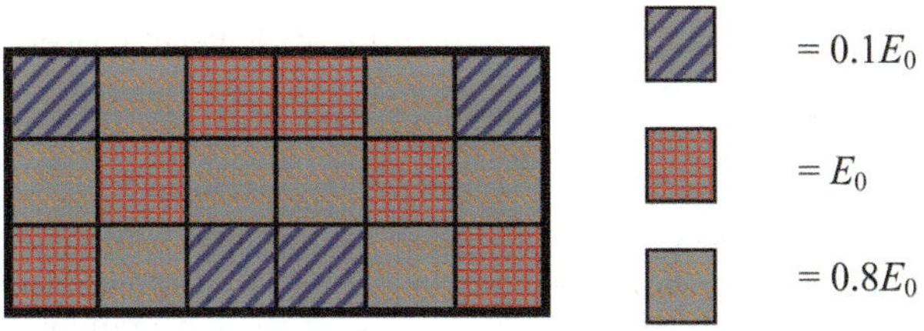

Fig. 19.3 Distribution of elasticity after topology optimization

19.1 Shape Optimization

An axial entry LP turbine blade having a fir tree root design in Figure 19.4 is used for shape optimization. A sector of a rotor disc is modeled to make use of its cyclic symmetry condition. The Blade is pre-twisted with a height of 290 mm. There are 60 blades in this LP stage and they are placed on the disk with the bottom of the blade root at a radius of 248 mm from the axis of the rotor. Using mapped meshing options, a solid element mesh with eight nodes is generated, by capturing all the critical regions with a finer mesh. Mesh around the singularities, blade and disc dovetail root fillet regions at higher radius, where the peak stresses are expected are captured with two to three layers of elements with element size as low as 0.235 mm. The mesh consists of 305,524 elements and 344,129 nodes.

For the analysis the blade along with the disk effect is considered by modeling a 1/60 sector of the disk with one blade and using cyclic symmetry boundary conditions applied on both the partition surfaces as shown in Figure 19.4. The common nodes on the pressure faces at six positions, where the load transfer between blade and disc takes place, are joined together to make it act as a single entity. The blade and disc are assumed to be made of the same material with yield stress of 585 MPa, Young's modulus 210 GPa, density 7900 kg/m^3 and Poisson's ratio 0.3.

An elastic stress analysis is conducted for a centrifugal load at full speed 8500 RPM. The Von Mises elastic stress field near the root region is shown in Figure 19.5. The root fillet in the first landing area experiences a severe stress of 1825 MPa at node 153608 well beyond yield 585 MPa, with an average sectional stress 256 MPa. The stress contour beyond yield is shown to be spread across three elements over a depth of 1.22 mm.

The hardening property of the material in the plastic region is given in Figure 19.6. The elasto-plastic analysis result for the Von Mises stress is given in Figure 19.7. The stress region beyond yield is also defined in the same figure. The root fillet now experiences a peak Von Mises stress of 768 MPa at a node 176017 in the same region, which is beyond the yield value 585 MPa. From Figure 19.7, it is observed that the plastic region has not changed from the simple elastic analysis result in Figure 19.5. The peak stress value has dropped considerably from the elastic analysis result of 1825 MPa to a value of 768 MPa just above the yield.

The material in the root region around the stress raiser location flows, thus easing the stress and raising the strain in accordance to the hardening law given in Figure 19.6. The peak strain observed at the node 153608 in the same region closer to peak stress location is 0.0153 [7].

Shape variables are generated using a recently developed mesh morphing technique that decreases considerably the time in DOE studies. Using the baseline finite element model in Figure 19.4, a suitable number of shapes in the vicinity of a baseline consistent with the available design space is defined by modifying the grid point locations, which are saved as perturbation vectors. The shapes generated are combinations of parameters shown in Figure 19.8. Shapes are then defined as variables by assigning lower and upper bounds to it. Table 19.1 gives the minimum and maximum values adopted for defining the shape variables. Shape variables can then be assigned as indicated to perturbation vectors, which control the shape of the model within a given bound. This is helpful in generating the required shape bounds without re-meshing the model.

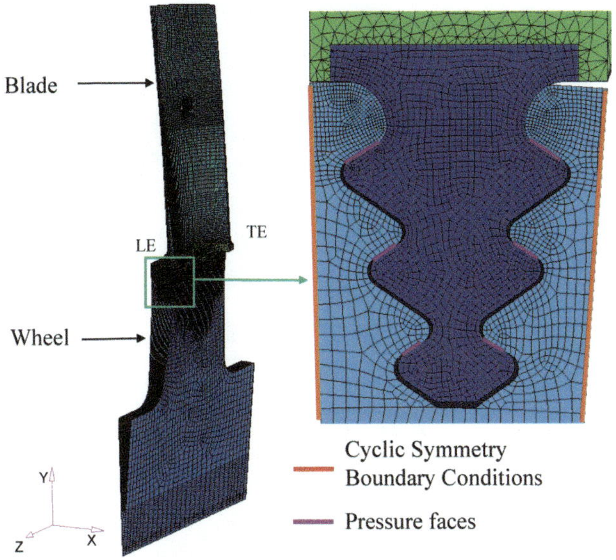

Fig. 19.4 Bladed disk model showing the FE mesh

Table 19.1 Shape variable definitions

Minimum Value (mm)	Maximum Value (mm)
$W_1 = 22.17$	$W_1 = 25.76$
$W_2 = 13.65$	$W_2 = 13.86$
$R_1 = 1.70$, $H = 5.67$, $V = 4.13$, $R_2 = 4.0$	$R_1 = 2.14$, $H = 4.85$, $V = 4.06$, $R_2 = 3.37$
$\theta = 29.86°$	$\theta = 16.25°$

HyperStudy is a solver-independent code that uses global optimization methods which are very general in that they can be used with any analysis code, including non-linear analysis codes. Global optimization methods use higher order polynomials to approximate the original structural optimization problem over a wide range of design variables. The polynomial approximation techniques are referred to as Response Surface methods. A sequential response surface method approach is used in which the objective and constraint functions are approximated in terms of design variables using a second-order polynomial. One can create a sequential response surface update by linear steps or by quadratic response surfaces. The process can also be used for non-linear physics and experimental analysis using wrap-around software, which can link with various solvers.

Shape optimization is carried by using the baseline model, having the cyclic symmetry boundary conditions imposed on the disc, with the objective to minimize the peak stresses. Shape variables generated in the previous section are used as design

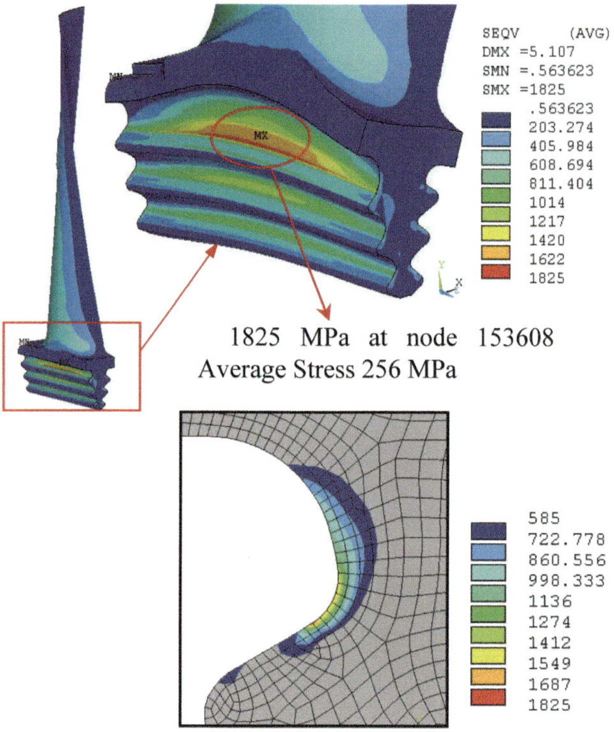

Fig. 19.5 Von Mises stress in elastic domain at 8500 RPM

variables. Figure 19.9a shows the objective value and Figure 19.9b shows the variables and as their variations during the iteration process.

The Von Mises stress distribution for the optimized shape is shown in Figure 19.10. Maximum stress has decreased marginally from 768 to 746 MPa by 22 MPa (2.86%) from baseline elasto-plastic analysis for 8500 RPM; however the peak plastic strains reduced from 0.0153 to 0.01126 by 26.4%. This is the major advantage in optimization for a blade root shape that can increase by four times or more. Table 19.2 shows the optimized shapes obtained from elasto-plastic analysis.

Many existing machines have roots designed by experience and there can be considerable margin in lowering peak strains and therefore enhanced life. This technology helps in true simulation and reduces considerably the testing and design development time.

19.2 Weight Optimization

Shape optimization was discussed in the previous section where the main aim was to increase life when the blades are subjected to local plastic conditions. If the local

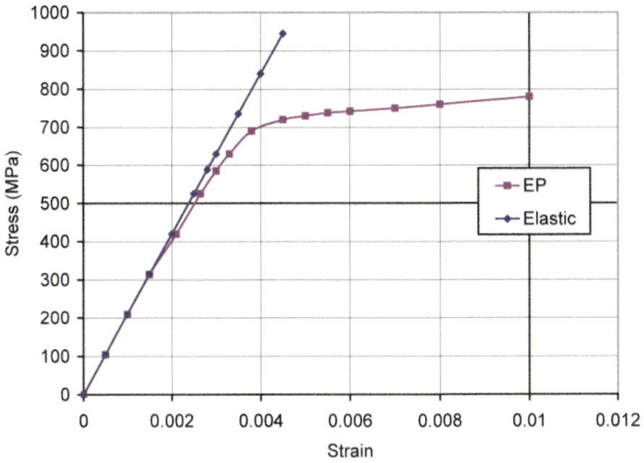

Fig. 19.6 Material hardening characteristic in the plastic region

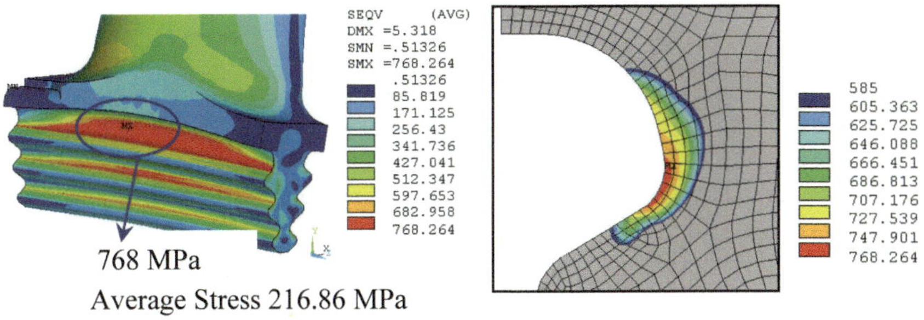

Fig. 19.7 Results of elasto-plastic analysis at 8500 RPM

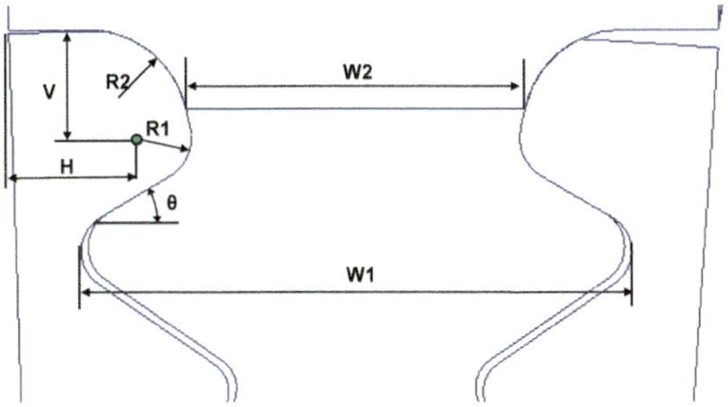

Fig. 19.8 Parameters used for defining the shape variables

plastic conditions are to be avoided, one may have to sacrifice the blade length so as to decrease the centrifugal loads with a corresponding loss in extraction of power from the turbine. The case of military aircraft engines, on the other hand, is different; here the life can be limited, but weight is an important criterion. Usually

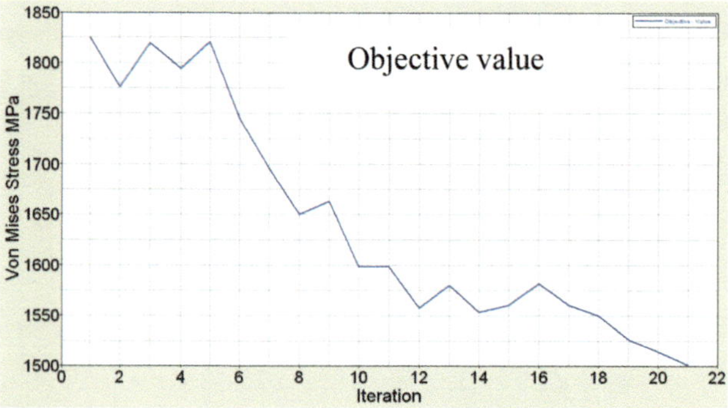

(a)

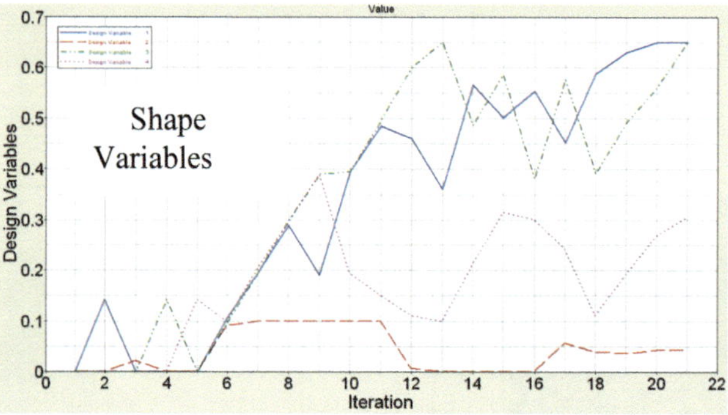

(b)

Fig. 19.9 (a) Variation of objective and shape variables during optimization. (b) Variation of objective and shape variables during optimization

considerable material sits near the platform region taking very little load and can be easily removed without endangering the structural integrity. Here, such a weight optimization problem is illustrated see [6].

Figure 19.11 shows the FE model of a typical aircraft engine LP compressor blade and disk made of a Ti-alloy; mass density is 4.42×10^{-9} Nsec^2/mm^4, Poisson's ratio is 0.3, Young's modulus is 102 GPa, yield strength is 820 MPa. Blade rotational speed is 1156.62 rad/sec. Cyclic symmetry is used in the analysis.

The disk has 106,066 Solid 45 elements with 121,948 nodes. The blade has 46,970 Solid 45 elements and 41,811 nodes. The nonlinear material property of the bladed-disk is shown in Figure 19.12.

The baseline results show the peak stress in the disk to be 787 MPa and the maximum stress in the blade is 721 MPa. HyperStudy is used to optimize the blade for weight reduction by limiting the peak stress to the yield value, 820 MPa in the blade-disk system.

The blade root and shank have considerable regions of stress well below yield and a baseline for optimization is chosen as given in Figure 19.13. As shown, eight holes with radius $R = 1.75$ mm are provided in the blade root and two cutouts in the shank are allowed to reduce the weight. The cutout proposed in the shank is also

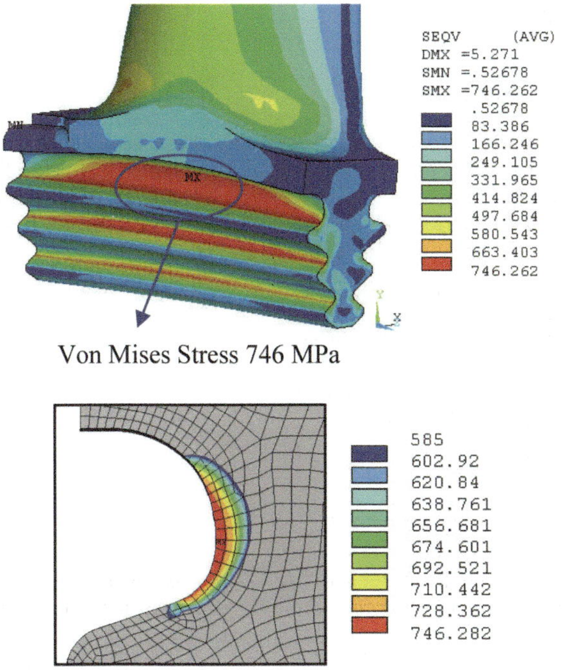

Fig. 19.10 Elasto-plastic analysis at 8500 RPM for optimized configuration

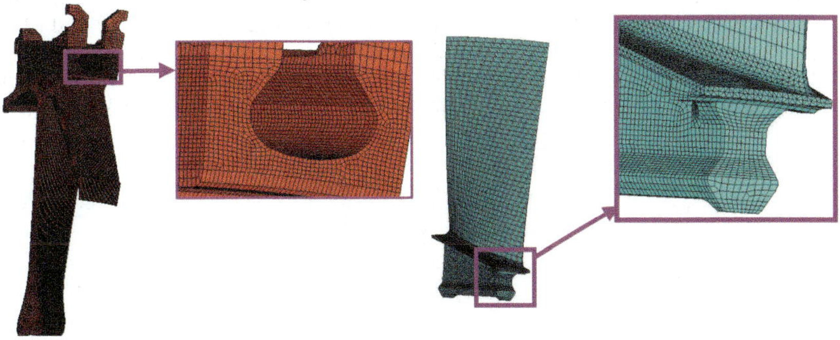

Fig. 19.11 FE model of disk and blade

given in Figure 19.13. The blade root originally without any cutouts was 7890.34 mm^3. The objective function is chosen to be this volume and it is minimized subject to the condition that the peak stress is limited to the yield value, namely 820 MPa.

The front shank has $D_1 = 1.1$, $W_{1a} = 4.75$, $W_{1b} = 13.94$ mm, while the rear shank $D_2 = 1.1$, $W_{2a} = 4.44$, $W_{2b} = 13.64$ mm. Table 19.2 gives the range of design variables allowed in optimization.

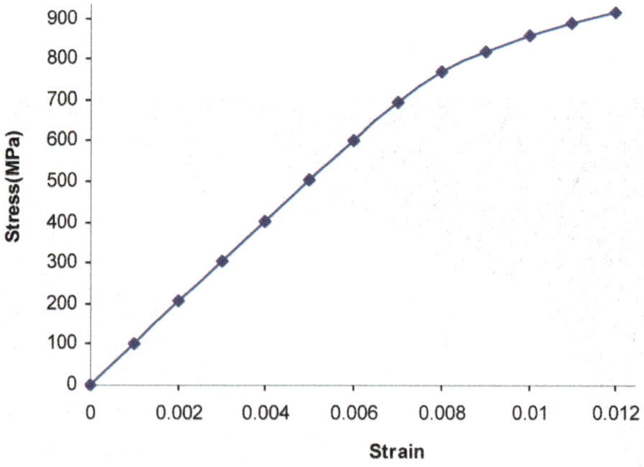

Fig. 19.12 Material stress-strain characteristics

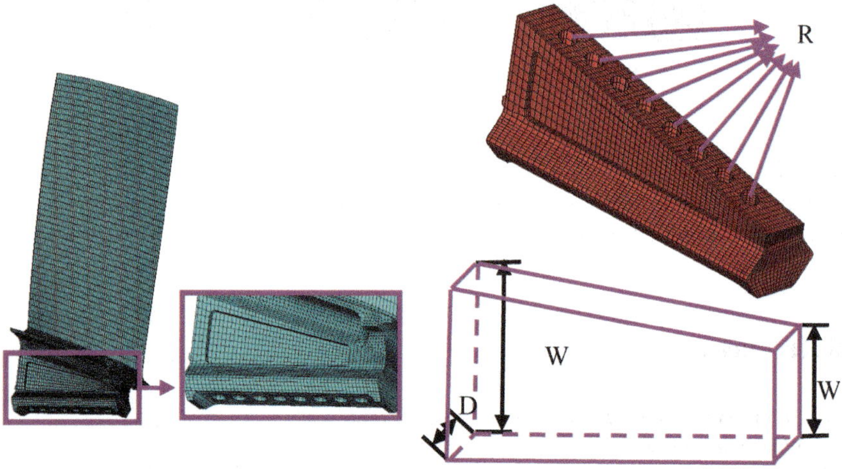

Fig. 19.13 Baseline for weight optimization and design variables on the front shank

In 16 steps the result was achieved. Optimum design variables are given in Table 19.2. The objective function value decreased from 7890.34 to 7098.93 mm^3, i.e., a reduction of 10.03%.

Because of rapid changes that have taken place due to the availability of commercial CAE tools, the design activity was accelerated and one was able to look at the nooks and crannies of every machine component and observe the state of stress. In considerable areas in a given structure, material was sitting idle in the conventional designs and questions were raised about reductions in weight, raising stiffness, etc. The optimization itself remained classical for a long time well into 1990s, but the scenario changed with Topological Optimization and the availability of commer-

Table 19.2 Range of design variables and optimized values

Variable	Range (mm)	Optimized Value (mm)
D_1	1.1 to 2.0	1.64
D_2	1.1 to 2.0	1.64
W_{1a}	4.75 to 6.5	5.09
W_{1b}	13.94 to 15.5	15.35
W_{2a}	4.44 to 6.0	5.57
W_{2b}	13.64 to 16.0	14.56
R	1.75 to 2.25	2.0

cial codes to deal with thousands of variables and multiple objective functions with several constraints.

References

1. Bendsøe, M.P. and Kikuchi, N. (1988) Generating Optimal Topologies in Optimal Design Using a Homogenization Method. *Comp. Meth. Appl. Mech. Eng.*, vol. 71, p. 197.
2. Bendsøe, M.P. and Sigmund, O. (2003) *Topology Optimization – Theory, Methods and Applications*, Springer, Dordrecht.
3. Bendsøe, M.P. et al. (1994) An Analytical Method to Predict Optimal Material Properties in the Context of Optimal Structural Design, *Journal of Applied Mechanics*, vol. 61, no. 4, p. 930.
4. Bendsøe, M.P. et al. (1995) Optimal Design of Material Properties and Material Distribution for Multiple Loading Conditions, *International Journal for Numerical Methods in Engineering*, vol. 38, no. 7, p. 1149.
5. Keng-Tuno, C. (1981) On Non-Smoothness in Optimal Design of Solid, Elastic Plates, *International Journal of Structures*, vol. 17, no. 8, p. 795.
6. Rao, J.S., Bhaskar Kishore, Ch. and Vasanthakumar, M. (2008) Weight Optimization of Turbine Blades, in *Proceedings 12th International Symposium on Transport Phenomena and Dynamics of Rotating Machinery*, Honolulu, Hawaii, ISROMAC12-2008-20020.
7. Rao, J.S. and Suresh, S. (2006) Blade Root Shape Optimization, in *Proceedings of Future of Gas Turbine Technology*, 3rd International Conference, Brussels, Belgium, 11–12 October 2006.
8. Rao, S.S. (1996) *Engineering Optimization – Theory and Practice*, New Age.

Chapter 20
Concluding Remarks

In just over three centuries, through a scientific revolution, mankind became scientists with an understanding of continuum phenomena, solids, fluids, thermodynamics and other manifestations of nature that existed around them.

Enabled by these sciences, a second revolution consisted of making stone tools to elevate man over the rest of the animal kingdom and to replace human labor with animal labor. The third revolution replaced animal labor by machines.

The industrial revolution began just over two centuries ago. We were for a time content with the reciprocating steam engine doing as many chores for us as possible. This miraculous machine began gradually to fade in importance over the last century and by now has practically disappeared.

The 20th century truly belonged to the rotating machines; steam and gas turbines and reciprocating internal combustion engines. Rotating machines were thought of as "vibration free engines" just about 100 years ago.

With ever increasing hunger for energy by humans, the steam and gas turbines were pushed to their design limits, despite having innumerable vibration problems with the so-called "vibration free" engines.

We changed our designs from rigid considerations to flexible considerations, still facing many vibration problems that are still with us. The 20th century truly belongs to Vibration Engineering, where we applied known science with the use of computational tools and revolutionalized our design methodologies.

We have more recently witnessed another revolution – Information Technology – which has rendered vacuum tubes and transistors useless in computation technologies. The integrated circuit has changed the way we work; engineers have quickly utilized this opportunity and, with advances in hardware, matched machines to software, developed in a commercial mode. Today we can solve accurately many design problems, thus paving the way for simulation and reduction of testing. We do not depend on the Strength of Materials approach that had evolved at the beginning of the 20th century – instead we depend on basic energy principles that were formulated during the period of science revolution that brought us finite element methods. We gained the ability to assess the projected life expectation of mechanical components in a routine manner and also create optimal designs to increase their utility.

The developments in just one century have been awe-inspiring. Many designers of today's generation may take it for granted that common present-day design practices have existed over a long period. This book has been written mainly to highlight the history of these developments and to illustrate the various uses that led to invention and improvement of rotating machinery, rotors, and blade dynamics. We believe that knowledge of this history will lead to a better understanding of methods in rotating machinery rotor and blade dynamics that have evolved in the past and a greater appreciation of the people who led their invention and implementation.

Index

GPSR Compliance

The European Union's (EU) General Product Safety Regulation (GPSR) is a set of rules that requires consumer products to be safe and our obligations to ensure this.

If you have any concerns about our products, you can contact us on ProductSafety@springernature.com

In case Publisher is established outside the EU, the EU authorized representative is:

Springer Nature Customer Service Center GmbH
Europaplatz 3
69115 Heidelberg, Germany

Batch number: 10816149

Printed by Printforce, the Netherlands